Biophysics and Biochemistry of Protein Aggregation

Experimental and Theoretical Studies on Folding, Misfolding, and Self-Assembly of Amyloidogenic Peptides

WORLD SCIENTIFIC LECTURE AND COURSE NOTES IN CHEMISTRY

ISSN: 1793-1363

Editor-in-charge: S. H. Lin

Vol. 1: Monte Carlo Methods in Ab *Initio* Quantum Chemistry
by B. L. Hammond, W. A. Lester, Jr. & P. J. Reynolds

Vol. 2: Quantum Chemistry Aided Design of Organic Polymers: An Introduction to the Quantum Chemistry of Polymers and Its Applications
by J.-M. André, J. Delhalle & J.-L. Brédas

Vol. 4: The Physical Chemistry of Biopolymer Solutions: Application of Physical Techniques to the Study of Proteins and Nuclei Acids
by R. F. Steiner & L. Garone

Vol. 5: Theoretical Heterogeneous Catalysis
by R. A. van Santen

Vol. 6: Density Matrix Method and Femtosecond Processes
by S. H. Lin, R. Alden, R. Islampour, H. Ma & A. A. Villaeys

Vol. 7: Spectroscopy and Dynamics of Orientationally Structured Adsorbates
by V. M. Rozenbaum & S. H. Lin

Vol. 8: Introduction to Dynamic Spin Chemistry: Magnetic Field Effects on Chemical and Biochemical Reactions
by H. Hayashi

Vol. 9: Biophysics and Biochemistry of Protein Aggregation: Experimental and Theoretical Studies on Folding, Misfolding, and Self-Assembly of Amyloidogenic Peptides
edited by J.-M. Yuan & H.-X. Zhou

World Scientific Lecture and Course Notes in Chemistry – Vol. 9

Biophysics and Biochemistry of Protein Aggregation

Experimental and Theoretical Studies on Folding, Misfolding, and Self-Assembly of Amyloidogenic Peptides

Editors

Jian-Min Yuan
Drexel University, USA

Huan-Xiang Zhou
Florida State University, USA

World Scientific

NEW JERSEY • LONDON • SINGAPORE • BEIJING • SHANGHAI • HONG KONG • TAIPEI • CHENNAI • TOKYO

Published by

World Scientific Publishing Co. Pte. Ltd.
5 Toh Tuck Link, Singapore 596224
USA office: 27 Warren Street, Suite 401-402, Hackensack, NJ 07601
UK office: 57 Shelton Street, Covent Garden, London WC2H 9HE

Library of Congress Cataloging-in-Publication Data
Names: Yuan, Jian-Min, 1944– editor. | Zhou, Huan-Xiang, editor.
Title: Biophysics and biochemistry of protein aggregation : experimental and theoretical studies on folding, misfolding, and self-assembly of amyloidogenic peptides / [edited by] Jian-Min Yuan, Huan-Xiang Zhou.
Other titles: World scientific lecture and course notes in chemistry ; v. 9.
Description: New Jersey : World Scientific, 2017. | Series: World scientific lecture and course notes in chemistry ; volume 9 | Includes bibliographical references and index.
Identifiers: LCCN 2017015029 | ISBN 9789813202375 (hardcover : alk. paper)
Subjects: | MESH: Protein Aggregation, Pathological--complications | tau Proteins--chemistry | Intrinsically Disordered Proteins | Protein Folding | Amyloidogenic Proteins--chemistry | Alzheimer Disease--etiology
Classification: LCC QP517.P76 | NLM WL 104 | DDC 572/.633--dc23
LC record available at https://lccn.loc.gov/2017015029

British Library Cataloguing-in-Publication Data
A catalogue record for this book is available from the British Library.

Typeset by Stallion Press
Email: enquiries@stallionpress.com

Printed in Singapore

Preface

Protein self-assembly is a common process in nature. It generates functional polymers, such as actin filaments and microtubules in the cytoskeleton of cells. However, the recent explosion of interest in protein aggregation arises from the recognition that many diseases are caused by the conversion of soluble, often intrinsically disordered, peptides and proteins into amyloid fibrils, which in turn form extracellular deposit or intra-cellular tangles. A few well-known examples among dozens of diseases are Alzheimer's disease (AD) associated with amyloid-β peptide (Aβ) and tau protein, Parkinson's disease with α-synuclein, type II diabetes with amylin, and Creutzfeldt–Jakob disease (mad cow disease) with prion. In recent years, the number of AD patients has risen with an increase in the ageing population and diabetes has become more wide-spread in developed and developing countries. These diseases pose a serious threat to the general health of the public and are a financial burden to patients, their families and society. Therefore, we cannot over-emphasize the importance of studying protein aggregation in order to curtail the spread of such diseases and, we hope, one day to cure them.

Even natively folded functional proteins under abnormal conditions can self-assemble into aggregated states; for example, the amyloid state. The amyloid state, also with a well-defined structure, is an alternative state to the native structure. To gain a complete picture of different states of a protein, one should understand under what conditions a functional protein can misfold and assemble into the amyloid state and how in general functional proteins avoid such conversion. The story is different for intrinsically disordered peptides and proteins, often responsible for neurodegenerative diseases, where "misfolding" refers to the peptides/proteins adopting aggregation-prone conformations. In general conformations of such peptides/proteins fluctuate among many alternatives; therefore, it is not surprising that their conversion into the amyloid state is more rapid and under milder conditions. Thus in this case it is even more critical to understand the aggregation and fibril formation processes. From a different, that is, materials science point of view, amyloids can be interesting materials to study with common cross-beta structures and well-defined widths in nanometer and lengths in micrometer. For example, amyloid fibers have been used as a template to make conducting nanowires and carbon dioxide absorbers. If we consider diseases as a failure of protein materials, based on the understanding of the mechanisms of their failure, materials science can provide a new perspective to prevent and cure diseases. These mechanisms of failure and toxicity are the focus of several chapters in this book.

Experimental and theoretical studies of the amyloid cascade processes provide ideas for drug targets, inhibitors and antibodies to toxic species. Such studies, however, are driven by the amyloid cascade hypothesis. In its original form this hypothesis assumes that the deposition of mature aggregated fibrils or plaques in the brain parenchyma is the crucial step that leads to AD. However, over the years this hypothesis has been modified due to the lack of linear correlation between dementia, or other cognitive alternations, and Aβ accumulation in the brain in the form of amyloid plaques. A more recent version of the hypothesis posits that the most toxic species are in some forms of soluble oligomers. However, the exact molecular composition and structures of the oligomers remain elusive. Although this is the working hypothesis of many studies in recent years, the transient nature of the oligomers and poorly defined mechanisms of their toxicity often lead to confusion. Now a newer modified version of the amyloid hypothesis is presented by Brigita Urbanc in Chapter 1 of this book. It involves longer-living natively cross-linked Aβ oligomers, which may form curvilinear protofibrils with pore-like structures. Many amyloid fibrils share common cross-β structures, which to some may imply similar aggregation pathways. However, in reality, even a slight difference in amino acid sequence as in Aβ40 and Aβ42 can cause significant differences in the structures of oligomers and self-assembly behaviors. This is also well illustrated in Chapter 1, where Urbanc presents a comprehensive review of the research done on the Aβ's and their relation to AD, starting with amyloid precursor protein (APP) and its mutations, leading to different isoforms of Aβ. She discusses the different mechanisms through which Aβ may cause loss of neuronal functions, and reviews the detailed molecular structural information on Aβ, its fragments, and the oligomers formed. This information is obtained by combining ingeniously discrete molecular dynamics simulations with all-atom molecular dynamics simulations. Particular attention is given to the roles played by N-termini and C-termini of Aβ40 and Aβ42 in their toxicity and inhibiting power, respectively. Overall, this chapter presents a succinct survey of the current status of the Aβ field, its complexity and urgency to find solutions.

Besides the presence of the Aβ plaque outside the neurons, another hallmark of AD is the neurofibrillary tangles formed by microtubule-binding tau protein inside the brain cells. Ruxi Qi, Guanghong Wei, Ruth Nussinov, and Buyong Ma present in Chapter 2 an up-to-date review of recent experimental as well as computational studies on tau protein. They emphasize the different properties of six human tau isoforms and two truncated tau constructs throughout the chapter, and review their own work on the conformations of the two constructs using all-atom explicit-solvent replica-exchange molecular dynamics simulations. Focusing on tau aggregation, they discuss the effects of cross-seeding with Aβ and other tau isoforms, mutations, temperature, phosphorylation, polyions, as well as osmolytes. The possibility of cross-seeding of tau by Aβ is of particular interest, because it is generally accepted that Aβ accumulation precedes tau pathology in AD. Studying all the effects mentioned above reveals the important role of extended conformations of tau as an intermediate state to fibrillary growth.

Regarding recent progresses in finding drugs that combat AD, Rita P.-Y. Chen in Chapter 3 talks about design principles of making anti-amyloid inhibitors that are peptide-based. It starts from the adoption of the Aβ sequences themselves to find segments that have anti-AD properties. A similar strategy is applied to designing drugs that combat the damage caused by tau protein. Chen also talks about *in vivo* and *in vitro* tests of the validity and efficacy of designed peptide drugs using animal models. She emphasizes that anti-amyloid drugs may not be anti-aggregation or anti-AD. Particularly useful is Table 1 presented in this chapter which lists the delivery, dosage and efficacy of all peptide drugs that have been tested so far *in vivo*.

The next three chapters turn to more theoretical approaches to protein aggregation, not restricted to Aβ, tau or AD. Liu Hong, Chiu Fan Lee, and Ya Jing Huang in Chapter 4 present a review of two different approaches: statistical mechanical and kinetic. In the statistical mechanical or thermodynamic approach, equilibrium properties of chemical species in aggregation processes are investigated. They review the study of a simple polymer model of chains, made of beads with sticky patches and bonds that can be stretched and bent, where analytical solutions to partition functions can be written down. They then delve into the studies of more and more complicated systems by going beyond the dilute limit and by taking in steps the polymer interactions, oligomers, internal structures of monomer, fibril morphologies, etc., into consideration. In this way they show that the same formalism can be expanded to treat more realistic systems. Important quantities coming out of this formalism are average length distributions (which are exponential) and critical fibril concentrations. In the second part of Chapter 4, Hong *et al.* give an extensive review of kinetic approaches to protein aggregation, covering essentially all aspects of the processes: elongation, primary nucleation, fragmentation and annealing, surface-catalyzed secondary nucleation, seeding, saturation, membrane toxic processes, etc. In these studies, the total mass concentration and number concentration of fibrils are treated as dynamic variables which often yield analytic or asymptotic scaling behaviors of the kinetic processes; for example, half-time or fibril growth as functions of monomer mass or time. This systematic presentation of kinetic properties of the amyloid formation is valuable to researchers interested in understanding the kinetic processes involved in amyloid formation and using the knowledge for therapeutic intervention.

In Chapter 5, John S. Schreck, John Bridstrup, and Jian-Min Yuan also review protein aggregation from the kinetic perspective. The studies focus on primary nucleation, monomer addition and subtraction, fragmentation and coagulation of fibrils, with size-dependent rate constants. Their approach is more computationally extensive, solving ten thousands of mass-action equations numerically on GPU. Results are obtained in the form of time evolution of length distribution, fibril growth, and contour maps. Since in and outside cells, the environments that proteins face are crowded with other proteins (DNA, lipids, etc.), with up to 40% of volume fraction, their aggregation properties can be very different from those obtained in dilute solutions or *in vitro* experiments. Schreck *et al.* review studies of the effects of crowders on protein aggregation properties using simple kinetic models.

In the spirit of Zimm and Bragg's statistical mechanical treatment of helix-coil transitions of proteins, two-state models have been applied to the studies of protein folding and aggregation in recent years. This subject is reviewed in Chapter 6 by Min-Yeh Tsai, Jian-Min Yuan, and Sheng Hsien Lin. In these treatments, the kinetic Ising model derived from master equations within mean-field approximations has been applied to the thermodynamics and kinetics of protein folding and aggregation. The analytic and numerical results obtained can be used to fit experimental data of prion Sup35, Aβ, insulin, and β-microglobulin to interpret the effects of concentration, temperature, seeding, etc. of these systems.

The last two chapters of the book present reviews on experimental approaches to the stability and oligomerization properties of Aβ. Yu-Jen Chang and Yun-Ru Chen in Chapter 7 review their studies on Aβ folding, misfolding, stability, fibrillation and toxicity, giving plenty of experimental details. Focusing on Aβ40, Aβ42, and many familial mutations, including the Taiwan mutation, they emphasize that folding stability is a determining factor for fibrillation. On the other hand, the instability caused by Zn^{2+} and Al^{3+} binding results in the exposure of hydrophobic core on Aβ, leading to the formation of annular protofibrils. Furthermore, cross-seeding of Aβ40 and Aβ42 with equal amount seems to produce oligomers that are particularly toxic, consistent with some cases found in AD patients. In the last chapter (Chapter 8), Jun Han, Erwen Mei, Mei-Ping Kung, Hank F. Kung, Jian-Min Yuan and Hai-Lung Dai review their study of dimer and oligomer formation of Aβ40 using the single-molecule fluorescence resonance energy transfer (smFRET) technique. Donor and acceptor dyes are covalently bonded to the N-terminus of Aβ40, efficiency of acceptor emission yields distance and kinetic information, when donor and acceptor dyes come together in the formation of dimers and oligomers. They give some specific results on dimers using the smFRET technique.

The stories of Aβ amyloid formation leading to AD and, more generally, protein aggregation to fibrils, are indeed very complicated. The contributions in this book look at this problem from different perspectives and provide an invaluable and insightful picture of the current status of the field.

We would like to thank all contributors for their efforts to make the publication of this book possible.

Jian-Min Yuan
Huan-Xiang Zhou

Contents

Perplexity of Amyloid β-Protein Oligomer Formation: Relevance to Alzheimer's Disease

1

Brigita Urbanc

Department of Physics, Drexel University,
Philadelphia, PA 19104, USA
Faculty of Mathematics and Physics,
University of Ljubljana,
1000 Ljubljana, Slovenia

Aberrant aggregation into oligomers and amyloid fibrils is a universal process shared across many proteins associated with human diseases. Alzheimer's disease (AD) is characterized by extracellular amyloid deposits, rich in amyloid β-protein (Aβ), intracellular neurofibrillary tangles, made of tau protein aggregates, and massive neuronal loss. Aβ is hypothesized to be the key protein that initiates AD pathology by forming soluble oligomers, which are posited to start the cascade of events leading to neurodegeneration. Aβ is generated from the amyloid precursor protein (APP) by sequential cleavages of β- (or α-) and γ-secretases. Although sporadic AD is the dominant form of the disease, a number of naturally occurring mutations in APP lead to early onset familial forms of AD (FAD) by either increasing Aβ production, altering Aβ assembly dynamics, or both. Studies of assembly dynamics of naturally occurring Aβ isoforms offer insights into interactions that drive Aβ assembly dynamics. This chapter highlights the perplexity of Aβ oligomer formation by comparing and contrasting its universal and specific aspects. The current understanding of Aβ oligomer formation and its role in AD is discussed in the light of recent findings that challenge the toxic oligomer paradigm.

1. Introduction

1.1. *Universality of protein aggregation*

Protein aggregation is ubiquitous in nature. Numerous *in vitro* studies have shown that under aggregation-promoting external conditions, such as high temperature or acidic pH, any protein can aggregate, suggesting a universal mechanism of protein self-assembly.[1] The process of protein aggregation is reminiscent of self-assembly of colloid materials, which display intriguing transitions between gas, liquid, solid, and liquid-crystalline phases.[2,3] Protein aggregation into amyloid fibrils and their subsequent deposition into extra or intracellular inclusions in the brain represents a hallmark of several neurodegenerative diseases. For amyloid β-protein (Aβ) and microtubule-associated protein tau in Alzheimer's disease (AD), α-synuclein in Parkinson's disease, amylin in diabetes II mellitus, prions, and many other proteins associated with human disorders, the process of aggregation involves a transition from a disordered statistical coil-like monomeric state to a highly structured amyloid fibril. While the aggregation process may not be fully understood, *the macromolecular structure of amyloid fibrils is universal* across different proteins and characterized by cross-β motif, comprising long, stacked β-sheets, stabilized by intermolecular hydrogen bonding that is parallel to the fibrillar axis.[4] Despite the common cross-β macromolecular structure, atomistic details and the molecular arrangement of individual peptides into a fibril depend on the amino acid sequence.[5] Moreover, the amino acid sequence does not uniquely determine the amyloid fibrillar structure. Under different fibril growth conditions, the same protein aggregates into fibrils with distinct molecular-level structures.[6] This molecular-level polymorphism of amyloid fibrillar structure was proposed to contribute to disease variations *in vivo*.[7]

1.2. *Intrinsically disordered proteins*

The process of aggregation from a monomeric statistical coil-like state into an ordered amyloid fibril is further complicated by the fact that many proteins associated with human diseases, including Aβ and tau, belong to a class of intrinsically disordered proteins (IDPs),[a] which makes their characterization difficult on all levels. IDPs challenge the classical concept of protein function arising from a rigid

[a] The term IDP describes a protein that lacks a well-defined three-dimensional (3D) structure when in a folded state. Intrinsic disorder applies to a spectrum of states from fully unstructured to partially structured.

three-dimensional (3D) structure.[8] Uversky *et al.* reported that IDPs differ from globular proteins by possessing a large net charge and low hydrophobicity,[9] resulting in hydrodynamic properties resembling a statistical coil in a poor solvent.[10] Intrinsic disorder is widespread among proteins as more than 30% of gene sequences in eukaryotic genomes encode proteins or protein regions that lack a well-defined 3D structure.[11,12] Their disordered and thus flexible nature allows them to easily adopt various conformations in the presence of natural ligands, which was proposed to be essential to their physiological function.[13]

1.3. *What role do Aβ oligomers play in AD?*

AD is the leading cause of dementia among elderly and despite an effort that spans more than a century, the puzzle of AD is still unsolved, causing increasing emotional and financial damages to humankind across the globe. This chapter summarizes several aspects of early events in self-assembly of Aβ, the principal protein that is posited to play the key role in triggering AD. Strong evidence suggests that amyloid fibril formation from monomeric states proceeds through formation of soluble, low-molecular-weight (LMW) assemblies, called oligomers, which are postulated to be the principal pathogenic species triggering the disease.[14,15] Although the noun "oligomer" is derived from the Greek word *oligos*, meaning "few", the usage of this noun in the current literature does not discriminate between LMW and other nonfibrillar or even fibrillar assemblies. A wealth of scientific findings supports their seminal role in AD and yet the term oligomer remains ill-defined.

The perplexity of Aβ oligomer formation dynamics stems from what appears to be a contradiction between its universality and specificity. Universality is evident considering that Aβ akin to numerous other proteins forms not only amyloid fibrils but also oligomers, which appear to share a common structural motif and perhaps also a common mechanism of toxicity.[16] Specificity manifests itself in *the multitude of naturally occurring Aβ variants*, which stem in part from genetic polymorphism and which produce a variety of *distinct AD-related pathologies*. An important lesson learned from numerous *in vitro*, *in silico*, and animal model studies of these mutations is that a single amino acid substitution can significantly alter assembly dynamics, assembly structure, and can strongly modify human pathology. Perhaps one of the major challenges is to figure out how to bridge a gap among *in vivo*, *in vitro*, and *in silico* findings, which sometimes appear contradictory. These difficulties originate in part in the intrinsically disordered nature of Aβ, which adopts a wide range of monomer conformations with structural characteristics that are strongly sensitive to external conditions.[17–19]

Despite a wealth of knowledge accumulated over past decades, there is no consensus on which of the many reported oligomeric species is the proximate neurotoxic species in AD, which structural aspects of Aβ oligomers might be mediating toxicity, and how to prevent or reverse the damage caused by these presumably toxic Aβ oligomers. Aβ oligomers are the central topic of this chapter. Aβ oligomer formation, their structural characteristics, and their role in AD are discussed from the experimentally driven computational biophysics perspective.

2. Aβ and AD: Cause and Effect?

Since the discovery of abnormal inclusions in the brain of the first AD patient examined by Alois Alzheimer in 1906,[20] substantial progress in the understanding of the molecular basis of the disease has been made.[21] Extracellular amyloid plaques, intracellular neurofibrillary tangles, and substantial loss of neurons have been identified as histopathological hallmarks of the disease. Aβ was reported to be the main proteinaceous component of amyloid plaques.[22,23] Aberrant Aβ aggregation into Aβ fibrils, which are subsequently deposited into amyloid plaques, emerged as one of the key processes associated with AD. Similar to Aβ deposition into extracellular amyloid plaques, misfolding and aberrant aggregation of tau protein and its splicing isoforms result in intraneuronal neurofibrillary tangles.[24] Tau is a microtubule-associated protein,[25,26] which forms amyloid fibrils despite its high solubility.[27] It was not immediately clear which of the two processes, Aβ or tau aggregation, triggers the massive neuronal loss in AD and whether the two processes are causally linked. Using stereological counting techniques on *post mortem* brain tissue, the number of neurons, senile plaques, and neurofibrillary tangles in a high-order association cortex, Gomez-Isla *et al.* reported that neuronal loss correlates with neurofibrillary tangles but not with the amyloid plaque load.[28] Only after familial AD (FAD) forms, which cause early onset of the disease, were identified and linked to naturally occurring mutations affecting Aβ production and/or primary structure,[29] it became clear that tau pathology must be secondary to Aβ pathology.

In 1992, Hardy and Higgins proposed the amyloid cascade hypothesis of AD, implicating Aβ and its deposition into amyloid plaques in triggering neurofibrillary tangle formation and neurodegeneration.[30] Despite several studies that found amyloid deposits to be spatially correlated with disruption of neuronal processes[31,32] and focal neuronal loss,[33] amyloid deposits were shown not to correlate with neuronal loss in AD.[28] A discovery of soluble Aβ in AD brain,[34] which correlated better with AD symptoms than the amyloid plaque load,[35–37] alongside the research using transgenic (Tg) mouse models, which revealed that cognitive impairment develops well before deposition of insoluble Aβ into amyloid plaques,[38–40] caused a significant paradigm shift in understanding the causes of AD pathology. Substantial evidence now implicates Aβ oligomers, which form at early aggregation stages prior to Aβ deposition into amyloid plaques, as the key neurotoxins in AD.[41–49] To date, the revised amyloid hypothesis,[50,51] which states that soluble oligomeric assemblies of Aβ are the proximate neurotoxins in the brain, remains arguably the most prominent working hypothesis of AD.

3. Aβ Isoforms and Their Relevance to AD

3.1. *C-terminal heterogeneity of Aβ isoforms*

The vast majority of Aβ in the brain is generated from a transmembrane APP through sequential cleavages by β-secretase and γ-secretase complex.[52,53] Of several APP isoforms, the most abundant are APP_{695}, APP_{751}, and APP_{770}, whereby APP_{695} is the predominant neuronal isoform.[54] β-Secretase cleavage of APP produces the C-terminal fragment, referred to as C99 (Fig. 1). C99 is subjected to γ-secretase activity, resulting in a number of C-terminally truncated Aβ species, the most abundant of which is 40 amino acids long $A\beta_{1-40}$. The far less abundant but significantly more aggregation prone is $A\beta_{1-42}$, which differs from $A\beta_{1-40}$ by additional two amino acids at the C-terminus, I41 and A42, respectively. Alongside $A\beta_{1-40}$ and $A\beta_{1-42}$, other C-terminally truncated species ranging from $A\beta_{1-38}$ through $A\beta_{1-43}$ are produced *in vivo*.[55] The naturally occurring C-terminal heterogeneity in secreted Aβ is a consequence of a peculiar activity of γ-secretase, which cleaves C99 multiple times. Initially, C99 is cleaved at a site close to the membrane/cytoplasmic boundary, a process referred to as the ε-cleavage, which produces two APP intracellular domains (AICD), AICD(49–99) and AICD(50–99), released together with $A\beta_{1-48}$ and $A\beta_{1-49}$ (Fig. 1).[56–59] Funamoto *et al.* showed that when $A\beta_{1-49}$ is transiently expressed in Chinese hamster ovary cells, these cells produce predominantly $A\beta_{1-40}$, whereas when $A\beta_{1-48}$ is expressed in these cells, they secrete predominantly $A\beta_{1-42}$.[60] Two groups of Aβ variants are generated through sequential cleavages of C99 by γ-secretase: $A\beta_{1-49} \rightarrow A\beta_{1-46} \rightarrow A\beta_{1-43} \rightarrow A\beta_{1-40}$ and $A\beta_{1-48} \rightarrow A\beta_{1-45} \rightarrow A\beta_{1-42} \rightarrow A\beta_{1-38}$, respectively.[61–63]

Whereas the majority of experimental and computational studies thus far explored $A\beta_{1-42}$, $A\beta_{1-43}$ was recently reported to be highly amyloidogenic and even more toxic than $A\beta_{1-42}$ in a Tg mouse model overexpressing $A\beta_{1-43}$.[64] Although $A\beta_{1-43}$ exists *in vivo* at significantly lower concentrations than $A\beta_{1-42}$, its level in cerebrospinal fluid, in combination with $A\beta_{1-42}$ and tau levels, may prove to be a

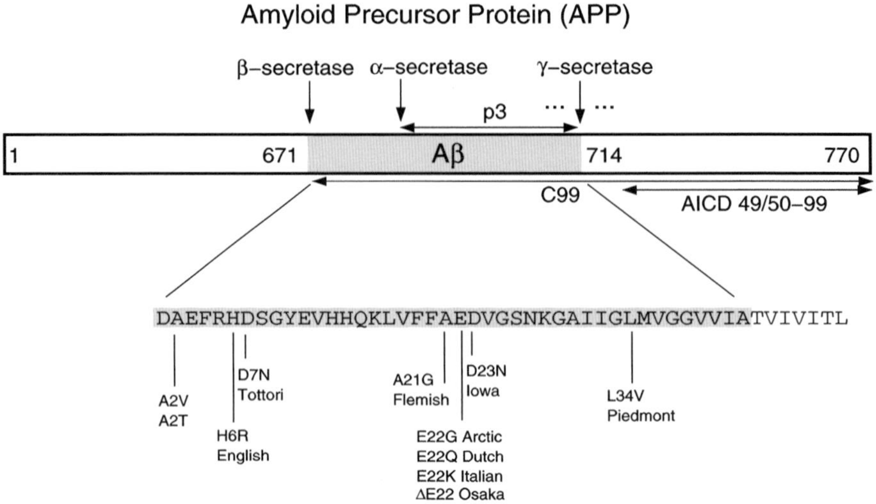

Figure 1: Schematic figure of APP processing by the three secretases and the resulting products. Along the amyloidogenic pathway, β-secretase cleavage produces C99, which is subjected to ε-cleavage by γ-secretase, releasing AICD (49–99) or AICD (50–99) alongside $A\beta_{1-49}$ or $A\beta_{1-48}$, respectively. Further sequential γ-secretase cleavages of $A\beta_{1-49}$ or $A\beta_{1-48}$ release the two predominant $A\beta$ alloforms, $A\beta_{1-40}$ and $A\beta_{1-42}$, respectively. $A\beta_{1-42}$ (gray-shaded region) spans the 672–713 region of APP. Along the nonamyloidogenic pathway, α- and γ-secretase cleavages produce the p3 fragment (sequences not drawn to scale). $A\beta_{1-49}$ and the gray-shaded $A\beta_{1-42}$ sequence within are displayed alongside the naturally occurring mutations within $A\beta_{1-42}$ that cause FAD.

useful biomarker for identifying patients with amnestic mild cognitive impairment[b] with an increased risk for AD.[65]

3.2. *N-terminally truncated and pyroglutamated Aβ isoforms*

The APP processing pathway through β-secretase and γ-secretase cleaving described above is referred to as the amyloidogenic pathway. During the alternative, nonamyloidogenic pathway of APP processing, α-secretase cleaves APP within the $A\beta$ region between K16 and L17, followed by γ-secretase processing, yielding a series of $A\beta$ fragments, including $A\beta_{17-40}$ and $A\beta_{17-42}$, referred to as p3 peptides (Fig. 1).

[b]A preclinical stage of AD is called mild cognitive impairment (MCI). When memory loss is the predominant symptom of MCI, it is referred to as amnestic MCI and viewed as a precursor of AD.

These hydrophobic p3 peptides have not received much attention in the literature because the α-secretase APP processing pathway is generally viewed as protective in the context of AD.[66]

In addition to cleaving APP at the β site, β-secretase was reported to cleave also at the β' site between Y10 and E11 within the Aβ region of C99.[67] The preference for the cleavage site may depend on whether the process occurs in the endoplasmic reticulum or trans-Golgi network.[68] Followed by γ-secretase cleavages, the β-secretase activity at the β' site gives rise to N-terminally truncated forms, $A\beta_{11-40}$ and $A\beta_{11-42}$, which are deposited into insoluble amyloid plaques in the brain alongside full-length Aβ.[69] Most other N-terminally truncated Aβ isoforms in the brain, including $A\beta_{3-40}$ and $A\beta_{3-42}$, arise from proteolysis of full-length Aβ peptides. N-terminally truncated forms of Aβ with glutamic acid as the N-terminal amino acid can be further modified through a cyclization process that involves an interaction of the N-terminal region with glutaminyl cyclases and produces pyroglutamated N-terminally truncated Aβ isoforms: $A\beta_{3(pE)-40}$, $A\beta_{3(pE)-42}$, $A\beta_{11(pE)-40}$, and $A\beta_{11(pE)-42}$, where "pE" stands for the pyroglutamated form of E.[70–74] Näslund *et al.* showed that $A\beta_{11(pE)-42}$ and $A\beta_{11-42}$ combined amounted to 20% of the total Aβ in the brain.[69] $A\beta_{3(pE)-42}$ and $A\beta_{3(pE)-43}$ were shown to contribute 25% to the total amount of $A\beta_{X-42}$ and $A\beta_{X-43}$ ($X \geq 1$), respectively, whereas $A\beta_{3-40}$ and $A\beta_{3-42}$ amounts are much smaller, suggesting that these N-terminally truncated isoforms are mostly pyroglutamated.[72,74] $A\beta_{3(pE)-42}$ was found to be the dominant pyroglutamated Aβ isoform,[72–74] existing in amounts equal or even surpassing the amount of $A\beta_{1-42}$.[71] The relevance of the pyroglutamated Aβ to the disease stems from their abundance in amyloid plaques and their increase (20–60%) with the progression of AD.[75–77]

4. Familial Forms of AD: APP Mutations

The most common form of AD has no specific family link and is referred to as a late-onset sporadic form, afflicting individuals older than ~65 years. In contrast, FAD forms run in specific families and are caused by mutations in four major genes that encode for APP, presenilin 1 (PSEN1), presenilin 2 (PSEN2), and/or apolipoprotein E (ApoE).[78,79] PSEN1 and PSEN2 comprise the catalytic core of the γ-secretase complex and the FAD-causing PSEN1 and PSEN2 mutations directly affect Aβ production by increasing the absolute amount of secreted Aβ and/or the $A\beta_{1-42}/A\beta_{1-40}$ ratio. ApoE is a polymorphic lipoprotein that functions as a major cholesterol carrier, supporting lipid transport and injury repair in the brain. Individuals carrying the $\varepsilon 4$ ($\varepsilon 2$) allele of ApoE are at increased (decreased) risk for

AD compared with those carrying the more common ε3 allele. ApoE is involved in Aβ clearance from the brain by promoting the proteolytic degradation of Aβ[80] yet the mechanism of the ApoE–Aβ interaction that leads to Aβ clearance remains unknown.[81]

Important clues about what causes the disease can be gleaned from the FAD forms associated with APP mutations.[82] There are ~58 APP mutations reported to date (see http://www.alzforum.org/mutations). A FAD phenotype is typically associated with a mutation that runs in a family and is named after the family's geographic location (Fig. 1). Strong genetic evidence in support of the seminal role Aβ and amyloid cascade hypothesis of AD stems from these naturally occurring APP mutations that affect either Aβ production or the primary structure of Aβ. The majority of FAD mutations lead to a considerably modified disease phenotype with an early onset and enhanced deposition of Aβ in the walls of small and medium-sized vessels of the cerebral cortex, a disorder known as cerebral amyloid angiopathy (CAA), which is often but not always associated with hemorrhagic strokes. FAD mutations can be divided into two groups: (1) APP mutations that are adjacent to the Aβ region and directly affect β- and γ-secretase processing and (2) mutations within the Aβ region of APP. Most FAD-associated Aβ mutants were reported to resist proteolytic degradation by neprilysin, which is a mechanism cells use to catabolize $A\beta_{1-42}$.[83] Such resistance to degradation may prolong the lifetime of Aβ mutants and contribute to their pathogenic nature.[84] The mutations within Aβ can be further classified with respect to the region, in which the mutation occurs. Although the majority of known Aβ mutations affect the region within or close to the central hydrophobic cluster (L17–A21), FAD-causing mutations within the N-terminal region of Aβ are particularly intriguing as discussed further in this chapter (Fig. 1). Another exception is the *Piedmont mutation* affecting the sequence in the C-terminal region (L34V), which leads to CAA-only pathology without amyloid plaques or neurofibrillary tangles.[85,86]

4.1. *Mutations affecting APP processing*

The **Swedish mutation** of APP was identified in two large Swedish families that are connected genealogically. This mutation is adjacent to the β-secretase cleavage site in APP at the N-terminus of Aβ and affects two consecutive amino acids, [K670N] and [M671L].[87] Swedish mutation increases production and secretion of $A\beta_{1-40}$, $A\beta_{1-42}$, and $A\beta_{1-43}$, which may be the lead cause of early onset AD in carriers of this mutation.[88–91]

Two mutations of the APP site adjacent to the γ-secretase cleavage site, at position 714, were reported to date. The **Austrian mutation** ([T714I]) was first discovered in an Austrian family and reported to produce a particularly aggressive form of AD with an average onset age of ~34 years.[92] The phenotype associated with the Austrian mutation exhibits extensive neuronal loss, mostly diffuse amyloid plaques, and neurofibrillary tangles. Interestingly, $A\beta_{1-40}$ was absent from amyloid deposits in the brain. This APP mutation appears to increase $A\beta_{1-42}$ production, resulting in an 11-fold increase in the $A\beta_{1-42}/A\beta_{1-40}$ ratio.[92] The same mutation was identified also in an African American family whose members exhibited an unusually early onset of AD (in their early 30s) with a rapidly progressing dementia.[93] The **Iranian mutation** ([T714A]) was identified in an Iranian family of nine documented individuals with an average onset age of ~55 years, which is significantly older than the average onset age of the Austrian mutation carriers, suggesting that strongly hydrophobic I may disrupt the APP structure and γ-secretase processing more than the less hydrophobic A.[94] The Iranian mutation was later found also in a Polish individual[95] and in another unrelated individual.[96] The neuropathology of the Iranian mutation is not well understood.

Two APP mutations in the proximity of γ-secretase cleavage sites, at position 715, were reported so far. The **French mutation** ([V715M]) was first discovered in an Italian family[97] and more recently in a Korean family.[98] This mutation distinguishes itself from the other APP mutations because it reduces the $A\beta_{1-40}$ production without affecting the $A\beta_{1-42}$ production and leads to an increase of N-terminally truncated $A\beta_{X-42}$ but not $A\beta_{X-40}$ ($X > 1$).[97] The **German mutation** ([V715])[99] was discovered in a German family and later also in unrelated individuals.[95] De Jonghe *et al.* demonstrated that several APP mutations at positions 714–717 affect γ-secretase processing in a way that increases the $A\beta_{1-42}/A\beta_{1-40}$ ratio.[100] Consistent with these findings, a nuclear magnetic resonance (NMR) spectroscopy study recently found that the French and Italian mutations alter the structure and dynamics of the transmembrane domain of APP, making it more susceptible to ε-cleavage, consequently shifting Aβ production toward $A\beta_{1-42}$.[101]

4.2. *Aβ mutations near the central hydrophobic cluster*

The **[K16N] mutation** at position 16 of Aβ is adjacent to the site of the α-secretase cleavage.[102] This mutation causes early onset dementia, most likely by suppressing the nonpathogenic APP processing pathway, which in turn makes APP more susceptible to β-secretase cleavage and increases Aβ production.[102] The resulting peptides appear to be less toxic than wild-type (WT) peptides, although they display increased resistance to neprilysin proteolysis.[102] Interestingly, heterotypic assemblies formed by both mutant and WT peptides appear to be highly toxic to cell cultures.[102]

The **Flemish mutation [A21G]** was first identified in a Flemish family with carriers suffering from early onset (~47 years) dementia and cerebral hemorrhage due to CAA.[103] Later on, additional individuals of British origin were reported to carry the same mutation with a similar disease phenotype.[104] In addition to prominent CAA, the Flemish Aβ mutation causes formation of unusually large amyloid plaque cores, containing predominantly Aβ_{1-40}, unlike the amyloid plaque composition in sporadic AD, which is dominated by Aβ_{1-42}.[105] Relative to WT peptides, Flemish mutants display increased resistance to neprilysin-mediated proteolysis, which may contribute to their pathogenic nature.[106]

Although the **Arctic mutation [E22G]** is clinically and neuropathologically indistinguishable from the late-onset sporadic AD, such as formation of compact plaques, it results in early onset pathology.[107,108] *In vitro* studies point at an increased generation of intermediate protofibrillar Aβ assemblies formed early during fibrillogenesis that may be responsible for the early onset.[108–110] Similar to Flemish Aβ mutants, [E22G]Aβ_{1-40} was demonstrated to confer resistance to neprilysin-mediated proteolysis.[84]

The **Dutch mutation [E22Q]**[111,112] and the **Iowa mutation [D23N]**[113,114] cause similar clinical, pathological, and *in vitro* characteristics, which are distinct from those of sporadic AD. Dutch mutation carriers suffer from CAA with massive accumulation of Aβ in cerebral vessel walls and diffuse plaques, leading to cerebral hemorrhages, strokes, and premature death.[111] Although the Iowa mutation carriers exhibit diffuse plaques with massive CAA similar to the Dutch mutation carriers, strokes and hemorrhages are rare.[113,115] The Iowa mutation carriers also display widespread neurofibrillary tangles and unusually extensive distribution of Aβ_{1-40} in amyloid plaques.[113] *In vitro* studies demonstrated that Dutch mutants form protofibrils and fibrils significantly faster than WT peptides.[116] Similarly, the Iowa mutation was shown to accelerate Aβ fibrillogenesis *in vitro*.[114]

Akin to the Dutch mutation, the **Italian mutation ([E22K])** causes hereditary cerebral hemorrhage with amyloidosis, leading to multiple strokes and cognitive decline, whereas parenchymal amyloid deposits are rare and neurofibrillary tangles are absent.[117,118] [E22K]Aβ_{1-42} aggregates *in vitro* at least twice as fast as Aβ_{1-42}.[119] Both [E22Q]Aβ_{1-42} and [E22K]Aβ_{1-42} peptides were shown to be potently neurotoxic to cell cultures.[119]

The **Osaka mutation [Δ E22]**, which is defined by a microdeletion of glutamic acid at position 22, was detected in several Japanese families with early onset AD-type dementia, which appears to be associated with low amyloid deposition in the brain as assessed by positron emission tomography scans using Pittsburgh compound B, suggesting that the Osaka mutants primarily assemble into soluble oligomers rather than forming amyloid fibrils.[120,121] This mutation markedly *reduces* secretion of total Aβ without affecting the Aβ_{1-42}/Aβ_{1-40} ratio; however, [ΔE22]Aβ peptides are more resistant to proteolytic degradation than WT peptides.[120] In contrast to what appears to be a dramatically reduced ability to form amyloid fibrils, *in vitro* studies reported a significantly higher rate of fibril formation by Osaka mutants with the resulting quaternary structure of fibrils that is distinct from the fibrillar structure of WT peptides.[122,123]

4.3. *Aβ mutations within the N-terminal region*

The **English mutation ([H6R])**, which is associated with an amino acid mutation in the N-terminal region of Aβ, was identified in an English family, causing early onset (45 years) AD; however, the pathology of this mutation is unclear.[124] Hori *et al.* reported that while this mutation does not alter Aβ production, it affects Aβ assembly kinetics *in vitro*.[125] English mutants display accelerated changes in the secondary structure during the assembly and form relatively large oligomers, which are significantly more toxic to cell cultures than oligomers formed by WT peptides.[126]

The **Tottori mutation ([D7N])** was identified in a Japanese family.[127] Unlike the majority of FAD mutations, the carriers of this mutation suffer from AD-like symptoms without cerebrovascular complications. Similar to the English mutation, the Tottori mutation does not alter Aβ production but affects the kinetics of Aβ assembly *in vitro*.[125] Relative to oligomers formed by WT peptides, Tottori mutants assemble into larger oligomers and exert increased toxicity to cell cultures.[126]

The **[A2V]** mutation differs from the other FAD mutations because it leads to early onset AD with CAA only in homozygous individuals, whereas heterozygous carriers are not affected.[128] Amyloid plaques of the homozygous carriers of the [A2V] mutation are unusually large, rich in Aβ_{1-40}, and peculiarly distributed in the brain.[129] This mutation was shown to increase the production of both Aβ_{1-40} and Aβ_{1-42}, presumably because the mutated amino acid position in Aβ is proximate to the β-secretase cleavage site. *In vitro*, this mutation does not alter Aβ_{1-42} aggregation; however, it accelerates Aβ_{1-40} aggregation.[130]

Perhaps the most inspiring of all APP mutations is the **Icelandic [A2T] mutation,** which was first found in a single aged individual without any cognitive decline[131] and relatively recently in a broader Icelandic population.[132] This mutation reportedly *protects against AD and cognitive decline in elderly*.[132] As the mutated amino acid site is close to the β-secretase cleavage site, the Icelandic mutation was reported to protect APP against the β-secretase cleavage, which results in a 40% decreased production of total Aβ.[132] In addition, the Icelandic mutants may display modified *in vitro* assembly dynamics. A recent *in vitro* study by Zheng *et al.* showed that this mutation inhibits formation of Aβ_{1-42} dodecamers, which were postulated to mediate Aβ_{1-42} oligomer-induced toxicity.[133]

4.4. *What do APP mutations tell us?*

The most obvious notion derived from APP mutations outside the Aβ region is that the disease pathology is enhanced by elevated Aβ_{1-42} levels or/and the elevated Aβ_{1-42}/Aβ_{1-40} ratio, which provides a strong support to the amyloid cascade hypothesis of AD. The reported effects of mutations within the Aβ region of APP, however, tell us that an increased amount of secreted Aβ may not be *always* associated with FAD pathology. Instead, the amino acid mutations that alter the sequence but do not affect APP processing (and do not cause increased Aβ production) must impart changes in Aβ folding and assembly, which profoundly affect the resulting disease pathology. The effect of sequence specificity on Aβ fibril formation is well documented. Early biochemical studies already demonstrated that *in vitro* Aβ_{1-42} and Aβ_{1-43} form fibrils more rapidly than Aβ_{1-39} and Aβ_{1-40}, indicating that a relatively small change in the Aβ sequence ($<$10%) can lead to accelerated or delayed assembly dynamics.[134,135] These studies show that the hydrophobic C-terminus of Aβ is strongly involved in aggregation. FAD mutations within Aβ, however, mostly affect amino acids in the central region of the sequence (A21, E22, and D23), suggesting that the central region is critically involved in formation of toxic oligomers. The most notable are the naturally occurring mutations in the N-terminal region, which demonstrate that the N-terminus of Aβ is not an inert peptide region but instead actively participates in Aβ assembly. Whereas the APP mutations place focus on Aβ, remaining questions are the following: which Aβ regions dominate Aβ oligomer formation and assembly pathways, which Aβ regions are critically involved in mediating toxicity, and how is Aβ toxicity mediated.

5. Aβ Oligomer Formation and Structure

Many important aspects of Aβ folding, oligomer formation, and fibrillization have been examined using various *in vitro* and *in silico* approaches, including molecular dynamics (MD) simulations.[136] Three major Aβ assembly states were identified: monomeric, oligomeric, and fibrillar. In addition to these, a range of intermediate assembly states between LMW oligomers and fibrils, hereby referred to as protofibrils, have been observed and studied. Following the definition by Walsh *et al.*,[137] the term protofibril is used here to describe <200-nm-long curvilinear assemblies, 6–8 nm in diameter, that appear to emerge from globular oligomers prior to developing into straight and significantly longer fibrils as observed by AFM or TEM. As described earlier, a paradigm shift from insoluble fibrils to soluble oligomers caused a substantial research effort aimed at identification and characterization of soluble Aβ oligomers. An Aβ oligomer is a soluble, noncovalently bonded assembly of two or more Aβ molecules, commonly referred to as a dimer, trimer, tetramer, pentamer, hexamer, etc, depending on its size (order).

5.1. *Identifying Aβ oligomers*

Oligomers formed *in vitro* by synthetic Aβ_{1-42} termed Aβ-derived diffusible ligands (ADDLs) were shown to kill mature neurons in cell cultures at nanomolar concentrations.[138] ADDLs were identified as globular LMW (17–42 kDa) assemblies, measuring 5–6 nm in diameter, which form even when fibril formation is suppressed.[138] Considering the molecular mass of Aβ_{1-42} (4.2 kDa), the reported ADDLs correspond to oligomers ranging from tetramers to decamers. Similarly, Chromy *et al.* observed formation of globular Aβ_{1-42} oligomers (trimers through 24-mers), which remained fibril-free at low temperatures and were toxic to cell cultures.[139] Barghorn *et al.* reported a highly stable Aβ_{1-42} oligomer, referred to as a globulomer, which was extracted from the brain of AD patients and formed also in Tg mice that over-expressed Aβ_{1-42}.[140] Globulomers, which were characterized as dodecamers, were shown to bind to neuronal dendritic processes and block long-term potentiation (LTP)[c] in hippocampal slices of rat brain.[140]

In juxtaposition to *in vitro*-generated oligomers formed by synthetic Aβ, Walsh *et al.* extracted "natural" oligomers of Aβ from human brain to demonstrate that addition of these *in vivo* oligomers (mostly dimers and trimers) to rats inhibited hippocampal LTP *in vivo*.[141] These natural oligomers may be a mixture of naturally occurring Aβ isoforms, possibly dominated by Aβ_{1-40} and Aβ_{1-42}. Gong *et al.* used oligomer-sensitive immunoassays to visualize soluble Aβ oligomers in human brain and showed that oligomers, which are highly elevated in AD relative to control brain, bind to dendritic surface of cultured hippocampal neurons.[142] Several studies thus reported that ADDLs or soluble Aβ oligomers from the brain inhibit hippocampal

[c]LTP, which refers to persistent strengthening of synapses induced by recent activity that produces a long-lasting increase in signal transmission between two neurons, is a paradigm for synaptic plasticity and memory storage mechanisms in the brain.

LTP and thereby cause impaired synaptic activity.[138,141,143] These findings lead to a proposition that AD is Aβ oligomer-induced synaptic failure.[144]

5.2. *In vitro characterization of Aβ oligomers*

The above mentioned research results showcase the multifaceted nature of Aβ oligomers that can adopt, depending on solvent conditions, different assembly states. *In vitro* studies of proteins allow for controlled conditions, under which oligomer formation can be characterized. There is no consensus on how Aβ should be prepared to study oligomer formation *in vitro*. To separate a mixture of macromolecules with different molecular masses, polyacrylamide gel electrophoresis (PAGE) with sodium dodecyl sulfate (SDS) can be used, whereby the species migrating in the gel separate into distinct bands such that the lowest/highest molecular mass-species populates the lowest/highest gel band. Noncovalently bonded oligomers, however, are dissolved by SDS into monomers, so SDS-PAGE cannot be directly applied to visualize and quantify gel bands corresponding to different oligomer orders. To solve this problem, Bitan *et al.* utilized photo-induced cross-linking of unmodified proteins (PICUP), a covalent cross-linking method previously described by Fancy and Kodadek,[145] to first covalently bond Aβ oligomers and then characterize oligomer populations by SDS-PAGE.[146,147]

Several important insights were derived through the application of PICUP/SDS-PAGE. Aβ_{1-40} and Aβ_{1-42} were shown to form oligomers with distinctly different oligomer size distributions.[148] Aβ_{1-40} and Aβ_{1-42} were reported to adopt dimers, trimers, and tetramers with relative abundances decreasing with oligomer order.[148] In addition to dimers through tetramers, Aβ_{1-42} also formed abundant pentamers and hexamers and their multiples, decamers to dodecamers, thereby displaying a multimodal oligomer size distribution.[148] In a follow-up study, Bitan *et al.* characterized oligomer size distributions of 34 Aβ isoforms, including naturally occurring Aβ mutants linked to FAD and N-terminally truncated Aβ peptides, which demonstrated that the oligomer size distribution is sensitive to small Aβ sequence modifications.[149] Akin to their distinct assembly pathways, Aβ_{1-40} and Aβ_{1-42} are affected by the same sequence modification in a unique and distinct manner.[149] Importantly, the hydrophobicity but not the size of I41 was shown to be critical for Aβ_{1-42} pentamer and hexamer formation.[149] These results highlight the important role of amino acid-specific hydropathy (hydrophobic and hydrophilic properties) in Aβ assembly dynamics. SDS-PAGE has its limitations. Artifacts such as SDS-induced Aβ_{1-42} trimers and other SDS-mediated Aβ_{1-42} assemblies may introduce bias into the Aβ_{1-42} oligomer size distribution by artificially increasing pentamer and hexamer abundances.[150] These difficulties aside, cross-linking combined with SDS-PAGE (or some other separation method, such as size-exclusion chromatography) remains to date one of a very few known methods for *in vitro* characterization of Aβ oligomer size distributions.

Aβ_{1-42} dimers, tetramers, hexamers, and dodecamers were detected also by the ion-mobility/mass-spectroscopy (IM-MS) teachnique.[151] The same technique was later used to detect and compare Aβ_{1-40} and Aβ_{1-42} oligomer formation, whereby in addition to Aβ_{1-40} and Aβ_{1-42} dimers and tetramers, hexamers, and dodecamers of Aβ_{1-42} were also identified.[152] Although the IM-MS technique involves electrospraying and vacuum conditions, which may affect the oligomer structure, this technique does not rely on either PICUP chemistry or SDS-PAGE. Bernstein *et al.*,[152] thus provided additional evidence that Aβ_{1-42} forms larger oligomers than Aβ_{1-40}, which is an indication of distinct assembly pathways, in agreement with PICUP/SDS-PAGE data.[148] Aβ_{1-42} dodecamers detected by Bernstein *et al.*[152] are particularly intriguing because they may correspond to the Aβ*56 species[d] detected in a human APP-expressing Tg mouse model, which induces memory deficits when administered to young rats.[153]

Recently, an alternative cross-linking method, copper and hydrogen peroxide induced cross-linking of unmodified proteins (CHICUP), combined with SDS-PAGE, was reported and shown to produce Aβ_{1-40} and Aβ_{1-42} oligomer size distributions that are very similar to PICUP/SDS-PAGE oligomer size distributions.[154] CHICUP utilizes copper, which is normally present in the brain, and hydrogen peroxide, which is the dominant reactive oxidative species associated with the AD brain.[154] Consequently, cross-linked Aβ oligomers can form *in vivo* and might be potentially more damaging than noncross-linked oligomers as cross-linking prolongs their lifetimes.[154] This proposition is supported by detection of dityrosine, which forms as a result of covalent cross-linking,[145,147,155] colocalized with intracellular Aβ and with extracellular amyloid plaques in AD patients, implying that dityrosine-rich Aβ oligomers can form in lysosomal compartments, trigger cell death,[156] and subsequently accumulate into extracellular amyloid plaques.[157]

Besides detection of Aβ_{1-40} and Aβ_{1-42} oligomers and characterization of their relative abundances, there is very limited experimental information on the oligomer structure. Ahmed *et al.* observed *in vitro* formed disc-shaped Aβ_{1-42} pentamers, which were significantly more toxic to mouse cortical neurons than fibrils, did not have the β-sheet structure of a fibril, and had C-termini protected from the solvent.[158] Ono *et al.* extracted cross-linked Aβ_{1-40} monomers, dimers,

[d]Here, "56" refers to the molar mass of 56 kDa and thus Aβ*56 corresponds to Aβ dodecamers.

trimers, and tetrames in pure form and then examined and compared their β-sheet content and toxicity, which were shown to increase with the oligomer order.[159]

5.3. *In silico characterization of Aβ oligomers*

Crystallization of full-length Aβ oligomers, if realized, would provide the coveted atomic structure of these presumably toxic entities. So far, pure oligomers of full-length Aβ have not been crystallized. The heterogeneous and short-lived nature of Aβ oligomers are serious obstacles to crystallization. Significant structural insights into early assembly event of Aβ_{1-40} and Aβ_{1-42} emerged from *in silico* studies.[160] MD of proteins using fully atomistic details and explicit solvent has become a standard method for studying protein dynamics. Aβ belongs to the class of IDPs with complex free-energy landscapes that need to be thoroughly sampled during simulations. For this reason, enhanced sampling techniques, such as the replica exchange protocol, coupled with MD (REMD)[161,162] are typically employed.[136,163–166] Despite significant understanding of Aβ dynamics derived from fully atomistic MD studies, assembly of a large number of initially separated and unstructured peptides into a steady-state population of monomers and oligomers is still computationally too demanding to be examined by fully atomistic explicit-solvent MD. To make simulations of protein populations more computationally accessible, lattice and off-lattice coarse-grained (CG) models of proteins have been developed. Numerous applications of CG protein models demonstrated that eliminating nonessential degrees of freedom increases the computational efficiency and allows for studies of processes that occur on long time scales, which helps identify, through a reductionist approach, the interactions that drive protein folding and assembly.[167–173]

To study protein assembly, discrete MD (DMD)[174,175] has been utilized by several research groups.[176–181] In DMD, continuous potentials are replaced by square-well potentials, rendering numerical calculations involved at each simulation step more computationally efficient. Computational efficiency can be further increased if the number of atoms in a simulation is reduced. This can be achieved (1) by replacing a fully atomistic protein model with a CG protein model and (2) introducing an implicit solvent force field to avoid modeling solvent molecules explicitly.[173,182,183] Aβ_{1-40} and Aβ_{1-42} dimer formation was first examined by DMD in combination with a four-bead protein model and the force field with backbone hydrogen bonding interactions,[184] which resulted in various β-hairpin dimer structures.[185] With backbone hydrogen bonding as the only attractive interaction, these dimers further assembled into a single planar β-sheet structure, which is inconsistent with experimental observations of quasi-spherical oligomers and elongated fibrils made of stacked β-sheets.[182]

Urbanc *et al.* developed the DMD4B-HYDRA approach by introducing amino acid-specific implicit solvent interactions due to hydropathy[186] into the four-bead model with backbone hydrogen bonding, which enabled simulations of $A\beta_{1-40}$ and $A\beta_{1-42}$ oligomer formation starting from spatially separated unstructured peptides.[187] The oligomer formation was observed as a fast collapse of initially monomeric populations into quasi steady-state ensembles of globular monomers and oligomers of various orders. In the quasi steady state, the oligomer size distribution did not significantly change over time.[188] The DMD4B-HYDRA approach captured distinct $A\beta_{1-40}$ and $A\beta_{1-42}$ oligomer size distributions, which were consistent with those determined *in vitro* by PICUP/SDS-PAGE.[148] DMD4B-HYDRA conformations of $A\beta_{1-40}$ and $A\beta_{1-42}$ monomers and oligomers resemble collapsed statistical coils with several turns and a prominent loop within the central folding region A21–A30 (Fig. 2).[187] The collapsed coil monomer conformation is consistent with experimentally determined $A\beta_{25-35}$ monomer conformation in aqueous solution.[18,189] Unlike $A\beta_{1-40}$ monomers, $A\beta_{1-42}$ monomers derived by DMD4B-HYDRA display a turn structure centered at G37-G38 (Fig. 2),[187] a structural difference that is consistent with experimental and explicit-solvent MD findings.[189–196] The β-strand content of $A\beta_{1-40}$ and $A\beta_{1-42}$ conformations derived by DMD4B-HYDRA increases with oligomer order,[188] in agreement with the β-strand content of the experimentally characterized cross-linked $A\beta_{1-40}$ oligomers.[159] Because the amino acid sequences of $A\beta_{1-40}$ and $A\beta_{1-42}$ differ at the C-terminus, it is intriguing that $A\beta_{1-40}$ and $A\beta_{1-42}$ oligomers obtained by DMD4B-HYDRA display distinct structural features in the N-terminal region, whereby $A\beta_{1-42}$ oligomers exhibit significantly more flexible and solvent-exposed N-termini than $A\beta_{1-40}$ oligomers.[187] This structural difference at N-termini was proposed to underlie increased toxicity of $A\beta_{1-42}$ relative to $A\beta_{1-40}$ oligomers (Fig. 2).[187,188]

The 3D *in vitro* structure of $A\beta_{1-40}$ and $A\beta_{1-42}$ oligomers has not been determined thus far. Streltsov *et al.* crystallized an $A\beta_{18-41}$ tetramer, trapped in the oligomeric state by four shark *Ig new antigen receptor* (IgNAR) proteins.[198] Although the interactions between $A\beta_{18-41}$ and the ligand IgNAR may have affected the resulting Aβ(18-41) tetramer structure, the reported *in vitro* conformation of the $A\beta_{18-41}$ tetramer is in a qualitative agreement with DMD4B-HYDRA oligomer conformations. Solution NMR study conducted by Pauwels *et al.* showed increased solvent accessibility of several amino acids within the N-terminal region D1-D7 of $A\beta_{1-42}$ relative to $A\beta_{1-40}$ assemblies (Supplementary Fig. S2 in Ref.[199]), an indication that N-terminal structural differences between $A\beta_{1-40}$ and $A\beta_{1-42}$ assemblies are present *in vitro*, consistent with DMD4B-HYDRA predictions.

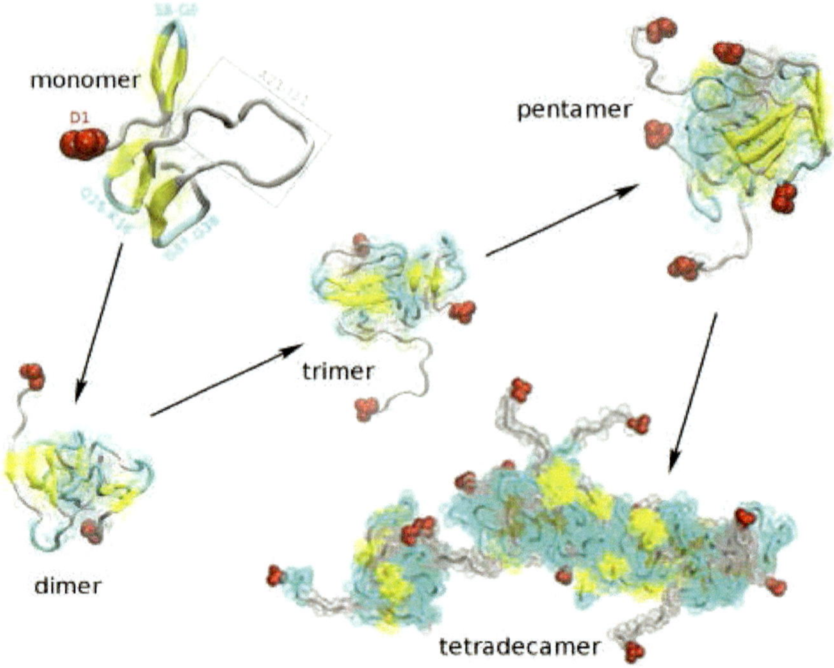

Figure 2: In DMD4B-HYDRA, unstructured $A\beta_{1-42}$ folds into a partially folded monomer (the top left) prior to forming a dimer (bottom left) and a trimer (center). Dimers and trimers assemble into pentamers (top right) and hexamers. Decamers, undecamers, tridecamers, and tetradecamers (bottom right) formed by modular assembly of LMW oligomers (dimers through hexamers) display elongated protofibril-like structure. The N-terminal amino acid D1 is represented by red spheres. $A\beta_{1-42}$ monomer (the leftmost conformation) is characterized by turns centered at S8-G9, Q15-K16, and G37-G38 (cyan) and an extended loop region at A21-I31 (gray rectangle). Conformations were generated using *Visual Molecular Dynamics* software package.[197] β-Strand, turn, and coil structures are colored in yellow, cyan, and gray, respectively.

Because CG protein models can explore a broader ensemble of conformations, an alternative approach to overcome the sampling problem is to use a large ensemble of conformations derived from a CG model, convert them into fully atomistic conformations, and use them as initial conformations for all-atom MD. Such a multiscale approach that combines CG modeling and all-atom MD was recently shown by Cheung *et al.* to be even more efficient than all-atom REMD.[200] Barz and Urbanc used a large ensemble of $A\beta_{1-40}$ and $A\beta_{1-42}$ monomer and dimer conformations derived by DMD4B-HYDRA (40–50 per peptide and per assembly state), converted them into fully atomistic structures, whereby the backbone conformations remained almost identical to their CG representations, and examined them by all-atom MD in explicit solvent.[201] The resulting all-atom monomers and dimers were slightly larger and less compact than their initial DMD4B-HYDRA conformations with comparable turn but lower β-strand propensities.[201] Importantly, the N-termini of fully atomistic $A\beta_{1-42}$ dimers were more flexibile and solvent exposed than the N-termini of fully atomistic $A\beta_{1-40}$ dimers, in agreement with DMD4B-HYDRA predictions.[201]

5.4. *Sequence modifications alter Aβ oligomer formation and structure*

Oligomer formation of **N-terminally truncated** $A\beta$ peptides: $A\beta_{3-40}$, $A\beta_{3-42}$, $A\beta_{11-40}$, and $A\beta_{11-42}$ was probed *in vitro* by PICUP/SDS-PAGE[149] and *in silico* by DMD4B-HYDRA.[202,203] PICUP/SDS-PAGE data showed that deletion of the first two amino acids in $A\beta_{1-40}$ resulted in a shift of oligomer size distributions to larger sizes, up to octamers, whereas the oligomer size distribution of $A\beta_{3-42}$ was similar to that of $A\beta_{1-42}$.[149] PICUP/SDS-PAGE data for $A\beta_{11-40}$ and $A\beta_{11-42}$ were complicated due to decreased cross-linking efficiency in the absence of Y10 and the corresponding size distributions appeared to be irregular.[149] DMD4B-HYDRA simulations revealed an increased oligomerization propensity of the four N-terminally truncated peptides relative to the full-length peptides, consistent with *in vitro* findings.[202] Importantly, $A\beta_{3-40}$ and $A\beta_{3-42}$ oligomer conformations displayed significantly more flexible and solvent-exposed N-termini than their full-length counterparts, whereas $A\beta_{11-40}$ and $A\beta_{11-42}$ oligomers were significantly more compact.[202] This compaction was mediated by hydrophobic V12 as verified by reducing and abolishing hydrophobicity at position 12, which resulted in significantly increased N-terminal to C-terminal distances of individual peptides within oligomers.[202,203]

As determined by PICUP/SDS-PAGE, the **Arctic mutation ([E22G])** alters *in vitro* oligomer size distributions of $A\beta_{1-40}$ and $A\beta_{1-42}$ in distinct ways.[149] [E22G]$A\beta_{1-40}$ forms more whereas [E22G]$A\beta_{1-42}$ forms less pentamers and hexamers than their respective WT peptides.[149] In addition, unlike for $A\beta_{1-42}$, no [E22G]$A\beta_{1-42}$ oligomers in the range of decamers to dodecamers were detected.[149] *In silico*, these observations were captured by the DMD4B-HYDRA approach.[188] The N-terminal region A2-F4 contributed to $A\beta_{1-40}$ oligomerization but was not involved in oligomerization of $A\beta_{1-42}$ or the two Arctic mutants.[188] The structure of [E22G]$A\beta_{1-40}$ and [E22G]$A\beta_{1-42}$ oligomers resembled the oligomer structure of $A\beta_{1-42}$ more than that of $A\beta_{1-40}$.[188] Because the shorter alloform is the more abundant alloform in the brain, [E22G]Aβ_{1-40} oligomers that are structurally similar to Aβ_{1-42} oligomers may represent the predominant toxic species in the Arctic form of FAD. These findings are consistent with experimental study of Päiviö *et al.*, who suggested that [E22G]Aβ_{1-40} plays an important role in the pathology of the Arctic form of AD.[109]

The **Flemish mutation ([A21G])** did not significantly affect Aβ_{1-40} oligomer size distribution as determined by PICUP-SDS/PAGE but enhanced formation of Aβ_{1-42} pentamers and hexamers at the expense of higher-order oligomers.[149] Interestingly, all naturally occurring mutations at positions 22 and 23 (Fig. 1) extended the *in vitro* Aβ_{1-40} oligomer size distribution to larger oligomers (up to octamers) and inhibited formation of higher-order Aβ_{1-42} oligomers (decamers through dodecamers).[149] Although no *in silico* data on oligomer formation by the Flemish, Dutch, Italian, Iowa mutants exist, it is plausible that all these mutations affect the oligomer structure of the more abundant Aβ_{1-40} in a way that increases toxicity of mutant Aβ_{1-40} oligomers. MD simulations of full-length Aβ mutants are sparse and mainly restricted to monomer folding dynamics. Lin *et al.* examined and contrasted the extent to which the Italian, Dutch, Arctic, and Iowa mutations alter ensembles of Aβ_{1-42} monomer conformations,[194] yet it is unclear how these distinct monomer characteristics affect oligomer formation.

In search for inhibitors of Aβ oligomer toxicity, Sinha *et al.* identified Lys-specific molecular tweezers, which bound to Aβ at an early assembly stage and inhibited Aβ toxicity *in vitro*, presumably by binding to lysine residues.[204] Although Aβ oligomers that assembled in the presence of molecular tweezers were of the same size and morphology as oligomers formed in the absence of this inhibitor, they appeared to be nontoxic and were not recognized by the oligomer-specific antibody.[204] To further examine the role of two lysines at positions 16 and 28 in Aβ aggregation dynamics, Sinha *et al.* studied the effect of two **single-amino-acid substitutions, [K16A] and [K28A]**, on Aβ_{1-40} and Aβ_{1-42} assembly and toxicity.[205] The results of this study revealed that the [K16A] substitution strongly inhibits toxicity of Aβ_{1-40} and Aβ_{1-42}, whereas the [K28A] substitution exerts only a moderate effect.[205] This result is intriguing because the two substitutions exert the same effect on the overall charge of Aβ_{1-40} and Aβ_{1-42}, and the amino acid content of the two mutant peptides, [K16A] Aβ and [K28A] Aβ, is the same despite distinct amino acid sequences.

Žganec *et al.* examined the effect of [K16A] and [K28A] substitutions on Aβ_{1-40} and Aβ_{1-42} oligomer formation *in silico* using the DMD4B-HYDRA approach.[206] DMD4B-HYDRA simulations revealed that structural changes to oligomers are both substitution- and alloform-specific, whereby [K28A] increased solvent exposure of hydrophobic residues and the intrapeptide N-terminal to C-terminal distance within oligomers more than [K16A].[206] Unlike [K28A], [K16A] decreased the β-strand content and enhanced tertiary contacts, resulting in more compact [K16A] $A\beta_{1-40}$ and [K16A] $A\beta_{1-42}$ oligomers relative to oligomers formed by [K28A] $A\beta_{1-40}$ and [K28A] $A\beta_{1-42}$, suggesting that oligomer compaction may reduce Aβ-oligomer mediated toxicity.[206] These findings are relevant to the recently discovered FAD-inducing Aβ mutation, **[K16N]**, which is located at the α-secretase cleavage site and affects APP processing by lowering nonpathogenic p3 and increasing $A\beta_{1-40}$ and $A\beta_{1-42}$ production.[102] Namely, *in vitro* [K16N] $A\beta_{1-40}$ and [K16N] $A\beta_{1-42}$ oligomers appear to be *less toxic* than oligomers formed by WT peptides,[102] which is consistent with reduced toxicity of [K16A] Aβ peptides,[205] possibly due to increased structural compactness of the resulting oligomer conformations.[206] Interestingly, mixtures of WT and mutant peptides produce oligomers that are even more toxic than $A\beta_{1-40}$ and $A\beta_{1-42}$ oligomers,[102] which opens up new venues for future studies.

5.5. *Aβ assembly pathways and polymorphism*

The assembly pathways of amyloid forming and disease-causing proteins need to be well understood in order to interfere with the assembly dynamics through efficient therapeutic approaches. For example, if the processes of oligomer and fibril formation are causally linked, inhibiting fibril formation may actually lead to an increased pool of toxic oligomers, which could have adverse effects on human pathology. Protein self-assembly into ordered amyloid cross-β fibrils,[4] polymers,[207,208] and microcrystals[209] are nucleation-dependent processes. This view is based on the existence of a lag time prior to the emergence of an ordered structure, and the fact that the lag phase disappears if preformed seeds are added to the protein solution. According to the classical nucleation theory, downhill polymerization is triggered by formation of the smallest ordered assembly that can template the ordered structure, a nucleus of a critical size, which is difficult to characterize both experimentally and computationally. A reevaluation of the the classical nucleation theory was proposed.[210] An alternative, two-step mechanism of protein nucleation, in which mesoscopic metastable protein clusters serve as precursors to the ordered nuclei, has been proposed.[211,212] One of the first computational studies in support of this two-step nucleation mechanism was reported by ten Wolde and Frenkel who examined homogeneous nucleation in a Lennard-Jones system of short-range attraction by Monte Carlo simulations.[213]

Consistent with a two-step nucleation mechanism, various studies employing AFM or TEM, observed formation of quasi-spherical Aβ oligomers that evolve into more elongated curvilinear beaded assemblies, protofibrils, which convert over time into long and straight fibrils.[137,146,214–217] These observations suggest that a conversion from mostly unstructured to an ordered cross-β motif occurs from within oligomers or protofibrils.[146,148] Experimental study by Ahmed *et al.* showed that Aβ_{1-42} oligomers, which lacked β-sheet content, underwent a structural conversion into β-sheets.[158] A more direct evidence for nucleated structural conversion was provided by Lee *et al.*, who used fluorescein arsenical helix binder (FlAsH), which fluoresces upon binding to tetracysteine tag, to monitor *in vitro* aggregation of Aβ_{1-40} modified by addition of two cysteine residues at the N-terminus.[218] This modified Aβ_{1-40} peptide was demonstrated to rapidly and efficiently form spherical oligomers that slowly convert into amyloid fibrils via a nucleated structural conversion mechanism.[218]

DMD4B-HYDRA simulations of Aβ self-assembly captured initial oligomer formation into a quasi-stable heterogeneous population of monomers and oligomers, which further assembled into elongated protofibril-like assemblies under aggregation-promoting conditions,[188,202] thus capturing the first stage of the proposed two-stage nucleation mechanism (Fig. 2). The second stage of the two-stage nucleation would correspond to a nucleated structural conversion within an elongated protofibril, which is by definition a rare event and thus difficult to detect in simulations. Barz and Urbanc recently examined assembly pathways using a minimal self-assembly model, in which a protein is represented by two hydrophobic and two hydrophilic beads arranged into a tetrahedron.[219] Self-assembly proceeded through formation of small globular oligomeric states, followed by formation of elongated protofibril-like assemblies, which grew larger or decreased in size either through monomer or oligomer attachment and detachment, thereby accounting for multiple coexisting assembly pathways.[219] Importantly, Barz and Urbanc detected a few rare events of a structural conversion from within a disordered elongated assembly into several distinctly ordered domains, showcasing polymorphism both in terms of distinct morphologies and multiple assembly pathways.[219] Computer simulations conducted by Pellarin *et al.* also revealed fibrillar polymorphism in a simplified model of an amyloid polypeptide, which was shown to be under kinetic control.[171] Computer simulation findings are supported by the results of a theoretical study by Schreck and Yuan, who applied a generalized version of Smoluchowski kinetics to describe protein aggregation by accounting for assembly growth and fragmentation of all possible assembly sizes, resulting in a prediction of multiple fibril formation pathways.[220]

Substantial evidence from experimental and theoretical studies thus supports the view that fibril formation is inevitably preceded by oligomer formation,[221] whereby oligomers may or may not be toxic, and that nucleated structural conversion from disordered to ordered structure can be initiated locally within a large assembly. The observation of concomitantly occurring multiple assembly pathways in simulations described above also helps explain the emergence of structural polymorphism of Aβ fibrils formed *in vitro*.[222,223]

5.6. *Are flexible N-termini key to Aβ oligomer toxicity?*

As discussed earlier, substantial evidence indicates that Aβ oligomers and not fibrils mediate toxicity. It appears that the phenomenon of oligomer toxicity extends beyond Aβ oligomers. Kayed *et al.* proposed that oligomer across a broad class of amyloid proteins share a common structure and mechanism of toxicity.[16] Aβ oligomers, particularly those formed by Aβ_{1-42}, have been shown to be at least 10-fold more toxic to neuronal cell cultures than fibrils.[224,225] Increased toxicity and propensity to deposit into amyloid plaques of Aβ_{1-42} relative to Aβ_{1-40} was established also *in vivo*, in a *Drosophila* and Tg mouse models of AD.[226,227]

Because of a strong sequence similarity (95%) between the two predominant Aβ alloforms, it is puzzling that oligomers formed by Aβ_{1-42} are significantly more toxic than Aβ_{1-40} oligomers. Which structural aspects of Aβ oligomers might be involved in mediating toxicity? One of the strategies that might help address this question is to compare and contrast ensembles of oligomer conformations formed by Aβ_{1-40} and Aβ_{1-42}. DMD4B-HYDRA simulations predicted globular Aβ_{1-40} and Aβ_{1-42} oligomer conformations with hydrophobic C-termini comprising a core and hydrophilic N-termini exposed to the solvent, whereby the N-termini are more flexible and solvent exposed in Aβ_{1-42} oligomers than in Aβ_{1-40} oligomers.[187] As such, Aβ_{1-42} oligomers might be able to interact with the membrane more avidly than Aβ_{1-40} oligomers.

Bitan *et al.* explored a series of Aβ C-terminal fragments (CTFs) as potential inhibitors of Aβ_{1-42}-induced cell culture toxicity.[228–230] DMD4B-HYDRA simulations of Aβ_{1-42} assembly in the presence of CTFs revealed that CTFs coassemble with Aβ_{1-42} to form Aβ42/CTF heterooligomers.[228,231] Aβ_{1-42} assembly was examined in the presence of three effective CTFs (Aβ_{31-42}, Aβ_{39-42}, and Aβ_{30-40}) and a control peptide Aβ_{21-30}, which did not inhibit Aβ_{1-42}-induced toxicity in cell cultures.[231] The three CTFs coassembled with Aβ_{1-42} into initially smaller heterooligomers that eventually formed a single large Aβ_{1-42}/CTF heterooligomer,

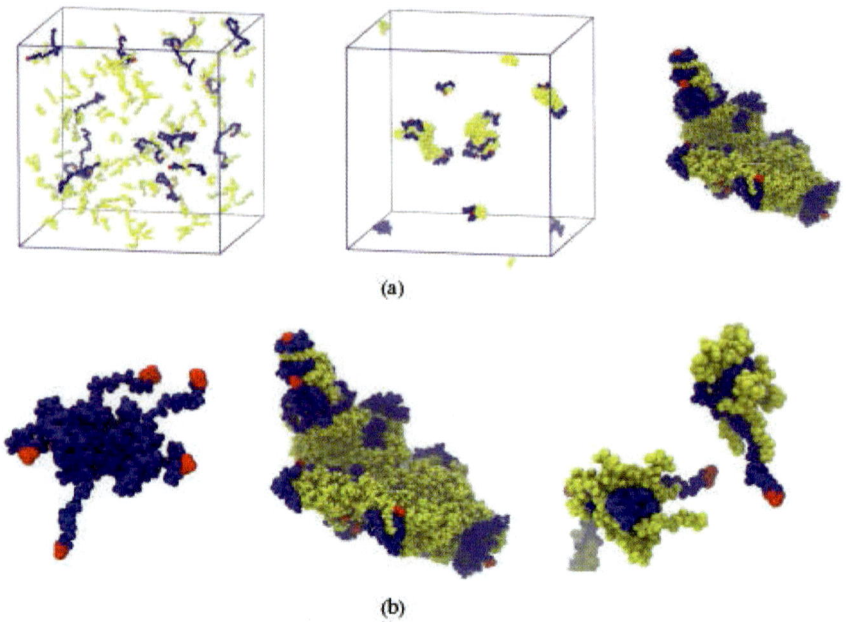

Figure 3: (a) $A\beta_{1-42}$ assembly in the presence of efficient toxicity inhibitors, $A\beta_{31-42}$, as simulated by DMD4B-HYDRA. Initially mixed ensembles of $16A\beta_{1-42}$ peptides and 128 $A\beta_{31-42}$ peptides (left panel) self-assembled into single hetero-oligomer populations (middle and right panels).[231] (b) $A\beta_{1-42}$ hexamer formed in the absence of peptide inhibitors (left panel), heterotypic assembly formed by Aβ1−42 and efficient toxicity inhibitor, Aβ31−42 (middle panel), and heterotypic assembly formed by Aβ1−42 and the control peptide Aβ21−30, which did not inhibit Aβ1−42 toxicity (right panel). The N-terminal D1 is represented by red spheres.

whereas the control peptide $A\beta_{21-30}$ coassembled with $A\beta_{1-42}$ into a population of smaller heterotypic assemblies.[231] CTFs and the control peptide displayed high binding affinities to the hydrophobic regions of $A\beta_{1-42}$, but only CTFs were able to bind the A2-F4 region of $A\beta_{1-42}$, thereby reducing the flexibility and solvent accessibility of the N-terminal region of $A\beta_{1-42}$ (Fig. 3).[231] These findings are in accord with predictions of a study by Gessel *et al.*, who examined interactions between $A\beta_{1-42}$ and CTFs using all-atom MD in conjunction with *in vitro* data obtained by IMS-MS.[166]

Several experimental findings are consistent with the proposition that flexible and solvent exposed N-termini mediate Aβ oligomer toxicity. Anti-Aβ monoclonal antibodies, which show promise in passive immunization treatment against AD, inhibit Aβ toxicity by binding to the N-terminal D1-S8 region.[232–234] Shankar *et al.*

demonstrated that soluble Aβ dimers, extracted from AD brain, disrupted the memory of learned behaviour in rats and that these deficits were rescued by coadministration of antibodies that bind to the N-terminus of Aβ.[235] Similarly, Jin *et al.* showed that Aβ dimers, derived from AD brain, mediate toxicity by inducing tau phosphorylation and that the toxic function was inhibited in the presence of monoclonal antibody that binds to the N-terminal D1, but not the antibody that binds to the C-terminus of Aβ.[236] The involvement of the N-terminal region in mediating Aβ toxicity is supported by *in vivo* studies of Luheshi *et al.*, who showed that amino acid substitutions within the N-terminal region strongly affected $A\beta_{1-42}$-mediated toxicity in a *Drosophila* model of AD.[237] Importantly, recently reported Aβ mutations, A2V and A2T, that result in either early onset AD or protect against AD, respectively, indicate a possibly critical role of the N-terminal region of Aβ in mediating toxicity.[128,132]

5.7. *How do Aβ oligomers mediate toxicity?*

5.7.1. *Synaptic failure*

If Aβ oligomers mediate toxicity, we need to know where and how they exert their toxic function. Substantial evidence supports the proposition that Aβ oligomers cause *synaptic failure*, suggesting that processes leading to eventual neurodegeneration are initiated at neuronal synapses.[144] Indeed, in Tg mouse models, density of presynaptic terminals decreased before any amyloid deposition occured[38] and was inversely correlated with soluble Aβ amounts.[39] Soluble Aβ was also shown to correlate with the loss of excitatory synapses in the proximity of amyloid plaques.[238] Low concentrations of synthetic $A\beta_{1-40}$, $A\beta_{1-42}$, $A\beta_{25-35}$, and ADDLs injected into rat hippocampal slices inhibited LTP and thus impaired synaptic plasticity, which was postulated to be the cause of memory deficits in the absence of neurodegeneration.[143,239] Aβ oligomers, extracted from the AD brain and injected into rats, inhibited LTP,[141] reduced dendritic spine density, and disrupted the memory of a learned behavior.[235] Shankar *et al.* showed that the observed spine loss was associated with a modulation of signaling pathways of the N-methyl D-aspartate (NMDA) receptor,[235,240] which also mediated sustained calcium influx in hippocampal neurons caused by oligomeric Aβ.[241]

5.7.2. *Disruption of cellular membranes*

There is no consensus with respect to the mechanism through which Aβ oligomers exert their toxic function, although Aβ-membrane interactions represent an underlying theme across the proposed mechanisms.[242] Aβ oligomers are believed to mediate their toxicity by interacting with cellular membranes, causing *increased membrane permeability*, which disrupts cellular homeostasis. The effect of Aβ on membrane permeability can be quantified by a calcein release assay, which measures the amount of fluorescent dye, calcein, released from the interior of the large unilamellar vesicles (LUVs) due to a reduced membrane permeability caused by extraneous Aβ.[243] Williams *et al.* showed that the ability to cause membrane damage was the largest for soluble Aβ at zero incubation time and then decreased with incubation time as Aβ started to form fibrils, whereby mature Aβ fibrils lacked the ability to disrupt the membrane.[243] Incorporation of negatively charged lipids (GM1 ganglioside)[244] and cholesterol[245,246] into the membrane was instrumental for Aβ binding to lipid vesicles[247] and mediating Aβ-induced membrane damage.[243,247] The ability of Aβ_{1-42} to cause membrane disruption was inhibited in the presence of europium III chloride ions, which were shown to strongly bind to GM1 ganglioside-containing LUVs and prevent their interaction with Aβ_{1-42}.[248] Importantly, soluble Aβ_{1-42} caused significantly more damage to biomimetic lipid vesicles than soluble Aβ_{1-40},[154] in agreement with distinct toxicities of Aβ_{1-40} and Aβ_{1-42} determined in cell culture[224,225] and animal model studies.[226,227]

5.7.3. *Potential toxicity mechanisms*

At least *three mechanisms of Aβ-induced membrane damage* have been proposed: (1) a "carpeting" mechanism, (2) detergent-like action, or (3) formation of ion channels or pores. In the carpeting model, the peptide adsorbs to one leaflet of the membrane surface, causing an asymmetrical pressure on the membrane that results in the leakage of small molecules.[249] In the detergent-like model, a peptide forming micelle-like structures acts as a surfactant and reduces the surface tension of the membrane by removing lipids from the bilayer, causing membrane thinning, disruption of the membrane curvature, hole formation,[249] and membrane disintegration.[250] The ion channel or pore formation models have garnered the most attention. The ion channel hypothesis of AD proposes that Aβ forms ion channels within the membrane, which disrupt calcium homeostasis, thereby damaging and/or killing neurons.[251–256] Calcium ion channels were reported to form as a result of the incorporation of Aβ_{1-40} within planar lipid membranes.[257,258] Bhatia *et al.* showed that nanomolar concentrations of globular (nonfibrillar) Aβ_{1-40}, Aβ_{1-42}, and Aβ_{25-35} induced morphological changes leading to cellular degeneration of cultured endothelial cells within minutes and that these cells were the most sensitive to Aβ_{1-42} globular oligomers.[259] The mechanism was linked to elevated calcium inside cells, presumably caused by formation of calcium-permeable $A\beta_{1-42}$ pores.[259]

5.7.4. *The ion-channel hypothesis*

Demuro *et al.* demonstrated that oligomeric (but not monomeric or fibrillar) forms of several amyloid proteins, including $A\beta_{1-42}$, applied to cell cultures, caused elevated intracellular calcium concentrations due to an increase in membrane permeability,[260] which could be explained by *the ion channel mechanism.* Similarly, Kawahara *et al.* reported formation of ion channels in neurons that resulted in elevated intracellular calcium as a generic mechanism underlying several diseases associated with aberrant protein aggregation.[261] Quist *et al.* applied AFM to visualize pore structures formed by several amyloidogenic proteins, including $A\beta_{1-40}$ and α synuclein.[262] A donut-shaped $A\beta_{1-40}$ pore with a diameter of 8–12 nm was shown to be embedded into a bilayer, whereby six hexamers were proposed to be organized into the 36-stranded β-barrel structure with the N-termini forming the pore lining.[262] These findings provide some evidence that an unregulated flux of ions, in particular calcium ions, through such ion channels (pore) into the cell may be the cause of oligomer-mediated toxicity not only in AD but in many other diseases associated with toxic oligomer formation.[263]

5.7.5. *Theoretical considerations*

Several *theoretical aspects of the Aβ-membrane* interaction have been studied by means of computer simulations. A theoretical MD study conducted by Strödel *et al.* revealed that $A\beta_{1-42}$ ion channels or pores may consist of tetrameric and hexameric β-sheet subunits.[264] Explicit-solvent MD simulations of Jang *et al.* provided evidence of a β-barrel topology of $A\beta_{1-42}$ ion channels,[265] in agreement with findings of Strödel *et al.*[264] These observations are consistent with biophysics intuition as there is no water within the lipid bilayer and thus no hydrophobic effect, which increases the importance of peptide-peptide hydrogen bonding and electrostatic interactions, resulting in increased β-strand propensity.[182,185,266] It is important to realize that *the structure of oligomers formed in water may be considerably different from the oligomer structure in a lipid bilayer*. For example, pores formed by amyloidogenic protein stefin B as captured by DMD4B-HYDRA solvent parameters, optimized for aqueous solutions, did not display significant β-strand, β-sheet, or β-barrel content.[267] Clearly, the process of insertion of Aβ oligomers into the membrane needs to be examined alongside the concomitant conformational changes within the oligomers to understand how Aβ-membrane interactions might contribute to cell toxicity. MD simulations by Jang *et al.* demonstrate that the interaction between Aβ oligomers and membrane is nonspecific and that several membrane-inserting pathways might coexist, depending on oligomer order and pre-existing structural motifs.[268] It is plausible that Aβ-membrane interactions lead to different toxicity mechanisms, depending on the membrane composition, cell type, and local solvent conditions, which govern Aβ assembly dynamics and the resulting oligomer conformations.[242,269]

6. Emerging Understanding of Aβ Assembly Dynamics and Its Role in AD

Aβ oligomers have been synonymous with toxicity and regarded as highly undesirable for several decades. Is Aβ oligomer formation and subsequent fibrillization always pathological? To find the answer to this question, we need to establish whether or not Aβ has a physiological function. It is convenient to view Aβ simply as an undesirable by-product of APP processing without any physiological function and hope that inhibition of Aβ production would abolish neurodegeneration associated with AD. Recent research findings challenge such a view. Synthetic Aβ was shown to exhibit potent antimicrobial activity *in vitro*.[270] In addition to antimicrobial activity, Aβ_{1-40} and even more so Aβ_{1-42} protected against replication of seasonal and pandemic strains of influenza A virus in cell cultures.[271] A similar antiviral activity of Aβ_{1-40} and Aβ_{1-42} against herpes simplex virus, which colocalizes with amyloid plaques *in vivo*, was recently reported.[272] Using Tg *Cenorhabditis elegans* and Tg mouse models, Kumar *et al.* recently showed that Aβ mediates its protective activity against microbial cells by forming oligomers, which bind to the surface of microbes and prevent their attachment to host cells, and fibrils, which encapsulate and trap microbial cells until immune cells kill them off.[273] These novel findings offer a complete change of perspective with regard to Aβ aggregation, which may be, at least to some extent, a normal process in the brain. Importantly, if this is correct, ongoing therapeutic efforts targeting Aβ oligomers need to be re-assessed, making the problem of finding a cure against AD even more challenging. Although Aβ oligomers may exert their normal function by preventing intruding microbial cells to attach to host cells,[273] a rapid increase in their abundance may cause a saturation of their normal activity and cause their presumably toxic interaction with synapses, thereby triggering AD pathology.

A rapid increase can be caused by any process that stabilizes Aβ oligomers and inhibits their conversion into less toxic fibrils. Williams *et al.* recently reported that *in vitro* cross-linking stabilizes Aβ_{1-40} and Aβ_{1-42} oligomers and significantly delays their conversion into fibrils.[154] The same study showed that cross-linked oligomers cause prolonged damage to biomimetic lipid vesicles,[154] consistent with the more toxic nature of oligomers relative to fibrils. Importantly, cross-linking process by CHICUP is facilitated by copper ions and hydrogen peroxide and, unlike PICUP, does not require any photo activation. Copper ions are normally present in the brain and hydrogen peroxide is the predominant reactive oxidative species, linked to the oxidative stress hypothesis of AD.[274–277] $A\beta_{1-42}$ was reported to have a particularly strong affinity to bind copper,[278] and copper metabolism is known to be strongly affected in AD,[279] causing copper to accumulate at high concentrations of up to $\sim$400 μM in amyloid plaques.[280–282] When in a complex with Aβ, Cu^{2+} reduction to Cu^{+} in the presence of molecular oxygen can catalyze production of hydrogen peroxide through a Fenton-like reaction, resulting in a highly reactive hydroxyl radical with the ability to modify carbonyl groups in proteins,[283] which leads to formation of dityrosine bonds.[155,156,284] It is thus reasonable to posit that cross-linking of Aβ oligomers through CHICUP can occur in human brain. These findings require a *modified amyloid hypothesis* stating that an accumulation of natively cross-linked Aβ oligomers, which are "trapped" in the oligomeric state and cannot rapidly convert into fibrils, represent the proximate toxic entities that initiate the AD pathology (Fig. 4). The modified amyloid hypothesis is a synthesis of three prevalent hypotheses of AD: (1) the amyloid cascade hypothesis that implicates toxic Aβ oligomers in triggering AD neurodegeneration,[50] (2) the metal hypothesis stating that interactions of Aβ with metals, such as copper and iron, cause the disease,[285] and (3) oxidative stress hypothesis positing that free radical formation initiates AD pathology.[286,287]

This chapter reviews the multifaceted aspects of Aβ oligomer formation and its role in AD. Yet the story of what causes neurodegeneration in AD does not end with Aβ oligomers. In a recent review, Musiek and Holtzman referred to the "tri-dimensional" nature of the amyloid cascade hypothesis, noting that Aβ acts primarily as a trigger of other downstream processes, particularly tau aggregation, and that Aβ, while necessary, is not sufficient to produce massive neurodegeneration observed in AD.[288] It is well known that aberrant aggregation of tau into intracellular neurofibrillary tangles is associated tau hyperphosphorylation.[289] The causal link between Aβ and tau was observed by Jin *et al.* who found that Aβ dimers, derived from the AD brain, induced tau hyperphosphorylation in hippocampal neurons, eventually causing neurodegeneration.[236] Using a Tg mouse models, Ittner *et al.* reported that Aβ oligomer toxicity is mediated by a dendritic function of tau and that tau-deficient Tg mice do not display memory deficits.[290] Aβ oligomers were argued to initiate tau missorting through a specific signaling event in neurons, an activation of the Src kinase Fyn.[290] The missing link between Aβ oligomer-induced

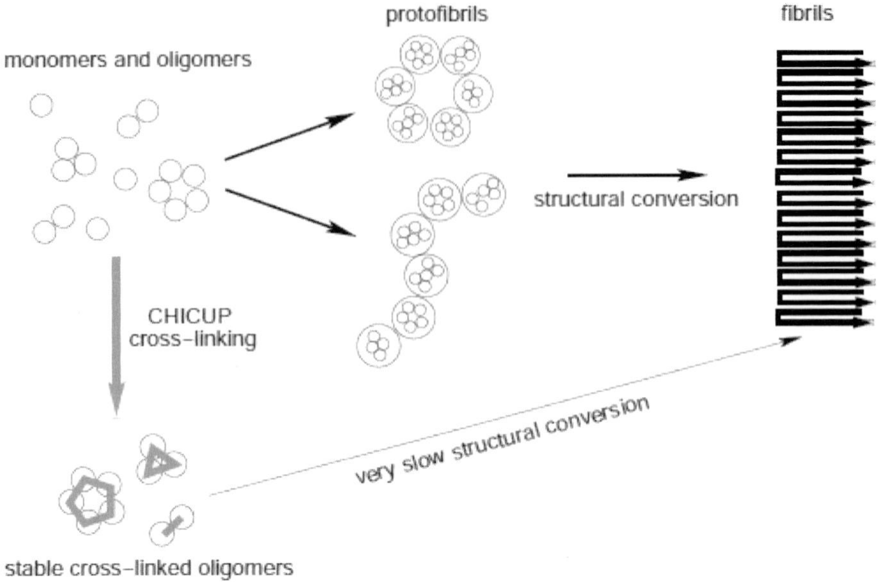

Figure 4: Schematics of Aβ assembly pathways. Soluble Aβ, which is a heterogeneous mixture of monomers and LMW oligomers, forms elongated curvilinear protofibrils of different morphologies, including pore-like structures, which undergo structural conversion into cross-β fibrils. Covalent cross-linking in the presence of copper and hydrogen peroxide stabilizes LMW oligomers and mitigates their conversion into fibrils.

synaptic and the dendritic function of tau was proposed to be the NMDA receptor, which acts as a substrate for Fyn activation.[290] Although several studies implicate Aβ oligomers in triggering aberrant tau aggregation, the causal link has not been adequately demonstrated in humans thus far. Despite their multifaceted nature with respect to their polymorphic structure and normal versus toxic function, Aβ oligomers remain the culprits underlying the molecular basis of AD.

References

1. Chiti, F. and Dobson, C. Protein misfolding, functional amyloid, and human disease. *Annu. Rev. Biochem.* **75**, 333–366 (2006).
2. Anderson, V. and Lekkerkerker, H. Insights into phase transition kinetics from colloid science. *Nature*. **416**(6883), 811–815 (2002).
3. Stradner, A., Sedgwick, H., Cardinaux, F., Poon, W., Egelhaaf, S., and Schurtenberger, P. Equilibrium cluster formation in concentrated protein solutions and colloids. *Nature*. **432**(7016), 492–495 (2004).

4. Serpell, L. C., Fraser, P. E., and Sunde, M. X-ray fiber diffraction of amyloid fibrils. *Methods. Enzymol.* **309**, 526–36 (1999).
5. Tycko, R. Solid-State NMR Studies of Amyloid Fibril Structure. Eds. Leone, S. R. and Cremer, P. S. and Groves, J. T. and Johnson, M. A., *Annu. Rev. Phys. Chem.* **62**, 279–299 (2011).
6. Tycko, R. and Wickner, R. B. Molecular structures of amyloid and prion fibrils: Consensus versus controversy. *Acc. Chem. Res.* **46**(7), 1487–1496 (2013).
7. Tycko, R. Amyloid polymorphism: Structural basis and neurobiological relevance. *Neuron.* **86**(3), 632–645 (2015).
8. Wright, P. and Dyson, H. Intrinsically unstructured proteins: Re-assessing the protein structure-function paradigm. *J. Mol. Biol.* **293**(2), 321–331 (1999).
9. Uversky, V., Gillespie, J., and Fink, A. Why are "natively unfolded" proteins unstructured under physiologic conditions? *Proteins: Struct., Funct., Genet.* **41**(3), 415–427 (2000).
10. Uversky,V. What does it mean to be natively unfolded? *Eur. J. Biochem.* **269**(1), 2–12 (2002).
11. Dunker, A., Lawson, J., Brown, C., Williams, R., Romero, P., Oh, J., Oldfield, C., Campen, A., Ratliff, C., Hipps, K., Ausio, J., Nissen, M. Reeves, R., Kang, C., Kissinger, C., Bailey, R., Griswold, M., Chiu, M.,Garner, E., and Obradovic, Z. Intrinsically disordered protein. *J. Mol. Graph. & Mod.* **19**(1), 26–59 (2001).
12. Dyson,H. and Wright, P. Intrinsically unstructured proteins and their functions. *Nat. Rev. Mol. Cell Biol.* **6**(3), 197–208 (2005).
13. Uversky, V. Natively unfolded proteins: A point where biology waits for physics. *Protein Sci.* **11**(4), 739–756 (2002).
14. Walsh, D. M., and Selkoe, D. J. Oligomers on the brain: The emerging role of soluble protein aggregates in neurodegeneration, *Protein Pept. Lett.* **11**(3), 213–228 (2004).
15. Fändrich, M. Oligomeric intermediates in amyloid formation: Structure determination and mechanisms of toxicity. *J. Mol. Biol.* **421**(4–5, 2, SI), 427–440 (2012).
16. Kayed, R., Head, E., Thompson, J. L., McIntire, T. M., Milton, S. C., Cotman, C. W., and Glabe, C. G. Common structure of soluble amyloid oligomers implies common mechanisms of pathogenesis. *Science.* **300**, 486–489 (2003).
17. Coles, M., Bicknell, W., Watson, A. A., Fairlie, D. P., and Craik,D. J. Solution structure of amyloid β-peptide(1-40) in a water-micelle environment. Is the membrane-spanning domain where we think it is? *Biochemistry.* **370**(31), 11064–11077 (1998).
18. Zhang, S., Iwata, K., Lachenmann, M. J., Peng, J. W., Li, S., Stimson, E. R., Lu, Y. Felix, A. M., Maggio, J. E., and Lee, J. P. The Alzheimer's peptide Aβ adopts a collapsed coil structure in water. *J. Struct. Biol.* **130**(2–3), 130–141 (2000).
19. Crescenzi, O., Tomaselli, S., Guerrini, R., Salvadori, S., D'Ursi, A. M., Temussi, P. A., and Picone, D. Solution structure of the Alzheimer amyloid β-peptide (1-42) in an apolar microenvironment: Similarity with a virus fusion domain. *Eur. J. Biochem.* **269**(22), 5642–5648 (2002).
20. Cipriani, G., Dolciotti, C., Picchi, L., and Bonuccelli, U. Alzheimer and his disease: A brief history, *Neurol. Sci.* **32**(2), 275–279 (2011).
21. Selkoe, D. J. Alzheimer's disease: Genes, proteins, and therapy. *Physiol. Rev.* **81**(2), 741–766 (2001).

22. Glenner, G. G. and Wong, C. W. Alzheimer's disease and Down's syndrome: Sharing of a unique cerebrovascular amyloid fibril protein. *Biochem. Biophys. Res. Commun.* **122**, 1131–1135 (1984).
23. Glenner, G. G. and C. W. Wong, C. W. Alzheimer's disease: Initial report of the purification and characterization of a novel cerebrovascular amyloid protein. *Biochem. Biophys. Res. Commun.* **20**, 885–890 (1984).
24. Goedert, M., Wischik, C., Crowther, R., Walker, J., and Klug, A. Cloning and sequencing of the cDNA-encoding a core protein of the paired helical filament of Alzheimer-disease: Identification as the microtubule-associated protein tau. *Proc. Natl. Acad. Sci. USA.* **85**(11), 4051–4055 (1988).
25. Ballatore, C., Lee, V. M. Y., and Trojanowski, J. Q. Tau-mediated neurodegeneration in Alzheimer's disease and related disorders. *Nat. Rev. Neurosci.* **8**(9), 663–672 (2007).
26. Serrano-Pozo, A., Frosch, M. P., Masliah, E., and Hyman, B. T. Neuropathological alterations in Alzheimer Disease. *Cold Spring Harb. Perspect. Med.* **1**(1), a006189 (2011).
27. Mandelkow, E. M. and Mandelkow, E. Biochemistry and Cell Biology of Tau Protein in Neurofibrillary Degeneration. *Cold Spring Harb. Perspect. Med.* **2**(7), a006247 (2012).
28. Gómez-Isla, T., Hollister, R., West, H., Mui, S., Growdon, J., Petersen, R., Parisi, J., and Hyman, B. Neuronal loss correlates with but exceeds neurofibrillary tangles in Alzheimer's disease. *Ann. Neurol.* **41**(1), 17–24 (1997).
29. Price, D. and Sisodia, S. Mutant genes in familial Alzheimer's disease and transgenic models. *Annu. Rev. Neurosci.* **21**, 479–505 (1998).
30. Hardy, J. A. and Higgins, G. A. Alzheimer's disease: The amyloid cascade hypothesis., *Science.* **256**, 184–185 (1992).
31. Knowles, R. B., Wyart, C., Buldyrev, S.V., Cruz, L., Urbanc, B., Hasselmo, M. E., Stanley, H. E., and Hyman B. T. Plaque-induced neurite abnormalities: implications for disruption of neural networks in Alzheimer's disease. *Proc. Natl. Acad. Sci. USA.* **96**(9), 5274–5279 (1999).
32. Le, R., Cruz, L., Urbanc, B., Knowles, R., Hsiao-Ashe, K., Duff, K., Irizarry, M., Stanley, H., and Hyman, B. T. Plaque-induced abnormalities in neurite geometry in transgenic models of Alzheimer disease: Implications for neural system disruption. *J. Neuropath. Exp. Neuro.* **60**, 753–758 (2001).
33. Urbanc, B., Cruz, L., Le, R., Sanders, J., Ashe, K. H., Duff, K., Stanley, H. E., Irizarry, M. C., and Hyman, B. T. Neurotoxic effects of thioflavin S-positive amyloid deposits in transgenic mice and Alzheimer's disease. *Proc. Natl. Acad. Sci. U. S. A.* **99**(22), 13990–13995 (2002).
34. Kuo, Y. M., Emmerling, M. R., Vigo-Pelfrey, C., Kasunic, T. C., Kirkpatrick, J. B., Murdoch, G. H., Ball, M. J., and Roher, A. E. Water-soluble Aβ (N-40, N-42) oligomers in normal and Alzheimer disease brains. *J. Biol. Chem.* **271**(8), 4077–4081 (1996).
35. Lue, L. F., Kuo, Y. M., Roher, A. E., Brachova, L., Shen, Y., Sue, L., Beach, T., Kurth, J. H., Rydel, R. E., and Rogers, J. Soluble amyloid β peptide concentration as a predictor of synaptic change in Alzheimer's disease. *Am. J. Pathol.* **155**(3), 853–862 (1999).
36. McLean, C. A., Cherny, R. A., Fraser,F. W., Fuller, S. J., Smith, M. J., Beyreuther, K., Bush, A. I., and Masters C. L., Soluble pool of Aβ amyloid as a determinant of severity of neurodegeneration in Alzheimer's disease. *Ann. Neurol.* **46**(6), 860–866 (1999).

37. Wang, J., Dickson, D., Trojanowski, J., and Lee, V. The levels of soluble versus insoluble brain Aβ distinguish Alzheimer's disease from normal and pathologic aging, *Exp. Neurol.* **158**(2), 328–337 (1999).
38. Hsia, A. Y., Masliah, E., McConlogue, L., Yu, G. Q., Tatsuno, G., Hu, K., Kholodenko, D., Malenka, R. C., Nicoll, R. A., and Mucke, L. Plaque-independent disruption of neural circuits in Alzheimer's disease mouse models. *Proc. Natl. Acad. Sci. USA.* **96**, 3228–3233 (1999).
39. Mucke, L., Masliah, E., Yu, G. Q., Mallory, M., Rockenstein, E. M., Tatsuno, G., Hu, K., Kholodenko, D., Johnson-Wood, K., and McConlogue, L. High-level neuronal expression of Aβ(1-42) in wild-type human amyloid protein precursor transgenic mice: Synaptotoxicity without plaque formation. *J. Neurosci.* **20**(11), 4050–4058 (2000).
40. Dodart, J. C., Bales, K. R., Gannon, K. S., DeMattos, S. J. G. R. B., Mathis, C., DeLong, C. A., Wu, S., Wu, X., Holtzman, D. M., and Paul, S. M. Immunization reverses memory deficits without reducing brain Aβ burden in Alzheimer's disease model. *Nat. Neurosci.* **5**, 452–457 (2002).
41. Kirkitadze, M. D., Bitan, G., and Teplow, D. B., Paradigm shifts in Alzheimer's disease and other neurodegenerative disorders: The emerging role of oligomeric assemblies. *J. Neurosci. Res.* **69**(5), 567–577 (2002).
42. Klein, W. L. ADDLs & protofibrils—the missing links? *Neurobiol. Aging.* **23**, 231–233 (2002).
43. Hardy, J. Alzheimer's disease: Genetic evidence points to a single pathogenesis. *Ann Neuro.* **54**(2), 143–144 (2003).
44. Klein, W. L., Stine, W. B., and Teplow, D. B. Small assemblies of unmodified amyloid β protein are the proximate neurotoxin in Alzheimer's disease. *Neurobiol. Aging.* **25**(5), 569–580 (2004).
45. Glabe, C. G. Amyloid accumulation and pathogensis of Alzheimer's disease: Significance of monomeric, oligomeric and fibrillar Aβ. *Subcell. Biochem.* **38**, 167–177 (2005).
46. Walsh, D. and Selkoe, D. Aβ Oligomers—A decade of discovery. *J. Neurochem.* **101**, 1172–1184 (2007).
47. Selkoe, D. Soluble oligomers of the amyloid β-protein impair synaptic plasticity and behavior. *Behav. Brain Res.* **192**, 106–113 (2008).
48. Roychaudhuri, R., Yang, M., Hoshi, M. M., and Teplow, D. B. Amyloid β-protein assembly and Alzheimer disease. *J. Biol. Chem.* **284**, 4749–53. (2008).
49. Rahimi, A., Shanmugam, A., and Bitan, G. Structure–function relationships of prefibrillar protein assemblies in Alzheimer's disease and related disorders. *Curr. Alzheimer Res.* **5**, 319–341 (2008).
50. Hardy, J. and Selkoe, D. J. The amyloid hypothesis of Alzheimer's disease: Progress and problems on the road to therapeutics. *Science.* **297**(5580), 353–356 (2002).
51. Hardy, J. The amyloid hypothesis for Alzheimer's disease: a critical reappraisal. *J. Neurochem.* **110**,1129–1134 (2009).
52. Haass, C., Kaether, C., Thinakaran, G., and Sisodia, S. Trafficking and proteolytic processing of APP. *Cold Spring Harb. Perspect. Med.* **2**(5), a006270 (2012).
53. Zhang, Y. W., Thompson, R., Zhang, H., and Xu, H. APP processing in Alzheimer's disease. *Mol. Brain.* **4**, 3 (2011).

54. Nalivaeva, N. N. and Turner, A. J. The amyloid precursor protein: A biochemical enigma in brain development, function and disease. *FEBS Lett.* **587**(13, SI), 2046–2054 (2013).
55. Kang, J., Lemaire, H. G., Unterbeck, A., Salbaum, J. M., Masters, C. L., Grzeschik, K. H., Multhaup, G., Beyreuther, K., and Muller-Hill, B.The precursor of Alzheimer's disease amyloid A4 protein resembles a cell-surface receptor. *Nature.* **325**, 733–736 (1987).
56. Gu, Y., Misonou, H., Sato, T., Dohmae, N., Takio, K., and Ihara, Y. Distinct intramembrane cleavage of the β-amyloid precursor protein family resembling γ-secretase-like cleavage of notch. *J. Biol. Chem.* **276**(38), 35235–35238 (2001).
57. Sastre, M., Steiner, H., Fuchs, K., Capell, A., Multhaup, G., Condron, M., Teplow, D., and Haass, C. Presenilin dependent γ-secretase processing of β-amyloid precursor protein at a site corresponding to the S3 cleavage of notch. *EMBO Rep.* **2**(9), 835–841 (2001).
58. Weidemann, A., Eggert, S., Reinhard, F., Vogel, M., Paliga, K., Baier, G., Masters, C., Beyreuther, K., and Evin, G. A novel ε-cleavage within the transmembrane domain of the Alzheimer amyloid precursor protein demonstrates homology with notch processing. *Biochemistry.* **41**(8), 2825–2835 (2002).
59. Sato, T., Dohmae, N., Qi, Y., Kakuda, N., Misonoum, H., Mitsumori, R., Maruyama, H., Koo, E., Haass, C., Takio, K., Morishima-Kawashima, M., Ishiura, S., and Ihara, Y. Potential link between amyloid β-protein 42 and C-terminal fragment γ 49-99 of β-amyloid precursor protein. *J. Biol. Chem.* **278**(27), 24294–24301 (2003).
60. Funamoto, S., Morshima-Kawashima, M., Tanimura, Y., Hirotani, N., Saido, T.,and Ihara, Y. Truncated carboxyl-terminal fragments of β-amyloid precursor protein are processed to amyloid β-proteins 40 and 42. *Biochemistry.* **43**(42), 13532–13540 (2004).
61. Qi-Takahara, Y., Morishima-Kawashima, M., Tanimura, Y., Dolios, G., Hirotani, N., Horikoshi, Y., Kametani, F., Maeda, M., Saido, T., Wang, R., and Ihara, Y. Longer forms of amyloid β protein: Implications for the mechanism of intramembrane cleavage by γ-secretase. *J. Neurosci.* **25**(2), 436–445 (2005).
62. Kakuda, N., Funamoto, S., Yagishita, S., Takami, M., Osawa, S., Dohmae, N., and Ihara, Y. Equimolar production of amyloid β-protein and amyloid precursor protein intracellular domain from β-carboxyl-terminal fragment by β-secretase. *J. Biol. Chem.* **281**(21), 14776–14786 (2006).
63. Takami, M., Nagashima, Y., Sano, Y., Ishihara, S., Morishima-Kawashima, M., Funamoto, S., and Ihara, Y. γ-secretase: Successive tripeptide and tetrapeptide release from the transmembrane domain of β-carboxyl terminal fragment. *J. Neurosci.* **29**, 13042–13052 (2009).
64. Saito, T., Suemoto, T., Brouwers, N., Sleegers, K., Funamoto, S., Mihira, N., Matsuba, Y., Yamada, K., Nilsson, P., Takano, J., Nishimura, M., Iwata, N., Van Broeckhoven, C., Ihara, Y., and Saido, T. C. Potent amyloidogenicity and pathogenicity of Aβ43. *Nat. Neurosci.* **14**(8), 1023–U120 (2011).
65. Lauridsen, C., Sando, S., Shabnam, A., Moller, I., Berge, G., Grontvedt, G., Bakken, I., Salvesen, O., Brathen, G., and White, L. Cerebrospinal fluid levels of amyloid β1-43 in patients with amnestic mild cognitive impairment or early Alzheimer's disease: A 2-year follow-up study. *Front. Aging Neurosci.* **8**, 30. (2016).

66. De Strooper, B., Vassar, R., and Golde, T. The secretases: Enzymes with therapeutic potential in Alzheimer disease. *Nat. Rev. Neurol.* **6**(2), 99–107 (2010).
67. Vassar, R., Bennett, B., Babu-Khan, S., Kahn, S., Mendiaz, E., Denis, P., Teplow, D., Ross, S., Amarante, P., Loeloff, R., Luo, Y., Fisher, S., Fuller, L., Edenson, S., Lile, J., Jarosinski, M., Biere, A., Curran, E., Burgess, T., Louis, J., Collins, F., Treanor, J., Rogers, G., and Citron, M. β-Secretase cleavage of Alzheimer's amyloid precursor protein by the transmembrane aspartic protease BACE. *Science.* **2860**(5440), 735–741 (1999).
68. Huse, J., Liu, K., Pijak, D., Carlin, D., Lee, V., and Doms, R. β-Secretase processing in the trans-Golgi network preferentially generates truncated amyloid species that accumulate in Alzheimer's disease brain. *J. Biol. Chem.* **277**(18), 16278–16284 (2002).
69. Näslund, J., Schierhorn, A., Hellman, U., Lannfelt, L., Roses, A. D., Tjernberg, L. O., Silberring, J., Gandy, S. E., Winblad, B., and Greengard, P. Relative abundance of Alzheimer Aβ amyloid peptide variants in Alzheimer disease and normal aging. *Proc. Natl. Acad. Sci. U. S. A.* **91**(18), 8378–8382 (1994).
70. Mori, H., Takio, K., Ogawara, M., and Selkoe, D. J. Mass spectrometry of purified amyloid β protein in Alzheimer's disease. *J. Biol. Chem.* **267**(24), 17082–6 (1992).
71. Saido, T. C., Iwatsubo, T., Mann, D. M., Shimada, H., Ihara, Y., and Kawashima, S. Dominant and differential deposition of distinct β-amyloid peptide species, $A\beta_{N3(pE)}$, in senile plaques. *Neuron.* **14**(2), 457–466 (1995).
72. Saido, T. C., W. Yamao-Harigaya, W., T. Iwatsubo, T., and S. Kawashima, S. Amino- and carboxyl-terminal heterogeneity of β-amyloid peptides deposited in human brain. *Neurosci. Lett.* **215**(3), 173–176 (1996).
73. Russo, C., Saido, T. C., DeBusk, L. M., Tabaton, M., Gambetti, P., and Teller, J. K. Heterogeneity of water-soluble amyloid β-peptide in Alzheimer's disease and Down's syndrome brains. *FEBS Lett.* **409**(3), 411–416 (1997).
74. Harigaya, Y., Saido, T. C., Eckman, C. B., Prada, C. M., Shoji, M., and Younkin, S. G. Amyloid β protein starting pyroglutamate at position 3 is a major component of the amyloid deposits in the Alzheimer's disease brain. *Biochem. Biophys. Res. Commun.* **276**(2), 422–427 (2000).
75. Sergeant, N., Bombois, S., Ghestem, A., Drobecq, H., Kostanjevecki, V., Missiaen, C., Wattez, A., David, J. P., Vanmechelen, E., Sergheraert, C., and Delacourte, A. Truncated β-amyloid peptide species in pre-clinical Alzheimer's disease as new targets for the vaccination approach. *J. Neurochem.* **85**(6), 1581–1591 (2003).
76. Güntert, A., Döbeli, H., and Bohrmann, B. High sensitivity analysis of amyloid-β peptide composition in amyloid deposits from human and PS2APP mouse brain. *Neuroscience.* **143**(2), 461–475 (2006).
77. Liu, K., Solano, I., Mann, D., Lemere, C., Mercken, M., Trojanowski, J., and Lee, V. Characterization of Aβ1-40/42 peptide deposition in Alzheimer's disease and young Down's syndrome brains: Implication of N-terminally truncated Aβ species in the pathogenesis of Alzheimer's disease. *Acta Neuropathol.* **112**, 163–174 (2006).
78. Tanzi, R. E. The genetics of Alzheimer Disease. *Cold Spring Harb. Perspect. Med.* **2**(10) (2012).
79. Piaceri, I., Nacmias, B., and Sorbi, S. Genetics of familial and sporadic Alzheimer's disease. *Front. Biosci. (Elite Ed.).* **5**, 167–177 (2013).

80. Jiang, Q., Lee, C. Y. D., Mandrekar, S., Wilkinson, B., Cramer, P., Zelcer, N., Mann, K., Lamb, B., Willson, T. M., Collins, J. L., Richardson, J. C., Smith, J. D., Comery, T. A., Riddell, D., Holtzman, D. M., Tontonoz, P., and Landreth, G. E. ApoE promotes the proteolytic degradation of Aβ, *Neuron.* **58**(5), 681–693 (2008).
81. Liu, C. C., Kanekiyo, T., Xu, H., and Bu, G. Apolipoprotein E and Alzheimer disease: risk, mechanisms and therapy. *Nat. Rev. Neurol.* **9**(2), 106–118 (2013).
82. Attar, A., Meral, D., Urbanc, B., and Bitan, G. *Bio-nanoimaging: Protein Misfolding and Aggregation: Assembly of Amyloid β-Protein Variants Containing Familial Alzheimer's Disease-Linked Amino Acid Substitutions*, Chapter 38. Elsevier; 2014. Amsterdam, The Netherlands. pp. 429–442. .
83. Iwata, N., Tsubuki, S., Takaki, Y., Watanabe, K., Sekiguchi, M., Hosoki, E., Kawashima-Morishima, M., Lee, H. J., Hama, E., Sekine-Aizawa, Y., and Saido, T. C. Identification of the major Aβ(1–42)-degrading catabolic pathway in brain parenchyma: Suppression leads to biochemical and pathological deposition. *Nat. Med.* **6**(2), 143–150 (2000).
84. Tsubuki, S., Takaki, Y., and Saido, T. Dutch, Flemish, Italian, and Arctic mutations of APP and resistance of Aβ to physiologically relevant proteolytic degradation. *Lancet.* **361**(9373), 1957–1958 (2003).
85. Obici, L., Demarchi, A., de Rosa, G., Bellotti, V., Marciano, S., Donadei, S., Arbustini, E., Palladini, G., Diegoli, M., Genovese, E., Ferrari, G., Coverlizza, S., and Merlini, G. A novel AβPP mutation exclusively associated with cerebral amyloid angiopathy. *Ann. Neurol.* **58**(4), 639–644 (2005).
86. Fossati, S., Cam, J., Meyerson, J., Mezhericher, E., Romero, I. A., Couraud, P. O., Weksler, B. B., Ghiso, J., and Rostagno, A. Differential activation of mitochondrial apoptotic pathways by vasculotropic amyloid-β variants in cells composing the cerebral vessel walls. *FASEB J.* **24**(1), 229–241 (2010).
87. Mullan, M., Crawford, F., Axelman, K., Houlden, H., Lilius, L., Winblad, B., and Lannfelt, L. A pathogenic mutation for probable Alzheimer's-disease in the APP gene at the N-terminus of β-amyloid. *Nat. Genet.* **1**(5), 345–347 (1992).
88. Citron, M., Oltersdorf, T., Haass, C., McConlogue, L., Hung, A. Y., Seubert, P., Vigo-Pelfrey, C., Lieberburg, I., and Selkoe, D. J. Mutation of the β-amyloid precursor protein in familial Alzheimer's disease increases β-protein production. *Nature.* **360**(6405), 672–674 (1992).
89. Johnston, J., Cowburn, R., Norgren, S., Wiehager, B., Venizelos, N., Winblad, B., Vigo-Pelfrey, C., Schenk, D., Lannfelt, L., and ONeill, C. Increased β-amyloid release and levels of amyloid precursor protein (APP) in fibroblast cell-lines from family members with the Swedish Alzheimer's-disease APP670/671 mutation. *FEBS Lett.* **354**(3), 274–278 (1994).
90. Citron, M., C. Vigo-Pelfrey, C., D. Teplow, D., C. Miller, C., D. Schenk, D., J. Johnston, J., B. Winblad, B., N. Venizelos, N., L. Lannfelt, L., and D. Selkoe, D. Excessive production of amyloid β-protein by peripheral cells of symptomatic and presymptomatic patients carrying the Swedish familial Alzheimer-disease mutation. *Proc. Natl. Acad. Sci. U. S. A.* **91**(25), 11993–11997 (1994).
91. Scheuner, D., Eckman, C., Jensen, M., Song, X., Citron, M., Suzuki, N., Bird, T. D., Hardy, J., Hutton, M., Kukull, W., Larson, E., LevyLahad, E., Viitanen, M., Peskind, E., Poorkaj, P., Schellenberg, G., Tanzi, R., Wasco, W., Lannfelt, L.,

Selkoe, D., and Younkin, S. Secreted amyloid β-protein similar to that in the senile plaques of Alzheimer's disease is increased in vivo by the presenilin 1 and 2 and APP mutations linked to familial Alzheimer's disease. *Nat. Med.* **2**, 864–870 (1996).

92. Kumar-Singh, S., De Jonghe, C., Cruts, M., Kleinert, R., Wang, R., Mercken, M., De Strooper, B., Vanderstichele, H., Lofgren, A., Vanderhoeven, I., Backhovens, H., Vanmechelen, E., Kroisel, P., and Van Broeckhoven, C. Nonfibrillar diffuse amyloid deposition due to a γ_{42}-secretase site mutation points to an essential role for N-truncated $A\beta_{42}$ in Alzheimer's disease. *Hum. Mol. Genet.* **9**(18), 2589–2598 (2000).
93. Edwards-Lee, T., Ringman, J., Chung, J., Werner, J., Morgan, A., Hyslop, P., Thompson, P., Dutton, R., Mlikotic, A., Rogaeva, E., and Hardy, J. An African American family with early-onset Alzheimer disease and an APP (T714I) mutation. *Neurology.* **64**(2), 377–379 (2005).
94. Pasalar, P., Najmabadi, H., Noorian, A., Moghimi, B., Jannati, A., Soltanzadeh, A., Krefft, T., Crook, R., and Hardy, J. An Iranian family with Alzheimer's disease caused by a novel APP mutation (Thr714Ala). *Neurology.* **58**(10), 1574–1575 (2002).
95. Zekanowski, C., Styczynska, M., Peplonska, B., Gabryelewicz, T., Religa, D., Ilkowski, J., Kijanowska-Haladyna, B., Kotapka-Minc, S., Mikkelsen, S., Pfeffer, A., Barczak, A., Luczywek, E., Wasiak, B., Chodakowska-Zebrowska, M., Gustaw, K., Laczkowski, J., Sobow, T., Kuznicki, J., and Barcikowska, M. Mutations in presenilin 1, presenilin 2 and amyloid precursor protein genes in patients with early-onset Alzheimer's disease in Poland. *Exp. Neurol.* **184**(2), 991–996 (2003).
96. Lindquist, S. G., Nielsen, J. E., Stokholm, J., Schwartz, M., Batbayli, M., Ballegaard, M., Erdal, J., Krabbe, K., and Waldemar, G. Atypical early-onset Alzheimer's disease caused by the Iranian APP mutation. *J. Neurol. Sci.* **268**(1–2), 124–130 (15, 2008).
97. Ancolio, K., Dumanchin, C., Barelli, H., Warter, J., Brice, A., Campion, D., Frebourg, T., and Checler, F. Unusual phenotypic alteration of β amyloid precursor protein (β APP) maturation by a new Val-715 Met β APP-770 mutation responsible for probable early-onset Alzheimer's disease. *Proc. Natl. Acad. Sci. U. S. A.* **96**(7), 4119–4124 (1999).
98. Park, H. K., Na, D. L., Lee, J. H., Kim, J. W., and Ki, C. S. Identification of PSEN1 and APP gene mutations in Korean patients with early-onset Alzheimer's disease. *J. Korean Med. Sci.* **23**(2), 213–217 (2008).
99. Cruts, M., Dermaut, B., Rademakers, R., Van den Broeck, M., Stogbauer, F., and Van Broeckhoven, C. Novel APP mutation V715A associated with presenile Alzheimer's disease in a German family. *J. Neurol.* **250**(11), 1374–1375 (2003).
100. De Jonghe, C., Esselens, C., Kumar-Singh, S., Craessaerts, K., Serneels, S., Checler, F., Annaert, W., Van Broeckhoven, C., and De Strooper, B. Pathogenic APP mutations near the γ-secretase cleavage site differentially affect Aβ secretion and APP C-terminal fragment stability. *Hum. Mol. Genet.* **10**(16), 1665–1671 (2001).
101. Chen, W., Gamache, E., Rosenman, D. J., Xie, J., Lopez, M. M., Li, Y. M., and Wang, C. Familial Alzheimer's mutations within APPTM increase Aβ42 production by enhancing accessibility of ε-cleavage site. *Nat. Commun.* **5**, 3037 (2014).
102. Kaden, D., Harmeier, A., Weise, C., Munter, L. M., Althoff, V., Rost, B. R., Hildebrand, P. W., Schmitz, D., Schaefer, M., Lurz, R., Skodda, S., Yamamoto, R., Arlt, S., Finckh, U., and Multhaup, G. Novel APP/Aβ mutation K16N produces highly toxic heteromeric Aβ oligomers. *EMBO Mol. Med.* **4**(7), 647–659 (2012).

103. Hendriks, L., van Duijn, C. M., Cras, P., Cruts, M., Van Hul, W., van Harskamp, F., Warren, A., McInnis, M. G., Antonarakis, S. E., Martin, J. J., Hofman, A., and Van Broeckhoven, C. Presenile dementia and cerebral haemorrhage linked to a mutation at codon 692 of the β-amyloid precursor protein gene. *Nat. Genet.* **1**(3), 218–221 (1992).
104. Brooks, W., Kwok, J., Halliday, G., Godbolt, A., Rossor, M., Creasey, H.,. Jones, A., and Schofield, P. Hemorrhage is uncommon in new Alzheimer family with Flemish amyloid precursor protein mutation. *Neurology*. **63**(9), 1613–1617 (2004).
105. Kumar-Singh, S., Cras, P., Wang, R., Kros, J., van Swieten, J., Lubke, U., Ceuterick, C., Serneels, S., Vennekens, K., Timmermans, J., Van Marck, E., Martin, J., van Duijn, C., and Van Broeckhoven, C. Dense-core senile plaques in the Flemish variant of Alzheimer's disease are vasocentric. *Am. J. Pathol.* **161**(2), 507–520 (2002).
106. Betts, V., Leissring, M. A., Dolios, G., Wang, R., Selkoe, D. J., and Walsh, D. M. Aggregation and catabolism of disease-associated intra-Aβ mutations: reduced proteolysis of Aβ A21G by neprilysin. *Neurobiol. Dis.* **31**(3), 442–450 (2008).
107. Kamino, K., Orr, H. T., Payami, H., Wijsman, E. M., Alonso, E., Pulst, S. M., Anderson, L., O'dahl, S., Nemens, E., White, J. A., Sadovnick, A. D., Ball, M. J., Kayue, J., Warren, A., McInnis, M., Antonarakis, S. T., Korenberg, J. R., Sharma, V., Kukull, W., Larson, E., Heston, L. L., Martin, G. M., Bird, T. D., and Schellenberg, G. D. Linkage and mutational analysis of familial Alzheimer disease kindreds for the APP gene region. *Am. J. Hum. Genet.* **51**, 998–1014 (1992).
108. Nilsberth, C., Westlind-Danielsson, A., Eckman, C. B., Condron, M. M., Axelman, K., Forsell, C., Stenh, C., Luthman, J., Teplow, D. B., Younkin, S. G., Näslund, J., and Lannfelt, L. The 'Arctic' APP mutation (E693G) causes Alzheimer's disease by enhanced Aβ protofibril formation. *Nat. Neurosci.* **4**(9), 887–893 (2001).
109. Päiviö, A., Jarvet, J., Gräslund, A., Lannfelt, L., and Westlind-Danielsson, A. Unique physicochemical profile of β-amyloid peptide variant Aβ1-40 E22G protofibrils: Conceivable neuropathogen in Arctic mutant carriers. *J. Mol. Biol.* **339**, 0 145–59 (2004).
110. Whalen, B., Selkoe, D., and Hartley, D. Small non-fibrillar assemblies of amyloid β-protein bearing the Arctic mutation induce rapid neuritic degeneration. *Neurobiol. Dis.* **20**, 254–266 (2005).
111. Levy, E., Carman,M. D., Fernandez-Madrid, I. J., Power, M. D., Lieberburg, I., van Duinen, S. G., Bots, G. T. A. M., Luyendijk, W., and Frangione, B. Mutation of the Alzheimer's disease amyloid gene in hereditary cerebral hemorrhage, Dutch-type. *Science*. **248**, 1124–1126 (1990).
112. Van Broeckhoven, C., Haan, J., Bakker, E., Hardy, J. A., Van Hul, W., Wehnert, A., Vegter-Van der Vlis, M., and Roos, R. A. Amyloid β protein precursor gene and hereditary cerebral hemorrhage with amyloidosis (Dutch). *Science*. **248**(4959), 1120–2 (1990).
113. Grabowski, T. J., Cho, H. S., Vonsattel, J. P. G., Rebeck, G. W., and Greenberg, S. M. Novel amyloid precursor protein mutation in an Iowa family with dementia and severe cerebral amyloid angiopathy. *Ann. Neurol.* **49**(6), 697–705 (2001).
114. Van Nostrand, W., Melchor, J., Cho, H., Greenberg, S., and Rebeck, G. Pathogenic effects of D23N Iowa mutant amyloid β-protein, *J. Biol. Chem.* **276**, 32860–32866 (2001).

115. Tomidokoro, Y., Rostagno, A., Neubert, T. A., Lu, Y., Rebeck, G., Frangione, B., Greenberg, S., and Ghiso, J. Iowa variant of familial Alzheimer's Disease—Accumulation of posttranslationally modified Aβ D23N in parenchymal and cerebrovascular amyloid deposits. *Am. J. Pathol.* **176**, 1841–1854 (2010).
116. Wisniewski, T., Ghiso, J., and Frangione, B. Peptides homologous to the amyloid protein of Alzheimer's disease containing a glutamine for glutamic acid substitution have accelerated amyloid fibril formation. *Biochem. Biophys. Res. Commun.* **179**, 1247–1254 (1991).
117. Bugiani, O., Padovani, A., Magoni, M., Andora, G., Sgarzi, M., Savoiardo, M., Bizzi, A., Giaccone, G., Rossi, G., and Tagliavini, F. An Italian type of HCHWA. *Neurobiol. Aging.* **19**, 238 (1998).
118. Bugiani, O., Giaccone, G., Rossi, G., Mangieri, M., Capobianco, R., Morbin, M., Mazzoleni, G., Cupidi, C., Marcon, G., Giovagnoli, A., Bizzi, A., Di Fede, G., Puoti, G., Carella, F., Salmaggi, A., Romorini, A., Patruno, G. M., Magoni, M., Padovani, A., and Tagliavini, F. Hereditary cerebral hemorrhage with amyloidosis associated with the E693K mutation of APP. *Arch. Neurol.* **67**(8), 987–995 (2010).
119. Murakami, K., Irie, K., Morimoto, A., Ohigashi, H., Shindo, M., Nagao, M., Shimizu, T., and Shirasawa, T. Neurotoxicity and physicochemical properties of Aβ mutant peptides from cerebral amyloid angiopathy. *J. Biol. Chem.* **278**(46), 46179–46187 (2003).
120. Tomiyama, T., Nagata, T., Shimada, H., Teraoka, R., Fukushima, A., Kanemitsu, H., Takuma, H., Kuwano, R., Imagawa, M., Ataka, S., Wada, Y., Yoshioka, E., Nishizaki, T., Watanabe, Y., and Mori, H. A new amyloid β variant favoring oligomerization in Alzheimer's-type dementia. *Ann. Neurol.* **63**, 377–387 (2008).
121. Kutoku, Y., Ohsawa, Y., Kuwano, R., Ikeuchi, T., Inoue, H., Ataka, S., Shimada, H., Mori, H., and Sunada, Y. A second pedigree with amyloid-less familial Alzheimer's disease harboring an identical mutation in the amyloid precursor protein gene (E693delta). *Int. Med.* **540**(2), 0 205–208 (2015).
122. Ovchinnikova, O. Y., Finder, V. H., Vodopivec, I., Nitsch, R. M., and Glockshuber, R. The Osaka FAD mutation E22Δ leads to the formation of a previously unknown type of amyloid β fibrils and modulates Aβ neurotoxicity. *J. Mol. Biol.* **408**(4), 780–791 (2011).
123. Inayathullah M. and Teplow, D. B. Structural dynamics of the E22 (Osaka) familial Alzheimer's disease-linked amyloid β-protein. *Amyloid.* **18**(3), 98–107 (2011).
124. Janssen, J. C., Beck, J. A., Campbell, T. A., Dickinson, A., Fox, N. C., Harvey, R. J., Houlden, H., Rossor, M. N., and Collinge, J. Early onset familial Alzheimer's disease — mutation frequency in 31 families. *Neurology.* **60**(2), 235–239 (2003).
125. Hori, Y., Hashimoto, T., Wakutani, Y., Urakami, K., Nakashima, K., Condron, M. M., Tsubuki, S., Saido, T. C., Teplow, D. B., and Iwatsubo, T. The Tottori (D7N) and English (H6R) familial Alzheimer disease mutations accelerate Aβ fibril formation without increasing protofibril formation. *J. Biol. Chem.* **282**(7), 4916–4923 (2007).
126. Ono, K., Condron, M. M., and Teplow, D. B. Effects of the English (H6R) and Tottori (D7N) familial Alzheimer disease mutations on amyloid β-protein assembly and toxicity. *J. Biol. Chem.* **285**(30), 23184–23195 (2010).
127. Wakutani, Y., Watanabe, K., Adachi, Y., Wada-Isoe, K., Urakami, K., Ninomiya, H., Saido, T. C., Hashimoto, T., Iwatsubo, T., and Nakashima, K. Novel amyloid precursor

protein gene missense mutation (D678N) in probable familial Alzheimer's disease. *J. Neuro. Neurosur. Psychiatry*. **750**(7), 0 1039–1042 (2004).

128. DiFede, G., Catania, M., Morbin, M., Rossi, G., Suardi, S., Mazzoleni, G., Merlin, M., Giovagnoli, A., Prioni, S., Erbetta, A., Falcone, C., Gobbi, M., Colombo, L., Bastone, A., Beeg, M., Manzoni, C., Francescucci, B., Spagnoli, A., Cant, L., Favero, E. D., Levy, E., Salmona, M., and Tagliavini, F. A recessive mutation in the APP gene with dominant-negative effect on amyloidogenesis. *Science*. **323**, 1473–1477 (2009).
129. Giaccone, G., Morbin, M., Moda, F., Botta, M., Mazzoleni, G., Uggetti, A., Catania, M., Moro, M. L., Redaelli, V., Spagnoli, A., Rossi, R. S., Salmona, M., Di Fede, G., and Tagliavini, F. Neuropathology of the recessive A673V APP mutation: Alzheimer disease with distinctive features. *Acta Neuropathol.* **120**(6), 803–812 (2010).
130. Benilova, I., Gallardo, R., Ungureanu, A. A., Cano, V. C., Snellinx, A., Ramakers, M., Bartic, C., Rousseau, F., Schymkowitz, J., and De Strooper, B. The Alzheimer disease protective mutation A2T modulates kinetic and thermodynamic properties of amyloid-β (Aβ) aggregation. *J. Biol. Chem.* **289**(45), 30977–30989 (2014).
131. Peacock, M., Warren, J., Roses, A., and Fink, J. Novel polymorphism in the A4-region of the amyloid precursor protein gene in a patient without Alzheimer's-disease. *Neurology*. **43**(6), 1254–1256 (1993).
132. Jonsson, T., Atwal, J. K., Steinberg, S., Snaedal, J., Jonsson, P. V., Bjornsson, S., Stefansson, H., Sulem1, P., Gudbjartsson, D., Maloney, J., Hoyte, K., Gustafson, A., Liu, Y., Lu, Y., Bhangale, T., Graham, R. R., Huttenlocher, J., Bjornsdottir, G., Andreassen, O. A., Jonsson, E. G., Palotie, A., Behrens, T. W., Magnusson, O. T., Kong, A., Thorsteinsdottir, U., Watts, R. J., and Stefansson, K. A mutation in APP protects against Alzheimer's disease and age-related cognitive decline. *Nature*. **488**, 96–99 (2012).
133. Zheng, X., Liu, D., Roychaudhuri, R., Teplow, D. B., and Bowers, M. T. Amyloid β-protein Assembly: Differential effects of the protective A2T mutation and recessive A2V familial Alzheimer's disease mutation. *ACS Chem. Neurosci.* **6**(10), 1732–1740 (2015).
134. Jarrett, J. T., Berger, E. P., and Lansbury, P. T. The carboxy terminus of the β amyloid protein is critical for the seeding of amyloid formation: Implications for the pathogenesis of Alzheimer's disease, *Biochem.* **32**(18), 4693–4697 (1993).
135. Jarrett, J. T., Berger E. P., and Lansbury P. T. The C-terminus of the β protein is critical in amyloidogenesis. *Ann. N Y Acad. Sci.* **695**, 144–148 (1993).
136. Shea J. and Urbanc, B. Insights into Aβ aggregation: A molecular dynamics perspective. *Curr. Topics Med. Chem., Special Issue: Protein Misfolding in Conformational Disorders, Guest Ed. Cláudio M. Gomes*. **12**, 2596–2610 (2012).
137. Walsh, D. M., Lomakin, A., Benedek, G. B., Condron, M. M., and Teplow, D. B. Amyloid β -protein fibrillogenesis. Detection of a protofibrillar intermediate. *J. Biol. Chem.* **272**(35), 22364–22372 (1997).
138. Lambert, M. P., Barlow, A. K., Chromy, B. A., Edwards, C., Freed, R., Liosatos, M., Morgan, T. E., Rozovsky, I., Trommer, B., Viola, K. L., Wals, P., Zhang, C., Finch, C. E., Krafft, G. A., and W. L. Klein, W. L. Diffusible, nonfibrillar ligands derived from Aβ (1–42) are potent central nervous system neurotoxins. *Proc. Natl. Acad. Sci. USA*. **95**(11), 6448–6453 (1998).

139. Chromy, B. A., Nowak, R. J., Lambert, M. P., and Viola, K. L., Chang, L., Velasco, P. T., Jones, B. W., Fernandez, S. J., Lacor, P. N.,Horowitz, P., Finch, C. E., Krafft, G. A., and Klein, W. L. Self-Assembly of Aβ(1-42) into globular neurotoxins. *Biochemistry*. **42**(44), 12749–12760 (2003).
140. Barghorn, S., Nimmrich, V., Striebinger, A., Krantz, C., Keller, P., Janson, B., Bahr, M., Schmidt, M., Bitner, R. S., Harlan, J., Barlow, E., Ebert, U., and Hillen, H. Globular amyloid β-peptide$_{1-42}$ oligomer—a homogenous and stable neuropathological protein in Alzheimer's disease. *J. Neurochem.* **95**(3), 834–847 (2005).
141. Walsh, D. M., Klyubin, I., Fadeeva, J. V., Cullen, W. K., Anwyl, R., Wolfe, M. S., Rowan, M. J., and Selkoe, D. J. Naturally secreted oligomers of amyloid β-protein potently inhibit hippocampal long-term potentiation in vivo. *Nature*. **416**(6880), 535–539 (2002).
142. Gong, Y., Chang, L., Viola, K. L., Lacor, P. N., Lambert, M. P., Finch, C. E., Krafft, G. A., and Klein, W. L. Alzheimer's disease-affected brain: Presence of oligomeric Aβ ligands (ADDLs) suggests a molecular basis for reversible memory loss. *Proc. Natl. Acad. Sci. USA*. **100**(18), 10417–10422 (2003).
143. Wang, H. W., Pasternak, J. F., Kuo, H., Ristic, H., Lambert, M. P., Chromy, B., Viola, K. L., Klein, W. L., Stine, W. B., Krafft, G. A., and Trommer, B. L. Soluble oligomers of β amyloid (1-42) inhibit long-term potentiation but not long-term depression in rat dentate gyrus. *Brain Res*. **924**(2), 133–140 (2002).
144. Selkoe, D. Alzheimer's disease is a synaptic failure. *Science*. **298**(5594), 789–791 (2002).
145. Fancy, D. A. and Kodadek, T. Chemistry for the analysis of protein-protein interactions: Rapid and efficient cross-linking triggered by long wavelength light. *Proc. Natl. Acad. Sci. USA*. **96**(11), 6020–6024 (1999).
146. Bitan, G., Lomakin, A., and Teplow, D. B. Amyloid β-protein oligomerization: prenucleation interactions revealed by photo-induced cross-linking of unmodified proteins. *J. Biol. Chem.* **276**(37), 35176–35184 (2001).
147. Bitan, G. Structural study of metastable amyloidogenic protein oligomers by photo-induced cross-linking of unmodified proteins. *Methods. Enzymol.* **413**, 217–236 (2006).
148. Bitan, G., Kirkitadze, M. D., Lomakin, A., Vollers, S. S., Benedek, G. B., and Teplow, D. B. Amyloid β-protein (Aβ) assembly: Aβ 40 and Aβ 42 oligomerize through distinct pathways. *Proc. Natl. Acad. Sci. U. S. A.* **100**(1), 330–335 (2003).
149. Bitan, G., Vollers, S. S., and Teplow, D. B. Elucidation of primary structure elements controlling early amyloid β-protein oligomerization. *J. Biol. Chem.* **278**(37), 34882–34889 (2003).
150. Pujol-Pina, R., Vilaprinyo-Pascual, S., Mazzucato, R., Arcella, A., Vilaseca, M., Orozco, M., and Carulla, N. SDS-PAGE analysis of Aβ oligomers is disserving research into Alzheimer's disease: appealing for ESI-IM-MS, *Sci. Rep.* **5**, 14809 (2015).
151. Bernstein, S. L., Wyttenbach, T., Baumketner, A., Shea, J.-E., Bitan, G., Teplow, D. B., and Bowers, M. T. Amyloid β-protein: monomer structure and early aggregation states of Aβ42 and Its Pro19 alloform. *J. Am. Chem. Soc.* **127**(7), 2075–2084 (2005).
152. Bernstein, S. L., Dupuis, N. F., Lazo, N. D., Wyttenbach, T., Condron, M. M., Bitan, G., Teplow, G. B., Shea, J.-E., Ruotolo, B. T., Robinson, C. V., and Bowers, M. T. Amyloid-β protein oligomerization and the importance of tetramers and dodecamers in the aetiology of Alzheimer's disease. *Nat. Chem.* **1**, 326–331 (2009).

153. Lesné, S., Koh, M. H., Kotilinek, L., Kayed, R., Glabe, C. G., Yang, A., Gallagher, M., and Ashe, K. H. A specific amyloid-β protein assembly in the brain impairs memory. *Nature*. **440**(7082), 352–357 (2006).
154. Williams, T., Serpell, L., and Urbanc, B. Stabilization of native amyloid β-protein oligomers by copper and hydrogen peroxide induced cross-linking of unmodified proteins (CHICUP). *Biochim. Biophys. Acta*. **1864**, 249–259 (2016).
155. Atwood, C., Perry, G., Zeng, H., Kato, Y., Jones, W., Ling, K., Huang, X., Moir, R., Wang, D., Sayre, L., Smith, M., Chen, S., and Bush, A. Copper mediates dityrosine cross-linking of Alzheimer's amyloid-β. *Biochemistry*. **43**(2), 560–568 (2004).
156. Al-Hilaly, Y. K., Williams, T. L., Stewart-Parker, M., Ford, L., Skaria, E., Cole, M., Bucher, W. G., Morris, K. L., Sada, A. A., Thorpe, J. R., and Serpell, L. C. A central role for dityrosine crosslinking of amyloid-β in Alzheimer's disease. *Acta Neuropathol. Commun.* **1**, 83 (2013).
157. Friedrich, R. P., Tepper, K., Rönicke, R., Soom, M., Westermann, M., Reymann, K., Käther, C., and Fändrich, M. Mechanism of amyloid plaque formation suggests an intracellular basis of Aβ pathogenicity. *Proc. Natl. Acad. Sci. U. S. A.* **107**(5), 1942–1947 (2010).
158. Ahmed, M., Davis, J., Aucoin, D., Sato, T., Ahuja, S., Aimoto, S., Elliott, J., Nostrand, W. V., and Smith S. Structural conversion of neurotoxic amyloid-β (1–42) oligomers to fibrils. *Nat. Struct. Mol. Biol.* **17**, 561–567 (2010).
159. Ono, K., Condron, M. M., and Teplow, D. B. Structure-neurotoxicity relationships of amyloid β-protein oligomers. *Proc. Natl. Acad. Sci. U. S. A.* **106**, 14745–14750 (2009).
160. Urbanc, B., Cruz, L., Teplow, D. B., and Stanley, H. E. Computer simulations of Alzheimer's amyloid β-protein folding and assembly. *Curr. Alzheimer Res.* **3**, 493–504 (2006).
161. Mitsutake, A., Sugita, Y., and Okamoto, Y. Generalized-ensemble algorithms for molecular simulations of biopolymers. *Biopolymers*. **60**, 96–123 (2001).
162. Nymeyer, H., Gnanakaran, S., and Garcia, A. E. Atomic simulations of protein folding, using the replica exchange algorithm. *Methods. Enzymol.* **383**, 119–149 (2004).
163. Baumketner, A., Bernstein, S. L., Wyttenbach, T., Bitan, G., Teplow, D. B., Bowers, M. T., and Shea, J.-E. Amyloid β-protein monomer structure: a computational and experimental study. *Protein Sci.* **15**(3), 420–428 (2006).
164. Baumketner, A. and Shea, J.-E. The structure of the Alzheimer amyloid-β (10–35) peptide probed through replica exchange molecular dynamics simulations in explicit solvent. *J. Mol. Biol.* **366**, 275–285 (2007).
165. Yang, M. and Teplow D. B. Amyloid β-protein monomer folding: free-energy surfaces reveal alloform-specific differences. *J. Mol. Biol.* **384**, 450–464 (2008).
166. Gessel, M. M., Wu, C., Li, H., Bitan, G., Shea, J.-E., and Bowers, M. T. Aβ (39–42) modulates Aβ42 oligomerization but not fibril formation. *Biochemistry*. **51**, 108–117 (2011).
167. Onuchic, J., LutheySchulten, Z., and Wolynes, P. Theory of protein folding: the energy landscape perspective. *Annu. Rev. Phys. Chem.* **48**, 545–600 (1997).
168. Pellarin, R. and Caflisch, A. Interpreting the aggregation kinetics of amyloid peptides. *J. Mol. Biol.* **360**(4), 882–892 (2006).
169. Pellarin, R., Guarnera, E., and Caflisch, A. Pathways and intermediates of amyloid fibril formation. *J. Mol. Biol.* **374**(4), 917–924 (2007).

170. Auer, S., Dobson, C. M., Vendruscolo, M., and Maritan, A. Self-templated nucleation in peptide and protein aggregation. *Phys. Rev. Lett.* **101**(25), 258101 (2008).
171. Pellarin, R., Schuetz, P., Guarnera, E., and Caflisch, A. Amyloid fibril polymorphism is under kinetic control. *J. Am. Chem. Soc.* **132**(42), 14960–14970 (2010).
172. Yang, Y., Meyer, R. B., and Hagan, M. F. Self-limited self-assembly of chiral filaments. *Phys. Rev. Lett.* **104**(25), 258102 (2010).
173. Wu, C. and Shea, J. E. Coarse-grained models for protein aggregation. *Curr. Opin. Struc. Biol.* **21**, 209–220 (2011).
174. Alder, B. J. and Wainwright, T. E. Studies in molecular dynamics. I. General method. *J. Chem. Phys.* **31**, 459–466 (1959).
175. Alder, B. J. and Wainwright, T. E. Studies in molecular dynamics. II. Behaviour of a small number of elastic spheres. *J. Chem. Phys.* **33**, 1439–1451 (1960).
176. Zhou, Y. and Karplus, M. Folding thermodynamics of a model three-helix-bundle protein. *Proc. Natl. Acad. Sci. U. S. A.* **94**, 14429–14432 (1997).
177. Zhou, Y., Karplus, M., Wichert, J. M., and Hall, C. K. Equilibrium thermodynamics of homopolymers and clusters: molecular dynamics and Monte Carlo simulations of systems with square-well interactions. *J. Chem. Phys.* **107**, 10691–10708 (1997).
178. Dokholyan, N. N., Buldyrev, S. V., Stanley, H. E., and Shakhnovich, E. I. Discrete molecular dynamics studies of the folding of a protein-like model. *Fold. Des.* **3**(6), 577–587 (1998).
179. Smith, A. V. and Hall, C. K. Protein refolding versus aggregation: computer simulations on an intermediate-resolution protein model. *J. Mol. Biol.* **312**, 187–202 (2001).
180. Peng, S., Ding, F., Urbanc, B., Buldyrev, S. V., Cruz, L., Stanley, H. E., and Dokholyan, N. V. Discrete molecular dynamics simulations of peptide aggregation. *Phys. Rev. E.* **69**, 041908 (2004).
181. Ding, F., Tsao, D., Nie, H., and Dokholyan, N. *Ab initio* folding of proteins with all-atom discrete molecular dynamics. *Structure.* **16**, 1010–1018 (2008).
182. Urbanc, B., Borreguero, J. M., Cruz, J. M., and Stanley, H. E. *Ab initio* discrete molecular dynamics approach to protein folding and aggregation. *Methods. Enzymol.* **412**, 314–338 (2006).
183. Betnel, M., Dokholyan, N. V., and Urbanc, B. From disordered amyloid β-proteins to soluble oligomers and protofibrils using discrete molecular dynamics, in *Alzheimer's disease: molecular basis of amyloid-β protein aggregation and fibril formation — Insights into low molecular weight and cytotoxic aggregates from computer simulations,* vol. 7, Chapter 12. Imperial College Press, London, UK, 2013. pp. 333–358. .
184. Ding, F., Borreguero, J. M., Buldyrev, S. V., Stanley, H. E., and Dokholyan, N. V. Mechanism for the α-helix to β-hairpin transition. *Proteins: Struct., Funct., Genet.* **53**(2), 220–228 (2003)
185. Urbanc, B., Cruz, L., Ding, F., Sammond, D., Khare, S., Buldyrev, S. V., Stanley, H. E., and Dokholyan, N. V. Molecular dynamics simulation of amyloid β dimer formation. *Biophys. J.* **87**(4), 2310–2321 (2004).
186. Kyte, K. and Doolittle, R. F. A simple method for displaying the hydropathic character of a protein. *J. Mol. Biol.* **157**(1), 105–132 (1982).
187. Urbanc, B., Cruz, L., Yun, S., Buldyrev, S. V., Bitan, G., Teplow, D. B., and Stanley, H. E. *In silico* study of amyloid β-protein folding and oligomerization. *Proc. Natl. Acad. Sci. U. S. A.* **101**(50), 17345–17350 (2004)

188. Urbanc, B., Betnel, M., Cruz, L., Bitan, G., and Teplow, D. B. Elucidation of amyloid β-protein oligomerization mechanisms: discrete molecular dynamics study. *J. Am. Chem. Soc.* **132**, 4266–4280 (2010).
189. Lazo, N. D., Grant, M. A., Condron, M. C., Rigby, A. C., and Teplow, D. B. On the nucleation of amyloid β-protein monomer folding. *Protein Sci.* **14**(6), 1581–1596 (2005).
190. Riek, R.,Güntert, P., Döbeli, H., Wipf, B., and Wüthrich, K. NMR studies in aqueous solution fail to identify significant conformational differences between the monomeric forms of two Alzheimer peptides with widely different plaque-competence, A$\beta(1\text{–}40)^{ox}$ and Aβ $(1\text{–}42)^{ox}$. *Eur J Biochem.* **268**(22), 5930–5936 (2001).
191. Murakami, K., Irie, K., Ohigashi, H., Hara, H., Nagao, M., Shimizu, T., and Shirasawa, T. Formation and stabilization model of the 42-mer Aβ radical: implications for the long-lasting oxidative stress in Alzheimer's disease. *J. Am. Chem. Soc.* **127**, 0 15168–15174 (2005).
192. Yan Y. and Wang, C. Aβ42 is more rigid than Aβ 40 at the C terminus: implications for a β aggregation and toxicity. *J. Mol. Biol.* **364**, 853–862 (2006).
193. Sgourakis, N., Yan, Y., McCallum, S., Wang, C., and Garcia, A. The Alzheimer's peptides Aβ40 and 42 adopt distinct conformations in water: A combined MD/NMR Study. *J. Mol. Biol.* **368**, 1448–1457 (2007).
194. Lin, Y.-S., Bowman, G. R., Beauchamp, K. A., and Pande, V. S. Investigating how peptide length and a pathogenic mutation modify the structural ensemble of amyloid β monomer. *Biophys. J.* **102**, 315–324 (2012).
195. Roychaudhuri, R., Yang, M., Deshpande, A., Cole, G., Frautschy, S., Lomakin, A., Benedek, G., and Teplow, D. C-terminal turn stability determines assembly differences between Aβ40 and Aβ42. *J. Mol. Biol.* (2012).
196. Ball, K. A., Phillips, A. H., Wemmer, D. E., and Head-Gordon, T. Differences in β-strand populations of monomeric Aβ40 and Aβ42. *Biophys. J.* **104**(12), 2714–2724 (2013).
197. Humphrey, W., Dalke, A., and Schulten, K. VMD: visual molecular dynamics. *J. Mol. Graphics.* **14**(1), 33–38 (1996).
198. Streltsov, V., Varghese, J., Masters, C., and Nuttall, S. Crystal structure of the amyloid-β p3 fragment provides a model for oligomer formation in Alzheimer's disease. *J. Neurosci.* **31**, 1419–1426 (2011).
199. Pauwels, K., Williams, T. L., Morris, K. L., Jonckheere, W., Vandersteen, A., Kelly, G., Schymkowitz, J., Rousseau, F., Pastore, A., Serpell, L. C., and Broersen, K. Structural basis for increased toxicity of pathological Aβ42:Aβ40 ratios in Alzheimer disease. *J. Biol. Chem.* **287**, 5650–5660 (2012).
200. Samiotakis, A., Homouz, D., and Cheung, M. Multiscale investigation of chemical interference in proteins. *J. Chem. Phys.* **132**, 175101 (2010).
201. Barz, B. and Urbanc, B. Dimer formation enhances structural differences between amyloid β-protein (1–40) and (1–42): an explicit-solvent molecular dynamics study. *PLoS One.* **7**, e34345 (2012).
202. Meral, D. and Urbanc, B. Discrete molecular dynamics study of oligomer formation by N-terminally truncated amyloid β-protein. *J. Mol. Biol.* **425**, 2260–2275 (2013).

203. Meral, D. and Urbanc, B. Erratum to discrete molecular dynamics study of oligomer formation by N-terminally truncated amyloid β-protein. *J. Mol. Biol.* **427**, 2726–2729 (2015).
204. Sinha, S., Lopes, D. H. J., Du, Z., Pang, E. S., Shanmugam, A., Lomakin, A., Talbiersky, P., Tennstaedt, A., McDaniel, K., Bakshi, R., Kuo, P.-Y., Ehrmann, M., Benedek, G. B., Loo, J. A., Klärner, F.-G., Schrader, T., Wang, C., and Bitan, G. Lysine-specific molecular tweezers are broad-spectrum inhibitors of assembly and toxicity of amyloid proteins. *J. Am. Chem. Soc.* **133**(42), 16958–16969 (2011).
205. Sinha, S., Lopes, D. H. J., and Bitan, G. A key role for lysine residues in amyloid β-protein folding, assembly, and toxicity. *ACS Chem. Neurosci* .**3**(6), 473–481 (2012).
206. Žganec, M., Kruczek, N., and Urbanc, B. Amino acid substitutions [K16A] and [K28A] distinctly affect amyloid β-protein oligomerization. *J. Biol. Phys.* **42**(3), 453–476(2016).
207. Ferrone, F. Polymerization and sickle cell disease: a molecular view. *Microcirculation.* **11**(2), 115–128 (2004).
208. Galkin, O., Nagel, R. L., and Vekilov, P. G. The kinetics of nucleation and growth of sickle cell hemoglobin fibers. *J. Mol. Biol.* **365**(2), 425–439 (2007).
209. Sawaya, M. R., Sambashivan, S., Nelson, R., Ivanova, M. I., Sievers, S. A., Apostol, M. I., Thompson, M. J., Balbirnie, M., Wiltzius, J. J. W., McFarlane, H. T., Madsen, A. O., Riekel, C., and Eisenberg, D. Atomic structures of amyloid cross-β spines reveal varied steric zippers. *Nature.* **447**(7143), 453–7 (2007).
210. Erdemir, D., Lee, A. Y., and Myerson, A. S. Nucleation of crystals from solution: classical and two-step models. *Acc. Chem. Res.* **42**(5), 621–629 (2009).
211. Vekilov, P. G. Nucleation. *Crystal Growth Design.* **10**(12), 5007–5019 (2010).
212. Vekilov, P. G. Nucleation of protein condensed phases. *Rev. Chem. Eng.* **27**(1–2), 1–13 (2011).
213. Ten Wolde, P. R. and Frenkel, D. Enhancement of protein crystal nucleation by critical density fluctuations. *Science.* **277**(5334), 1975–1978 (1997).
214. Harper, J. D., Wong, S. S., Lieber, C. M., and Lansbury, P. T. Observation of metastable Aβ amyloid protofibrils by atomic force microscopy. *Chem Biol.* **4**(2), 119–125 (1997).
215. Walsh, D. M., Hartley, D. M., Kusumoto, Y., Fezoui, Y., Condron, M. M., Lomakin, A., Benedek, G. B., Selkoe, D. J., and Teplow, D. B. Amyloid β-protein fibrillogenesis. Structure and biological activity of protofibrillar intermediates. *J. Biol. Chem.* **274**(36), 25945–25952 (1999).
216. Serpell, L. C. Alzheimer's amyloid fibrils: structure and assembly. *Biochim. Biophys. Acta.* **1502**(1), 16–30 (2000).
217. Kirkitadze, M. D., Condron, M. M., and Teplow, D. B. Identification and characterization of key kinetic intermediates in amyloid β-protein fibrillogenesis. *J Mol Biol.* **3120**(5), 1103–1119 (2001).
218. Lee, J., Culyba, E., Powers, E., and Kelly, J. Amyloid-β forms fibrils by nucleated conformational conversion of oligomers. *Nat. Chem. Biol.* **7**, 602–609 (2011).
219. Barz, B., and Urbanc, B. Minimal model of self–assembly: emergence of diversity and complexity. *J. Phys. Chem. B.* **118**, 3761–3770 (2014).
220. Schreck, J., and Yuan, J. A kinetic study of amyloid formation: fibril growth and length distributions. *J. Phys. Chem. B.* **117**, 6574–6583 (2013).
221. Uversky, V. N. Mysterious oligomerization of the amyloidogenic proteins, *FEBS J.* **2770**(14), 2940–2953 (2010).

222. Petkova, A. T., Leapman, R. D., Guo, Z.,Yau, W.-M., Mattson, M. P., and Tycko, R. Self-propagating, molecular-level polymorphism in Alzheimer's β-amyloid fibrils. *Science*. **307**, 262–265 (2005).
223. Paravastu, A., Leapman, R., Yau, W.-M., and Tycko, R. Molecular structural basis for polymorphism in Alzheimer's β-amyloid fibrils. *Proc. Natl. Acad. Sci. U. S. A.* **105**, 18349–18354 (2008).
224. Dahlgren, K. N., Manelli, A. M., Stine, W. B., Baker, L. K., Krafft, G. A., and LaDu, M. J. Oligomeric and fibrillar species of amyloid-β peptides differentially affect neuronal viability. *J. Biol. Chem.* **277**(35), 32046–32053 (2002).
225. Maiti, P., Lomakin, A., Benedek, G., and Bitan, G. Despite its role in assembly, methionine 35 is not necessary for amyloid β-protein toxicity. *J. Neurochem.* **113**, 1252–1262 (2010).
226. Iijima, K., Liu, H., Chiang, A., Hearn, S., Konsolaki, M., and Zhong, Y. Dissecting the pathological effects of human Aβ40 and Aβ42 in Drosophila: a potential model for Alzheimer's disease. *Proc. Natl. Acad. Sci. U. S. A.* **101**, 6623–6628 (2004).
227. McGowan, E., Pickford, F., Kim, J., Onstead, L., Eriksen, J., Yu, C., Skipper, L., Murphy, M. P., Beard, J., Das, P., Jansen, K., Delucia, M., Lin, W.-L., Dolios, G., Wang, R., Eckman, C. B., Dickson, D. W., Hutton, M., Hardy, J., and Golde, T. Aβ42 is essential for parenchymal and vascular amyloid deposition in mice. *Neuron*. **47**(2), 191–199 (2005).
228. Fradinger, E., Monien, B. H., Urbanc, B., Lomakin, A., Tan, M., Li, H., Spring, S. M., Condron, M. M., Cruz, L., Xie, C.-W., Benedek, G. B., and Bitan, G. C-terminal peptides coassemble into Aβ42 oligomers and protect neurons against Aβ42-induced neurotoxicity. *Proc. Natl. Acad. Sci. U. S. A.* **105**, 14175–14180 (2008).
229. Li, H., Monien, B. H., Fradinger, E. A., Urbanc, B., and Bitan, G. Biophysicsl characterization of Aβ42 C-terminal fragments: inhibitors of Aβ42 neurotoxicity. *Biochemistry*. **49**, 1259–1267 (2010).
230. Li, H., Monien, B. H., Lomakin, A., Zemell, R., Fradinger, E. A., Tan, M., Spring, S. M., Urbanc, B., Xie, C.-W., Benedek, G. B., and Bitan, G. Mechanistic investigation of the inhibition of Aβ42 assembly and neurotoxicity by C-terminal Aβ42 fragments. *Biochemistry*. **49**, 6358–6364 (2010).
231. Urbanc, B., Betnel, M., Cruz, L., Li, H., Fradinger, E., Monien, B. H., and Bitan, G. Structural basis of $A\beta_{1-42}$ toxicity inhibition by Aβ C-terminal fragments: discrete molecular dynamics study. *J. Mol. Biol.* **410**, 316–328 (2011).
232. Gardberg, A., Dice, L., Ou, S., Rich, R., Helmbrecht, E., Ko, J., Wetzel, R., Myszka, D., Patterson, P., and Dealwis, C. Molecular basis for passive immunotherapy of Alzheimer's disease. *Proc. Natl. Acad. Sci. U. S. A.* **104**, 15659–15664 (2007).
233. Miles, L., Wun, K., Crespi, G., Fodero-Tavoletti, M., Galatis, D., Bagley, C., Beyreuther, K., Masters, C., Cappai, R., McKinstry, W., Barnham, K., and Parker, M. Amyloid-β — anti-amyloid-β complex structure reveals an extended conformation in the immunodominant B-cell epitope. *J. Mol. Biol.* **377**, 181–192 (2008).
234. Robert, R., Dolezal, O., Waddington, L., Hattarki, M., Cappai, R., Masters, C., Hudson, P., and Wark, K. Engineered antibody intervention strategies for Alzheimer's disease and related dementias by targeting amyloid and toxic oligomers. *Protein Eng. Des. Sel.* **22**, 199–208 (2009).

235. Shankar, G. M., Li, S. M., Mehta, T. H., Garcia-Munoz, A., Shepardson, N. E., Smith, I., Brett, F. M., Farrell, M. A., Rowan, M. J., Lemere, C. A., Regan, C. M., Walsh, D. M., Sabatini, B. L., and Selkoe, D.J. Amyloid-β protein dimers isolated directly from Alzheimer's brains impair synaptic plasticity and memory. *Nat. Med.* **14**(8), 837–842 (2008).
236. Jin, M., Shepardson, N., Yang, T., Chen, G., Walsh, D., and Selkoe, D. J. Soluble amyloid β-protein dimers isolated from Alzheimer cortex directly induce Tau hyperphosphorylation and neuritic degeneration. *Proc. Natl. Acad. Sci. U. S. A.* **108**, 5819–5824 (2011).
237. Luheshi, L. M., Tartaglia, G. G., Brorsson, A. C., Pawar, A. P., Watson, I. E., Chiti, F., Vendruscolo, M., Lomas, D. A., Dobson, C. M., and Crowther, D. C. Systematic in vivo analysis of the intrinsic determinants of amyloid β pathogenicity. *PLoS Biol.* **5**, e290 (2007).
238. Koffie, R. M., Meyer-Luehmann, M., Hashimoto, T., Adams, K. W., Mielke, M. L., Garcia-Alloza, M., Micheva, K. D., Smith, S. J., Kim, M. J., Lee, V. M., Hyman, B. T., and Spires-Jones, T. L. Oligomeric amyloid beta associates with postsynaptic densities and correlates with excitatory synapse loss near senile plaques. *Proc. Natl. Acad. Sci. U. S. A.* **106**(10), 4012–4017 (2009).
239. Chen, Q. S., Kagan, B. L., Hirakura, Y., and Xie, C. W. Impairment of hippocampal long-term potentiation by Alzheimer amyloid β-peptides. *J. Neurosci. Res.* **60**(1), 65–72 (2000).
240. Shankar, G. M., Bloodgood, B. L., Townsend, M., Walsh, D. M., Selkoe, D. J., and Sabatini, B. L. Natural oligomers of the Alzheimer amyloid-β protein induce reversible synapse loss by modulating an NMDA-type glutamate receptor-dependent signaling pathway. *J. Neurosci.* **27**, 2866–2875 (2007).
241. Kelly, B. L. and Ferreira, A. β-amyloid-induced dynamin 1 degradation is mediated by N-methyl-D-aspartate receptors in hippocampal neurons. *J. Biol. Chem.* **281**(38), 28079–28089 (2006).
242. Williams, T. and Serpell, L. Membrane and surface interactions of Alzheimer's Aβ peptide – insights into the mechanism of cytotoxicity. *FEBS J.* **278**, 3905–3917 (2011).
243. Williams, T., Day,I., and Serpell, L. The effect of Alzheimer's Aβ aggregation state on the permeation of biomimetic lipid vesicles. *Langmuir.* **26**, 17260–17268 (2010).
244. Kremer J. J. and Murphy, R. M. Kinetics of adsorption of β-amyloid peptide Aβ(1–40) to lipid bilayers. *J. Biochem. Biophys. Methods.* **57**, 159–169 (2003).
245. Subasinghe, S., Unabia, S., Barrow, C., Mok, S., Aguilar, M., and Small, D. Cholesterol is necessary both for the toxic effect of Aβ peptides on vascular smooth muscle cells and for Aβ binding to vascular smooth muscle cell membranes. *J. Neurochem.* **84**, 471–479 (2003).
246. Micelli, S., Meleleo, D., Picciarelli, V., and Gallucci, E. Effect of sterols on β-amyloid peptide (AβP 1–40) channel formation and their properties in planar lipid membranes. *Biophys. J.* **86**, 2231–2237 (2004).
247. Williams, T., Johnson, B., Urbanc, B., Jenkins, T., Connell, S., and Serpell, L. Alzheimer's Aβ42 oligomers but not fibrils simultaneously bind to and cause damage to ganglioside containing lipid membranes. *Biochem. J.* **439**, 67–77 (2011).

248. Williams, T. L., Urbanc, B., Marshall, K. E., Vadukul, D. M., Jenkins, A. T. A., and Serpell, L. C. Europium as an inhibitor of amyloid-β (1–42) induced membrane permeation. *FEBS Lett.* **589**(21), 3228–3236 (2015).
249. Hebda, J. and Miranker, A. The interplay of catalysis and toxicity by amyloid intermediates on lipid bilayers: insights from type II diabetes. *Ann. Rev. Biophys.* **38**, 125–152 (2007).
250. Shai, Y. Mechanism of the binding, insertion, and destabilization of phospholipid bilayer membranes by α-helical antimicrobial and cell non-selective membrane-lytic peptides. *Biochim. Biophys. Acta.* **1462**, 55–70 (1999).
251. Pollard H. B., Rojas E., and Arispe. N. A new hypothesis for the mechanism of amyloid toxicity, based on the calcium channel activity of amyloid β protein (AβP) in phospholipid bilayer membranes. In ed. Nitsch, R. M., Growdon, J. H., Corkin, S., and Wurtman, R. J. *Alzheimer's disease: Amyloid precusor proteins, signal transduction, and neuronal transplantation*, vol. 695, *Ann. N. Y. Acad. Sci.*, pp. 165–168 (1993).
252. Arispe, N. and Pollard, H. B. and Rojas, E. β-Amyloid Ca^{2+}-channel hypothesis for neuronal death in Alzheimer-disease. *Mol. Cell Biochem.* **140**(2), 119–125 (1994). 4th Fall Symposium of the Membrane-Biophysics-Subgroup of the Biophysical-Society, Beaufort, NC, Oct 1993.
253. Kawahara, M., Arispe, N., Kuroda, Y., Rojas, E. Alzheimers disease amyloid β-protein forms Zn2+-sensitive, cation-selective channels across excised membrane patches from hypothalamic neurons. *Biophys. J.* **73**(1), 67–75 (1997).
254. Lin, H., Bhatia, R., and Lal, R., Amyloid β protein forms ion channels: implications for Alzheimer's disease pathophysiology. *FASEB J.* **15**, 0 2433–2444 (2001).
255. Kagan, B. L., Hirakura, Y., Azimov, R., Azimova, R., and Lin, M. C. Channel hypothesis of Alzheimer's disease: current status. *Peptides.* **23**(7), 1311–1315 (2002).
256. Lal, R., Lin, H., and Quist, A. P. Amyloid β ion channel: 3D structure and relevance to amyloid channel paradigm. *Biochim. Biophys. Acta.* **1768**, 1966–1975 (2007).
257. Arispe, N., Rojas, E., and Pollard, H. B. Alzheimer disease amyloid β protein forms calcium channels in bilayer membranes: blockade by tromethamine and aluminum. *Proc. Natl. Acad. Sci. U. S. A.* **90**, 567–571 (1993).
258. Arispe, N., Pollard, H. B., and Rojas, E. Giant multilevel cation channels formed by Alzheimer disease amyloid β-protein [AβP-(1-40)] in bilayer membranes. *Proc. Natl. Acad. Sci. U. S. A.* **90**, 10573–10577 (1993).
259. Bhatia, R., Lin, H., and Lal, R. Fresh and globular amyloid β protein (1–42) induces rapid cellular degeneration: evidence for AβP channel-mediated cellular toxicity. *FASEB J.* **14**, 1233–1243 (2000).
260. Demuro, A., Mina, E., Kayed, R., Milton, S. C., Parker, I., and Glabe, C. G. Calcium dysregulation and membrane disruption as a ubiquitous neurotoxic mechanism of soluble amyloid oligomers. *J. Biol. Chem.* **280**(17), 17294–300 (2005).
261. Kawahara, M., Kuroda, Y., Arispe, N., and Rojas, E. Alzheimer's β-amyloid, human islet amylin, and prion protein fragment evoke intracellular free calcium elevations by a common mechanism in a hypothalamic GnRH neuronal Cell Line. *J. Biol. Chem.* **275**(19), 14077–14083 (2000).
262. Quist, A., Doudevski, I., Lin, H., Azimova, R., Ng, D., Frangione, B., Kagan, B., Ghiso, J., and Lal, R. Amyloid ion channels: a common structural link for protein-misfolding disease. *Proc. Natl. Acad. Sci. U. S. A.* **102**, 10427–10432 (2005).

263. Demuro, I., Parker, and Stutzmann, G. E. Calcium signaling and amyloid toxicity in Alzheimer disease. *J. Biol. Chem.* **285**(17), 12463–12468 (2010).
264. Strödel, J., Lee, C., Whittleston, and Wales, D. Transmembrane structures for Alzheimer's $A\beta_{1-42}$ oligomers.
265. Jang, H., Arce, F. T., Ramachandran, S., Capone, R., Lal, R., and Nussinov, R. β-barrel topology of Alzheimer's β-amyloid ion channels. *J. Mol. Biol.* **404**(5), 917–934 (2010).
266. Yun, S., Urbanc, B., Cruz, L., Bitan, G., Teplow, D. B., and Stanley, H. E. Role of electrostatic interactions in amyloid β-protein (Aβ) oligomer formation: a discrete molecular dynamics study. *Biophys. J.* **92**, 4064–4077 (2007).
267. Žganec, M., Žerovnik, E., and Urbanc, B. Self-assembly of globular protein stefin B into polymorphic oligomers probed by discrete molecular dynamics. *J. Chem. Theory Comput.* **11**, 2355–2366 (2015).
268. Jang, H., Connelly, L., Arce, F. T., Ramachandran, S., Kagan, B. L., Lal, R., and Nussinov, R. Mechanisms for the insertion of toxic, fibril-like β-amyloid oligomers into the membrane. *J. Chem. Theory Comput.* **9**(1), 822–833 (2013).
269. Deshpande, A., Mina, E., Glabe, C., and Busciglio, J. Different conformations of amyloid β induce neurotoxicity by distinct mechanisms in human cortical neurons. *J. Neurosci.* **26**(22), 6011–6018 (2006).
270. Soscia, S. J., Kirby, J. E., Washicosky, K. J., Tucker, S. M., Ingelsson, M., Hyman, B., Burton, M. A., Goldstein, L. E., Duong, S., Tanzi, R. R., and Moir, R. D. The Alzheimer's disease-associated amyloid β-protein is an antimicrobial peptide. *PLoS One.* **5**(3), e9505 (2010).
271. White, M. R., Kandel, R., Tripathi, S., Condon, D., Qi, L., Taubenberger, J., and Hartshorn, K. L. Alzheimer's associated β-amyloid protein inhibits influenza A virus and modulates viral interactions with phagocytes. *PLoS One.* **9**(7) (2014).
272. Bourgade, K., Garneau, H., Giroux, G., Le Page, A. Y., Bocti, C., Dupuis, G., Frost, E. H., and Fülöp Jr., T. β-Amyloid peptides display protective activity against the human Alzheimer's disease-associated herpes simplex virus-1. *Biogerontology.* **16**(1), 85–98 (2015).
273. Kumar, D.K., Choi, S. H., Washicosky, K. J., Eimer, W. A., Tucker, S., Ghofrani, J., Lefkowitz, A., McColl, G., Goldstein, L. E., Tanzi, R. E., and Moir, R.D. Amyloid-β peptide protects against microbial infection in mouse and worm models of Alzheimer's disease. *Sci. Transl. Med.* **8**(340), 340ra72.
274. Butterfield, A., Reed, T., Newman, S. F., and Sultana, R. Roles of amyloid β-peptide-associated oxidative stress and brain protein modifications in the pathogenesis of Alzheimer's disease and mild cognitive impairment. *Free Rad. Biol. Med.* **43**(5), 658–677 (2007).
275. Rival, T., Page, R. M., Chandraratna, D. S., Sendall, T. J., Ryder, E., Liu, B., Lewis, H., Rosahl, T., Hider, R., Camargo, L. M., Shearman, M. S., Crowther, D. C., and Lomas, D. A. Fenton chemistry and oxidative stress mediate the toxicity of the β-amyloid peptide in a Drosophila model of Alzheimer's disease. *Eur. J. Neurosci.* **29**(7), 1335–1347 (2009).
276. Chassaing, S., Collin, F., Dorlet, P., Gout, J., Hureau, C., and Faller, P. Copper and heme-mediated Aβ toxicity: redox chemistry, Aβ oxidations and anti-ROS compounds. *Curr. Top. Med. Chem.* **12**(22), 2573–2595 (2012).

277. Greenough, M. A., Camakaris, J., and Bush, A. I. Metal dyshomeostasis and oxidative stress in Alzheimer's disease. *Neurochem. Int.* **62**(5, SI), 540–555 (2013).
278. Atwood, R., Scarpa, X., Huang, R., Moir, W., Jones, D., Fairlie, R., Tanzi., and Bush, A. Characterization of copper interactions with Alzheimer amyloid β peptides: identification of an attomolar-affinity copper binding site on amyloid β 1–42. *J. Neurochem.* **75**(3), 1219–1233 (2000).
279. Squitti, R., Barbati, G., Rossi, L., Ventriglia, M., Dal Forno, G., Cesaretti, S., Moffa, F., Caridi, I., Cassetta, E., Pasqualetti, P., Calabrese, L., Lupoi, D., and Rossini, P. M. Excess of nonceruloplasmin serum copper in AD correlates with MMSE, CSF beta-amyloid, and h-tau. *Neurology.* **67**(1), 76–82 (2006).
280. Lovell, M., Robertson, J., Teesdale, W., Campbell, J., and Markesbery, W. Copper, iron and zinc in Alzheimer's disease senile plaques. *J. Neurol. Sci.* **158**(1), 47–52 (1998).
281. Dong, J., Atwood, C., Anderson, V., Siedlak, S., Smith, M., Perry, G., and Carey, P. Metal binding and oxidation of amyloid-β within isolated senile plaque cores: Raman microscopic evidence. *Biochemistry.* **42**(10), 2768–2773 (2003).
282. Miller, L. M., Wang, Q., Telivala, T. P., Smith, R. J., Lanzirotti, A., and Miklossy, J. Synchrotron-based infrared and X-ray imaging shows focalized accumulation of Cu and Zn co-localized with β-amyloid deposits in Alzheimer's disease. *J. Struct. Biol.* **155**(1), 30–37 (2006).
283. Opazo, C., Huang, X., Cherny, R., Moir, R., Roher, A., White, A., Cappai, R., Masters, C., Tanzi, R., Inestrosa, N., and Bush, A. Metalloenzyme-like activity of Alzheimer's disease β-amyloid-Cu-dependent catalytic conversion of dopamine, cholesterol, and biological reducing agents to neurotoxic H_2O_2. *J. Biol. Chem.* **277**(43), 40302–40308 (2002).
284. Roberts, B. R., Ryan, T. M., Bush, A. I., Masters, C. L., and Duce, J. A. The role of metallobiology and amyloid-β peptides in Alzheimer's disease. *J. Neurochem.* **120**(1, SI), 149–166 (2012).
285. Bush, A. I. and Tanzi, R. E. Therapeutics for Alzheimer's disease based on the Metal Hypothesis. *Neurotherapeutics.* **5**(3), 421–432 (2008).
286. Markesbery, W. Oxidative stress hypothesis in Alzheimer's disease. *Free Rad. Biol. Med.* **23**(1), 134–147 (1997).
287. Praticò, D. Oxidative stress hypothesis in Alzheimer's disease: a reappraisal. *Trends Pharmacol. Sci.* **29**(12), 609–615 (2008).
288. Musiek, S. and Holtzman, D. M., Three dimensions of the amyloid hypothesis: time, space, and 'wingmen'. *Nat. Neurosci.* **18**(6), 800–806 (2015).
289. Goetz, A., Juergen nd Gladbach., Pennanen, L., van Eersel, J., Schild, A., David, D., and Ittner, L. M. Animal models reveal role for tau phosphorylation in human disease. *Biochim. Biophys. Acta.–Molec. Basis Dis.* **1802**(10, SI), 860–871 (2010).
290. Ittner, L., Ke, Y., Delerue, F., Bi, M., Gladbach, A., van Eersel, J., Wölfing, H., Chieng, B., Christie, M., Napier, I., Eckert, A., Staufenbiel, M., Hardeman, E., and Götz, J. Dendritic function of Tau mediates amyloid-β toxicity in Alzheimer's disease mouse models. *Cell.* **142**, 387–397 (2010).

Exploring the Aggregation Mechanism of Intrinsically Disordered Tau Protein

2

Ruxi Qi, Guanghong Wei,* Ruth Nussinov†,‡ and Buyong Ma†*

**State Key Laboratory of Surface Physics, Key Laboratory for Computational Physical Sciences (MOE), and Department of Physics, Fudan University, Shanghai, P. R. China*
†Basic Science Program, Leidos Biomedical Research, Inc. Cancer and Inflammation Program, National Cancer Institute, Frederick, Maryland 21702, USA
‡Sackler Inst. of Molecular Medicine, Department of Human Genetics and Molecular Medicine, Sackler School of Medicine, Tel Aviv University, Tel Aviv 69978, Israel

One of the hallmarks of Alzheimer's disease is the neurofibrillary tangles formed by the aggregation of microtubule-binding protein tau. Many factors, such as phosphorylation and polyanions, influence the pathological self-assembly of tau proteins. In this chapter, we review recent progress in elucidating the mechanisms of tau aggregation obtained by experiments and computer simulations. Particularly, we focus on the conformational ensemble of tau and its fragments and discuss the impact of several factors, including cross seeding, mutations, phosphorylation, temperature, polyanions, and osmolytes. Emerging common features of tau aggregation are highlighted and discussed.

1. Introduction

Alzheimer's disease (AD) is characterized by the presence of the extracellular senile plaques of amyloid-β (Aβ) proteins and the intracellular neurofibrillary tangles (NFTs) of tau proteins in the brain of AD patients.[1,2] Tau dysfunctions, manifested as hyper-phosphorylation and aggregation, are major proximal causes of neuron loss in AD.[3] However, during the past decades, research focused more on the aggregation of Aβ, probably due to its small size, though accumulating evidence shows that tau plays an indispensable role in AD pathogenesis.[4–6] Recent experimental studies reported that tau depletion can protect against Aβ-associated neuron death.[7,8] Thus, with a growing body of evidence providing solid experimental support for Aβ-induced tau pathology, it is generally accepted that Aβ accumulation occurs prior to tau in AD pathogenesis and triggers the conversion of tau from normal to toxic states.[5,8–13] Tau proteins are microtubule-binding proteins. Six tau isoforms exist in human brain tissue, with different projection domains and different number of microtubule-binding repeats (Fig. 1). Three isoforms have three binding repeats (3R) and the other three have four binding repeats (4R).[15] Both

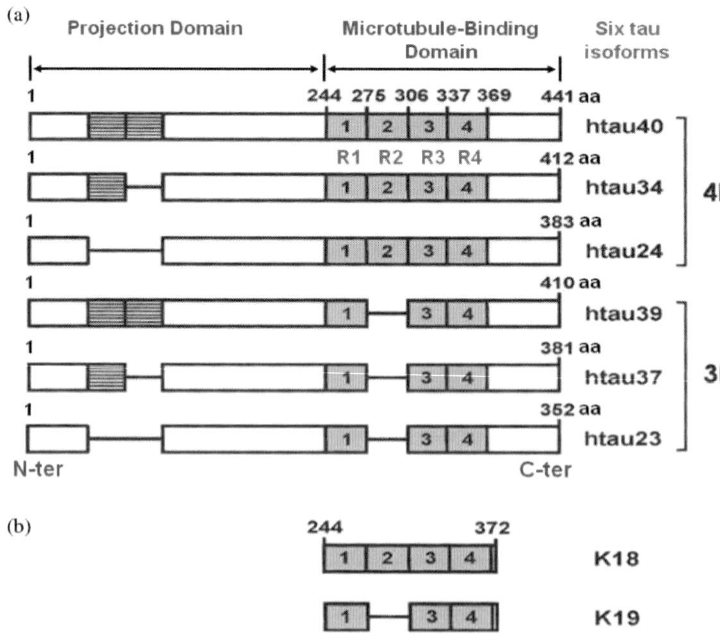

Figure 1: Schematic representation of the six human tau isoforms and two tau constructs. (a) Tau isoforms differ by the absence or presence of one or two 29-amino acid inserts in the amino-terminal part, in combination with either three (R1, R3, and R4) or four (R1–R4) repeat regions (black boxes) in the carboxy-terminal part. Isoform sizes range from 352 amino acids (aa) to 441 aa. (b) The truncated constructs K18 and K19, which contain the repeats (four for K18 and three for K19) plus three additional residues at the C termini. This figure is adapted with permission from Ref. 14

3R- and 4R-containing isoforms are found in the fibrils deposited in the brain of AD patients.[16] K18 and K19 are the truncated constructs of the 4R and 3R tau, consisting of only four (residues 243–372) and three (with repeat-two being deleted) repeats, respectively. The microtubule-binding repeats constitute the core of the tau fibrils,[17] whereas the remaining residues of tau form an unstructured fuzzy coat.[18]

A number of biophysical and spectroscopic studies revealed that both Aβ and tau belong to the category of intrinsically disordered proteins (IDPs),[19–23] which lack a well-defined three-dimensional structure. Although the monomeric tau appears a random polymer,[24,25] some regions exhibit a preference for short helix or β-strand conformations.[26] Because of their lack of a stable unique structure, IDPs such as tau constantly exchange between different conformations, making it challenging to capture the structural properties of tau.

In recent years, an increasing number of experimental and computational studies have focused on exploring the aggregation mechanism of tau proteins, which provide novel structural and mechanistic insights into the nature of tau aggregation. Here we overview these to elucidate the underlying mechanisms of tau aggregation. We focus on the conformational ensemble of monomeric tau constructs and the effects of several factors (such as cross seeding, mutations, phosphorylation, temperature, polyanions, and osmolytes) on the structural and aggregation properties of tau and its fragments.

2. Conformational Ensembles of Truncated Tau Constructs K18 and K19

The tau protein is an intrinsically disordered protein, thus it is experimentally challenging to characterize its three-dimensional conformations that are crucial for tau fibrillation. Complementing experiments, all-atom molecular dynamics (MD) simulations can provide detailed structural information of tau at the atomic level. By performing extensive all-atom explicit-solvent replica-exchange MD (REMD) simulations, Luo *et al.* investigated the conformational ensembles of K18 and K19, two truncated tau constructs from the microtubule (MT)-binding domain.[27] The top four most-populated conformations are presented in Fig. 2a and b for K18 and K19, respectively. The simulation results achieved quantitative agreement with experimental secondary chemical shifts (SCSs) of the C_α atom for both K18 and K19 (Fig. 2c and d). High β-sheet content and large hydrophobic surface exposure for the two core-nucleating motifs (^{275}VQIINK^{280} and ^{306}VQIVYK^{311}) were also

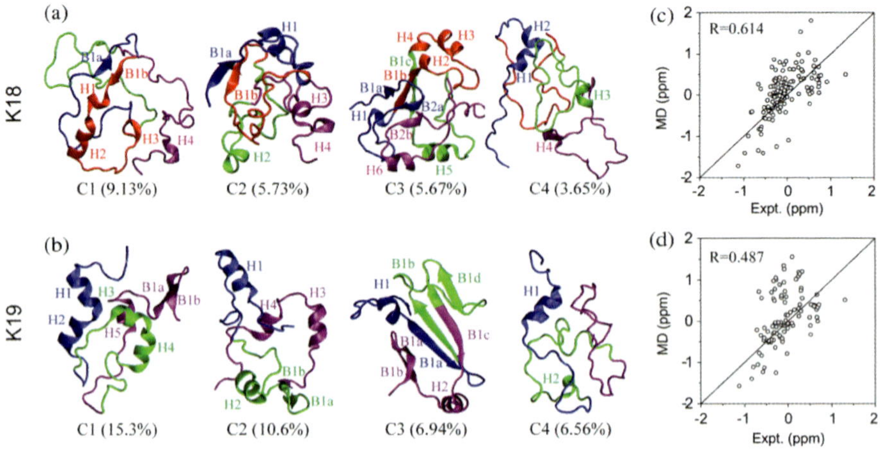

Figure 2: Conformational analysis of monomeric K18 and K19 in aqueous solution at 310 K. Representative conformations for the top four most-populated clusters (labeled by "Cn", n = 1 ~ 4) of K18 (a) and K19 (b) along with their corresponding probabilities. Secondary structures are displayed in new-cartoon representation, with different colors representing different repeats, blue for R1, red for R2, green for R3, and purple for R4, and the last three residues after R4. For each structure, helices are indicated with H1, H2,..., and β-sheets are labeled with B1, B2, ...; β-strands in the same sheet are labeled with B*n*a, B*n*b, B*n*c, ...(*n* = 1,2, ...). Two adjacent β-strands (for example, a and b) are labeled using two neighboring letters in the alphabet. Scatter plots comparing experimental (Expt.) and SPARTA-predicted (MD) SCSs of the C_α atom for K18 (c) and K19 (d). The Pearson correlation coefficients (R) between experimental and MD-generated SCSs are indicated. This figure is adapted with permission from Ref. 27

observed. These results show that K18 and K19 sample both ordered and disordered conformations although their ensemble average is disordered.

3. Cross-Seeding Effects on Tau Aggregation

Aβ amyloid plaques are usually deposited in the extracellular space and tau is produced in the neurons. However, recent studies showed that Aβ also accumulates intraneuronally[2,10] and binds to tau to form insoluble complexes within the same neurons.[28] Exposure of tau to Aβ oligomers[29] could lead to loss of microtubule integrity,[30] formation of new insoluble aggregates,[28] or an enhancement of NFT formation.[11] Although the mechanisms linking Aβ and tau-pathology have not been conclusively identified, several potential mechanisms have been postulated. One of the promising mechanisms that have drawn increasing attention is that Aβ oligomers directly trigger tau aggregation.[3,9,31–33]

Interestingly, the propagation of toxic misfolded Aβ and tau bears a striking resemblance to prion protein.[31,32] The injection of preaggregated synthetic amyloid peptides[11] or aggregated amyloid peptide enriched brain extracts[34] induced tau aggregation not only at the injection site but also in functionally connected brain regions that are remote from the site, indicating that amyloid peptides can initiate and propagate tau aggregation in these regions as well.[11,34] This prion-like behavior[35,36] of Aβ and tau underscores the need for an in-depth understanding of the cross-seeding mechanism through which Aβ induces tau pathology and tau aggregation gets propagation.

3.1. *How does Aβ cross-seed tau?*

An important issue to address in Aβ-induced tau pathology is how Aβ triggers tau aggregation, or, to put it differently, how Aβ cross-seeds tau. As a first step, Miller *et al.* investigated the structural stability of constructed U-shaped fibril-like Aβ-tau hetero-oligomers by performing all-atom MD simulations.[37] The Aβ-tau hetero-oligomers were constructed for tau's R2, R3, or R4 repeats exploiting solid-state nuclear magnetic resonance spectroscopy (NMR) derived Aβ fibril models and additional available experimental data.[20,22,38] The simulations showed that a fragment derived from repeat R2 can form stable hetero-oligomers with Aβ with better alignment in the turn region and the β-structure domain, while fragments from repeats R3 and R4 exhibit less stable oligomers when interacting with Aβ. Binding energy calculations using the generalized born method with molecular volume (GBMV) revealed that the Aβ–R2 hetero-oligomers is more stable energetically than the Aβ–Aβ and R2-R2 homo-oligomers. For repeats R3 and R4, however, the two homo-oligomers are more stable than the Aβ–tau hetero-oligomers.[37]

Recently, by using ion-mobility mass spectrometry (IM-MS), atomic force microscopy (AFM), and replica exchange MD (REMD) simulations, Do *et al.* probed the interactions between Aβ_{25-35} and tau$_{273-284}$, two fragments that contain important regions of the full-length Aβ and tau proteins.[39] Aβ_{25-35} has been shown to possess similar aggregation propensity and toxicity as the Aβ_{1-42},[40,41] and tau$_{273-284}$ fragment, which is located in the R2 repeat of tau, which encompasses one of the two core-nucleating motifs (^{275}VQIINK^{280} and ^{306}VQIVYK^{311}).[42–45] The results showed that tau$_{273-284}$ binds to Aβ_{25-35} with higher affinity to form hetero-oligomers, and formation of hetero-oligomers triggers conformational transitions from compact to extended forms for both peptides as early as the dimer stage.

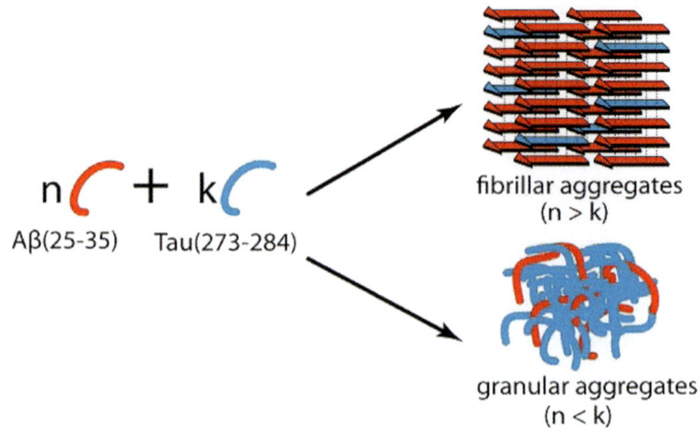

Figure 3: Illustration of how fibrillar and granular aggregates can be formed in the mixture of Ac-Aβ (25–35)-NH_2 and Ac-Tau(273–284)-NH_2. The morphologies of the aggregates are determined by the population of the individual peptides. This figure is adopted with permission from Ref. 39

Direct coaggregation of $tau_{273-284}$ and $A\beta_{25-35}$ peptides was observed, although hetero-oligomers with higher tau content are trapped at relatively small sizes and produce granular aggregates, and only those with higher Aβ content can grow into hetero-fibrils, as shown in Fig. 3. Compared to the self-aggregation of $A\beta_{25-35}$, the incorporation of $tau_{273-284}$ into β-rich $A\beta_{25-35}$ oligomers reduces the aggregation propensity of $A\beta_{25-35}$ but does not fully abolish fibril formation. $Tau_{273-284}$ acts as a modulator that drives $A\beta_{25-35}$ self-assembly into both granular and hetero-fibrillar aggregates.

Recent experiments reported that Aβ accumulation occurs prior to tau in AD pathogenesis, triggering tau-pathology, and aggregation of tau into toxic states.[46,47] How does Aβ protofibril directly cross-seed tau? To answer this question, we recently investigated the effects of a preformed $A\beta_{1-42}$ hexameric protofibril on tau monomers at the atomic level by employing all-atom REMD simulations.[48] We focused on the two truncated constructs of tau — K18 and K19. The amino acid sequences of $A\beta_{1-42}$ and of K18 are given in Fig. 4. Through a total of 19.2 μs simulations, we found that tau's conformations are regulated by the Aβ-protofibril template. Compared with isolated K18 and K19 monomers,[27] both K18 and K19 in the Aβ-tau complex are stretched into more extended conformations. Furthermore, Aβ protofibril drastically reduces the ordered secondary structure probability of tau monomers, impeding tau's intramolecular hydrogen bond and salt-bridge formation, and exposing the two buried fibril nucleating cores, ^{275}VQIINK^{280} in R2 and

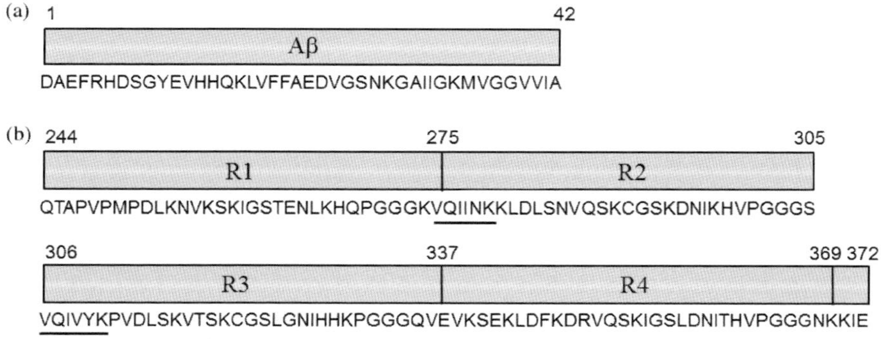

Figure 4: Amino acid sequences of (a) $A\beta_{1-42}$ and of (b) the truncated tau construct K18 which contains the four repeats (R1, R2, R3, and R4) plus three additional residues at the C terminal. The two core fibril-nucleating motifs (^{275}VQIINK^{280} and ^{306}VQIVYK^{311}) are underlined.

^{306}VQIVYK^{311} in R3, to water solvent. This observed behavior could be the first step of $A\beta$ cross-seeding tau into prion-like aggregation. Contrary to common intuition that $A\beta$ cross-seeding tau would first enhance tau's β-sheet content, it is more likely that $A\beta$ first stretches tau into more extended states, which facilitates intermolecular interactions favorable for tau fibrillation. On the basis of our findings, we proposed a "stretching-and-packing" mechanism of $A\beta$ cross-seeding tau to promote tau aggregation at the early stage. The mechanism is depicted in Fig. 5 which shows that the $A\beta$-protofibril template stretches tau into a more extended conformation by disrupting the intramolecular interactions of tau, exposing the fibril nucleation core to water. This may drive tau into β-sheet-rich conformations favorable for the next aggregation step, thus triggering prion-like tau spreading. This process would also alter the free energy landscape of tau by shifting the population of tau toward more extended conformations.

3.2. *How do tau's different isoforms cross seed each other?*

The previous section mainly focuses on $A\beta$-cross-seeded-tau aggregation mechanisms. It is also important to ask how tau's different isoforms cross-seed each other. We present some studies devoted to this issue.

Previous work has shown that asymmetric seeding barriers exist between different isoforms of tau. For example, filaments of 3R tau can recruit four-repeat-containing tau, whereas filaments of 4R tau cannot recruit 3R tau.[49] Conformational selection and population shift are two key mechanisms in biomolecular

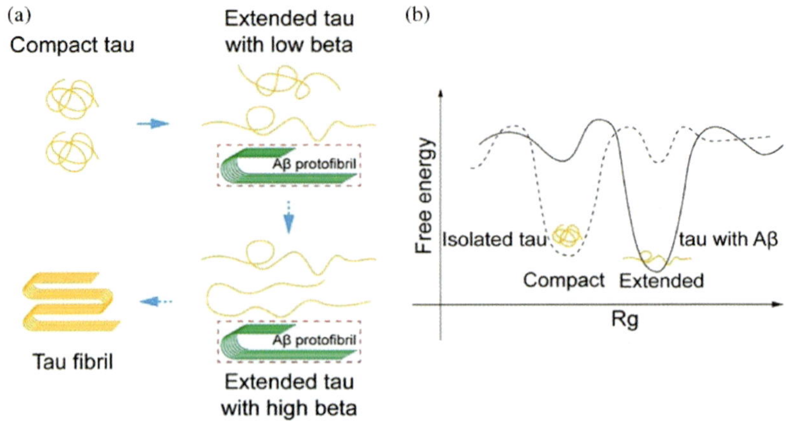

Figure 5: Schematic diagram of $A\beta$ cross-seeding tau with a "stretching-and-packing" mechanism. (a) In the first step, $A\beta$ protofibrils stretch compact tau into more extended state with reduced β-sheet content, hydrogen bond number, and salt-bridges; in the next step, the extended tau molecules form stacking conformation with regained beta content, gradually entering a prion-like growth pathway. Finally, tau aggregates into the fibril state. (b) The interaction of $A\beta$ protofibrils with tau shifts the free energy landscape of isolated tau and drives tau's population to more extended state. This figure is adopted with permission from Ref. 48

recognition.[50–52] Monomers and small oligomers binding to amyloid seeds in fibril growth is a molecular recognition event,[53] and conformational selection underlies all recognition events, including asymmetric barriers between three-repeat and four-repeat tau cross seeding. To understand the conformational selection mechanism involved in the fibril formation of three-repeat and four-repeat tau, Yu *et al.* constructed K18/K19 octamers with R3 as U shaped, L shaped or straight line (SL) shaped motifs based on the previous models by Miller *et al.*[37] They investigated the structural stabilities and related populations of different fibril-like oligomers of K18 and K19.[54] They showed that K18/K19 octamers with R3 in an L shape and those with R3 in an SL shape are highly populated and exhibit the highest structural similarity among all simulated K18 and K19 octamers, suggesting that similar folded states of K18/K19 may serve as the structural core for the K18–K19 coassembled hetero-filaments. By analyzing the structural and energetic contributions, they demonstrated that formation of stable R2 and R3 conformations is the critical step for K18 aggregation, and the formation of stable R3 conformation is critical for K19 fibrillization. The different core units in K18 and K19 may create a cross-seeding barrier for the K18 seed to trigger K19 fibril growth because R2 is not available for K19. The polymorphic nature of protein aggregation could be magnified in the cross-seeding process, which could also contribute to the asymmetric barrier between K18 and K19.

Using double electron–electron resonance (DEER) spectroscopy in combination with extensive computational analysis, Siddiqua *et al.* investigated the molecular basis of the seeding barrier.[55] They showed that filaments of three-repeat- and four-repeat-containing tau are distinct in conformation. Interestingly, the models by Yu *et al.*[54] provide a good interpretation of the experimental observations by Siddiqua *et al.* Distance measurements between spin labels in the third repeat (R3) reveal tau amyloid filaments as ensembles of known β-strand–turn–β-strand U-turn motifs (Fig. 6). Tau filaments seeded with K19 are structurally homogeneous, whereas

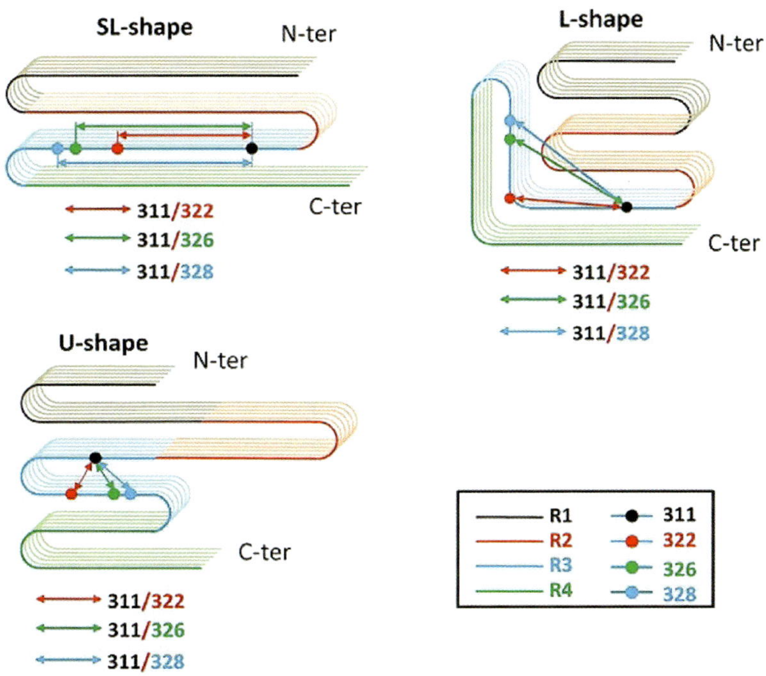

Figure 6: Structural models of tau filaments. The K18 models were optimized by extensive MD simulations and emphasize the three basic conformations of the third repeat (blue): SL-, L-, and U-shapes, which have also been modeled for K19 filaments.[54] The remaining tau repeats are represented by different colors. These repeats can assume additional conformations.[54] However, for clarity, those conformations are not depicted. Labeled residues (C_α) are indicated by colored dots. Pairs of labels for which distances were determined by DEER are labeled by arrows. Repeats 1–4 are abbreviated as R1–R4. This figure is adopted with permission from Ref. 55

filaments seeded with K18 are heterogeneous, composed of at least three distinct conformers.[55] tau-pathology is involved in prion-like transmission,[31,32,56] thus its structural heterogeneity may help explain the seeding barriers between different tau isoforms and the barriers in prion-like transmission.

4. Mutation Effects on Tau Aggregation

The primary effects of about 16 known tau mutations are reduction of the ability of tau proteins to interact with microtubules and increase of the propensity of tau to assemble into abnormal filaments.[57] Hence elucidating the influence of single point mutations at key positions on tau aggregation, especially on initial seeding process is important. Here we review some recent works that may provide new insights into the underlying mechanism.

It has been demonstrated that the K280 deletion (ΔK280) in the tau$_{273-284}$ fragment leads to increased aggregation rates.[48] Since ΔK280 is located in the core fibril-nucleating motif ^{275}VQIINK^{280} and is positively charged, tau's aggregation would be sensitive to this mutation. Larini *et al.* have investigated the aggregation of a fragment of wild-type R2 (R2/wt, with the sequence ^{273}GKVQIINKKLDL^{284}) and its ΔK280 mutant using IM-MS, transmission electron microscopy (TEM) and REMD simulations.[58] They found that R2/ΔK280 aggregates more readily than its wild-type counterpart. Simulations and experiments reveal that the R2/ΔK280 peptide adopts extended conformations to a greater extent than the R2/wt peptide. R2/wt monomer was shown to be stabilized by the presence of a salt bridge between K280 and D283, and to a lesser extent by the hydrogen bonding interaction. The removal of this constraint in R2/ΔK280 leads to more extended structures with a low content of intra-peptide hydrogen bonds, facilitating the aggregation. Addition of heparin dramatically accelerates fibril formation, particularly in the case of the mutant, which may be explained by the mechanism that heparin binds to extended species in aqueous solution and promotes formation of extended fibril-precursor species.[58] Interestingly, using the same techniques, a recent study by Ganguly *et al.* compared the aggregation properties of individual and binary mixtures of tau fragments R2/wt (^{273}GKVQIINKKLDL^{284}), R3/wt (^{306}VQIVYKPVDLSK^{317}) and R2/ΔK280. They found that R3/wt has a much stronger aggregation propensity than either R2/wt or R2/ΔK280. Hetero-dimers containing R3/wt are less stable than R3/wt homo-dimers but much more stable than homo-dimers of R2/wt and R2/ΔK280. R2/ΔK280 binds more strongly to R3/wt than R2/wt. These data indicate a dominant role of the core fibril-nucleating motif ^{306}VQIVYK^{311} in tau aggregation.[59]

As shown in Section 3.2, tau proteins show intriguing seeding properties as asymmetric seeding barrier between 3R and 4R tau.[49] To further investigate the underlying molecular basis, Margittai and coworkers have reexamined the seed selection of tau monomers into fibrils by a systematic single point mutation study using DEER spectroscopy.[60] They observed that when the mutated tau grows onto WT seeds, the mutations ΔK280, P301S, P312I, and D314I markedly change the distribution of fibril conformers after template-assisted growth, whereas other mutations I308M, S320F, G323I, G326I, and Q336R do not. These seed-selection effects can be explained by the conformational compatibilities of individual mutants with WT seeds. Consistent with the conformational selection concept,[50–52] these studies established conformational compatibility as an important driving force in tau fibril propagation.

5. Posttranslational Modification Effects on Tau Aggregation

As an intrinsically disordered protein, tau is enriched in phosphorylation sites. Indeed, the regulation of tau primarily involves posttranslational modifications (PTMs) including phosphorylation, truncation, nitration, glycation, acetylation, and methylation.[61–64] The hyperphosphorylation of tau leads to the dissociation of microtubules and the promotion of filament formation in AD.[65] Other PTMs can also regulate tau aggregation; for example, methylation has been shown to suppress tau's aggregation propensity, whereas glycation and acetylation promote pathological tau aggregation.[62,63] Here we focus on some studies of the regulation of tau aggregation by phosphorylation, and the molecular basis of tau phosphorylation and self-acetylation.

Antibodies have been developed to specifically recognize three phosphoepitopes in tau protein. The Fab fragment forms complex with the tau peptide ^{224}KKVAVVR(pT^{231})PPK-(pS^{235})PSSAKC^{241} (two phosphorylation sites at Thr-231 and Ser-235).[66] In the Fab-peptide cocrystal structure, 10 aa (^{225}KVAVVR(pT)PPK^{234}) are visible, of which six (^{225}KVAVVR(pT^{231})) interact directly with the Fab fragment, and the remaining eight residues of the peptide are disordered. The segment 224-241 is in the N-terminal side of the tau repeat domains, which starts at residue 243. Since the antibody recognizes the phosphorylated epitope in the intact molecule,[66] the segment 224–241 must be exposed in the full-length tau. The critical phosphorylation site (pThr-231) is exclusively recognized by CDR-H2, which forms a positively charged pocket to accommodate the phosphate. The highly specific phosphate recognition explains why the antibody does not bind to nonphosphorylated peptides with the same sequence.[66]

Using the time-resolved fluorescence resonance energy transfer (FRET) method, Chin *et al.* studied the conformational property of $tau_{173-183}$ (AKTP-PAPKTPP) peptide.[67] They showed that phosphorylation extends the end-to-end distance and increases the effective persistence length of this tau-derived peptide. However, the peptide extension is independent of salt concentration, indicative of a nonelectrostatic origin. These data indicate that geometric extension and stiffening at the peptide scale may be an important conformational consequence of phosphorylation in disordered proteins.

In a recent simulation study, Xu *et al.* explored the conformational consequences of hyperphosphorylation on tau proteins.[68] By combining hybrid-resolution, implicit solvent all-atom MD simulations and GBMV-based energy calculation methods, they demonstrated that the presence of the phosphorylated terminal domains alters the relative stabilities of conformations in the K18 ensemble, thus shifting the populations of full-length tau filaments. The hyperphosphorylation of the two terminal domains decreases the attractive interactions among the N- and C-terminus and repeat domains. However, the structure with the straight repeats in the core region is still the most stable, and the exposure of the repeat domains upon hyperphosphorylation could enhance tau filament aggregation.[68]

Except for the phosphorylation effect, tau has been found to possess intrinsic enzymatic activity capable of catalyzing self-acetylation mediated by a pair of catalytic cysteine residues residing within the microtubule-binding domain.[69] The acetylation of tau was shown to be able to inhibit its function and promote its pathological aggregation.[62,69,70] To understand the self-acetylation activity and aggregation propensity of tau, Luo *et al.* investigated the conformational ensembles of K18 and K19 using extensive 4.8 μs REMD simulations.[27] The simulation results revealed dynamically ordered conformations with close lysine–cysteine distances essential for tau self-acetylation (Fig. 7). The "order in disorder" property in tau conformations provides the structural basis for tau self-acetylation.

6. Polyanion and Osmolyte Effects on Tau Aggregation

It has long been known that tau is very sensitive to the presence of polyanions, such as heparin.[71] Various cellular polyanions (e.g., RNA, acidic peptides, or micelles) drastically accelerate aggregation of tau and they were found together with tau aggregates in the brains of AD patients.[72,73] *In vitro*, tau proteins do not aggregate spontaneously and adopt predominantly disordered conformations, but they

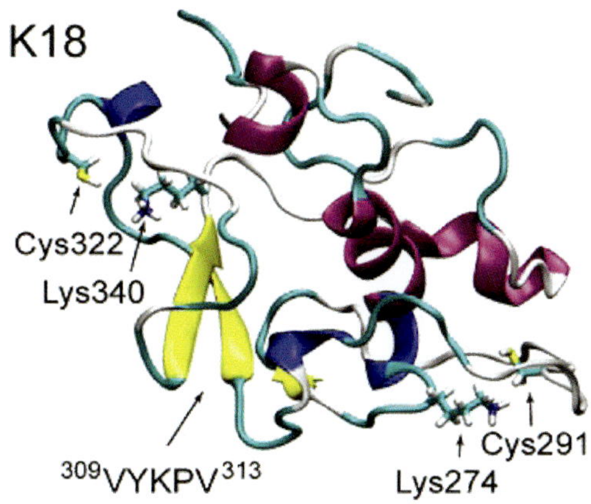

Figure 7: The representative snapshot showing close lysine-cysteine distances in the simulation.

undergo rapid aggregation in the presence of heparin.[14,20,72,74] In the work by Larini *et al.*,[58] the addition of heparin dramatically accelerates fibril formations by R2/wt, and particularly by the ΔK280 mutant. Also, the R3 peptide in the presence of heparin self-assembles into typical fibrillar structures that exhibit similar structural and morphological properties as those assembled from the full-length tau protein.[75] In tau aggregation studies, heparin is often used for accelerating the aggregation process. However, recently, using TEM and AFM, in combination with circular dichroism (CD) spectroscopy, thioflavin T (ThT) fluorescence, and X-ray scattering characterization methods, Adamcik *et al.* demonstrated that in the absence of heparin, $tau_{306-327}$ R3 fragment is able to self-assemble into large, flat, multistranded ribbons consisting of up to 45 laterally assembled protofilaments.[76] These results indicate that the tau protein and its fragments have an intrinsic ability to assemble into amyloid structures of large dimensions in the absence of aggregation-enhancing species, such as heparin or alternative polyanions.

In addition to heparin, polyglutamic acid has also been employed to study the structural consequences of tau aggregation. Different from heparin, polyglutamic acid does not immediately convert tau into oligomers.[20] Tau is predominantly monomeric in the presence of polyglutamic acid at low temperature and only aggregates into oligomers and fibrils at higher temperature and longer incubation time. Utilizing this feature, through a combined NMR spectroscopy and molecular ensemble calculation method, Akoury *et al.* examined the conformational ensembles of K18 in the presence and absence of the aggregation enhancer polyglutamic acid.[77]

They showed that binding of polyglutamic acid to tau remodels the conformational ensemble of tau. The temperature effects on the tau aggregation are consistent with previous computational and experimental studies. Using REMD simulations and polar sensitive fluorescence assays, Luo *et al.* studied the tau aggregation property upon temperature change.[78] Their REMD simulations predicted that tau proteins could form amyloid fibrils at a high temperature of 343 K, and the tau amyloid fibrils may cold dissociate at 275 K. This intriguing feature was then confirmed by fluorescence experiments and it was observed that K18 fibrils cold dissociate when cooled. They also found that heparin locks the tau fibril and prevents its reversion. The temperature-dependent stabilization of the fibril is coupled with its temperature-sensitive monomeric properties, driven by both hydrophobic and electrostatic interactions.[78]

As an IDP, the tau protein is highly soluble due to its large number of polar and charged residues. An interesting question is what the effect of solvents on tau aggregation is. In a recent study, Shea and coworkers investigated osmolytes in regulating tau structural and aggregation properties.[79] Their REMD simulation results showed that the osmolyte urea promotes the formation of extended R2/wt monomeric conformations, whereas another osmolyte trimethylamine N-oxide (TMAO) stabilizes compact R2/wt monomers. Their ThT and TEM experiments revealed that in the presence of heparin, urea suppresses R2/wt aggregation, whereas TMAO accelerates tau aggregation by decreasing the lag phase. Further analyses showed that osmolyte hydration affects peptide conformations, and osmolyte–water interactions can significantly affect the formation of both intrapeptide hydrogen bonds and salt bridges (Fig. 8). A new mechanism of redistribution of water around the perimeter of the peptide was proposed to explain the promotion of tau aggregation by TMAO.

Later, using TEM, ThT, IM-MS, electron paramagnetic resonance (EPR), Overhauser dynamic nuclear polarization (ODNP) techniques, and a N-terminus-spin-labeled R2/wt, Eschmann *et al.* studied how the peptide conformational shifts by osmolytes urea or TMAO before peptide aggregation influence the downstream peptide propensity to form fibrillary aggregates.[80] They found that varying solution conditions by adding the osmolyte urea or TMAO, produces significant differences in early monomer/dimer populations and conformations, and finally affect the aggregation pathways and end products. These results led the authors to conclude that the driving forces that accelerate R2/wt aggregation, when heparin is added, do not override the subtle intra- or inter-protein interactions induced by the initial solvent conditions.[80]

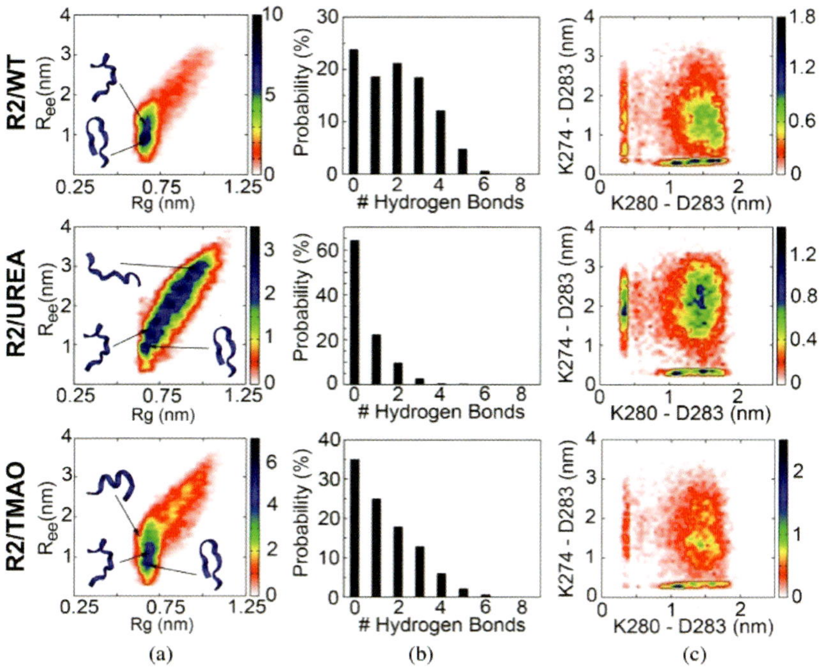

Figure 8: (a) Population densities (in normalized arbitrary units) of peptide conformations at room temperature over 150 ns (R2/wt, R2/urea) or 100 ns (R2/TMAO). R_{ee} is a measure of the peptide end-to-end distance (C terminus to N terminus), and R_g is the peptide radius of gyration. Highly populated conformations are indicated in blue, whereas less populated conformations are indicated in red and yellow. (b) Probabilities of forming intramolecular hydrogen bonds (C=O to *N*–H). (c) Population densities of monomeric salt bridges are indicated by the relative proximities between lysine and aspartic acid. This figure was adapted with permission from Ref. 80.

7. Conclusions

Recent computational and experimental studies revealed that many factors control the conformational ensemble and aggregation properties of the tau protein and its fragments, including but not limited to the cross seeding, mutations, phosphorylation, temperature, polyanions and osmolytes. Even though these studies provided in-depth insights into the nature of tau's aggregation and its pathological role in AD, controversial outputs and related questions remain to be addressed to fully understand tau aggregation.

IDPs such as tau typically contain multiple charges and few hydrophobic residues. Despite these characteristics, IDPs are not typically found in extended states, but rather populate compact states due to hydrogen bonds and salt bridges.[81,82] In our studies discussed in the previous section, we have noticed that extended conformations are frequently implicated in tau aggregation. Of course, this conformational shift can be driven by various of factors, such as the ΔK280 mutation in $tau_{273-284}$,[58] $A\beta_{25-35}$ binding to $tau_{273-284}$,[39] the urea effect on $tau_{273-284}$,[80] $A\beta_{1-42}$ cross-seeding tau K18 and K19,[48] and phosphorylation of $tau_{173-183}$.[67] These processes may either promote[39,58,67] or hinder tau aggregation,[79] and it may only reflect the initial oligomerization process of tau that needed for fibrillary growth.[48] Regarding the complexity of IDP's role in amylodic oligomerization, these conformation elongation events may reveal an important common feature that is intrinsically embedded in tau aggregation, i.e., that extended conformations are potential intermediate states. It is plausible that intermediate states will facilitate prion-like growth of tau amyloid.

Acknowledgments

Guanghong Wei acknowledges the financial support from the MOST of China (Grant No.: 2016YFA0501702) and the NSF of China (Grant No.: 11674065). Ruth Nussinov and Buyong Ma thank the financial support from NCI, NIH, under contract number HHSN261200800001E. This research was supported (in part) by the Intramural Research Program of the NIH, National Cancer Institute, Center for Cancer Research.

References

1. Alzheimer, A. Über eine eigenartige Erkankung der Hirnrinde. *Allg. Z. Psychiatr. Ihre. Grenzgeb.* **64**, 146–148 (1907).
2. LaFerla, F. M., Green, K. N., Oddo, S. Intracellular amyloid-[beta] in Alzheimer's disease. *Nat. Rev. Neurosci.* **8**(7), 499–509 (2007).
3. Bloom, G. S. Amyloid-beta and tau: the trigger and bullet in Alzheimer disease pathogenesis. *JAMA.* **71**(4), 505–508 (2014).
4. Sperling, R. A., Laviolette, P. S., O'Keefe, K., O'Brien, J., Rentz, D. M., Pihlajamaki, M., *et al.* Amyloid deposition is associated with impaired default network function in older persons without dementia. *Neuron.* **63**(2), 178–188 (2009).
5. St George-Hyslop, P. H., Morris, J. C. Will anti-amyloid therapies work for Alzheimer's disease?. *Lancet.* **372**(9634), 180–182 (2008).
6. Holmes, C., Boche, D., Wilkinson, D., Yadegarfar, G., Hopkins, V., Bayer A., *et al.* Long-term effects of Aβ42 immunisation in Alzheimer's disease: follow-up of a randomised, placebo-controlled phase I trial. *Lancet.* **372**(9634), 216–223 (2008).

7. Leroy, K., Ando, K., Laporte, V., Dedecker, R., Suain, V., Authelet, M., *et al.* Lack of tau proteins rescues neuronal cell death and decreases amyloidogenic processing of APP in APP/PS1 mice. *Am. J. Pathol.* **181**(6), 1928–1940 (2012).
8. Nussbaum, J. M., Schilling, S., Cynis, H., Silva, A., Swanson, E., Wangsanut, T., *et al.* Prion-like behaviour and tau-dependent cytotoxicity of pyroglutamylated amyloid-beta. *Nature.* **485**(7400), 651–655 (2012).
9. Stancu, I. C., Vasconcelos, B., Terwel, D., Dewachter, I. Models of beta-amyloid induced tau-pathology: the long and "folded" road to understand the mechanism. *Mol. Neurodegener.* **9**(1), 51 (2014).
10. Forny-Germano, L., Lyra, ESNM, Batista, A. F., Brito-Moreira, J., Gralle, M., Boehnke, S. E., *et al.* Alzheimer's disease-like pathology induced by amyloid-beta oligomers in nonhuman primates. *J. Neurosci.* **34**(41), 13629–13643 (2014).
11. Gotz, J., Chen, F., van Dorpe, J., Nitsch, R. M. Formation of neurofibrillary tangles in P3011 tau transgenic mice induced by Abeta 42 fibrils. *Science.* **293**(5534), 1491–1495 (2001).
12. Vossel, K. A., Zhang, K., Brodbeck, J., Daubm A. C., Sharma, P., Finkbeiner, S., *et al.* Tau reduction prevents Abeta-induced defects in axonal transport. *Science.* **330**(6001), 198 (2010).
13. Nussbaum, J. M., Seward, M. E., Bloom, G. S. Alzheimer disease: a tale of two prions. *Prion.* 7(1), 14–19 (2013).
14. Zhu, H. L., Fernandez, C., Fan, J. B., Shewmaker, F., Chen, J., Minton, A. P., *et al.* Quantitative characterization of heparin binding to tau protein: implication for inducer-mediated tau filament formation. *J. Biol. Chem.* **285**(6), 3592–3599 (2010).
15. Buee, L., Bussiere, T., Buee-Scherrer, V., Delacourte, A., Hof P. R. Tau protein isoforms, phosphorylation and role in neurodegenerative disorders. *Brain Res. Rev.* **33**(1), 95–130 (2000).
16. Goedert, M, Spillantini, M. G., Cairns, N. J., Crowther, R. A. Tau proteins of Alzheimer paired helical filaments: abnormal phosphorylation of all six brain isoforms. *Neuron.* **8**(1), 159–168 (1992).
17. Novak, M., Kabat, J., Wischik, C. M. Molecular characterization of the minimal protease resistant tau unit of the Alzheimer's disease paired helical filament. *EMBO J.* **12**(1), 365–370 (1993).
18. Wegmann, S., Medalsy, I. D., Mandelkow, E., Muller, D. J. The fuzzy coat of pathological human Tau fibrils is a two-layered polyelectrolyte brush. *Proc. Natl. Acad. Sci. U. S. A.* **110**(4), E313–E3121 (2013).
19. Uversky, V. N. Intrinsic disorder in proteins associated with neurodegenerative diseases. In ed. Ovádi, J. and Orosz, F. *Protein Folding and Misfolding: Neurodegenerative Diseases.* Dordrecht, The Netherlands: Springer; 2009. pp. 21–75.
20. Mukrasch, M. D., Biernat, J., von Bergen, M., Griesinger, C., Mandelkow, E., and Zweckstetter, M. Sites of Tau important for aggregation populate β-structure and bind to microtubules and polyanions. *J. Biol. Chem.* **280**(26), 24978–24986 (2005).
21. Jeganathan, S., von Bergen, M., Mandelkow, E. M., Mandelkow, E. The natively unfolded character of tau and its aggregation to Alzheimer-like paired helical filaments. *Biochemistry.* **47**(40), 10526–10539 (2008).

22. Mukrasch, M. D., Bibow, S., Korukottu, J., Jeganathan, S., Biernat, J., Griesinger C., *et al.* Structural polymorphism of 441-residue tau at single residue resolution. *PLoS Biol.* 7(2), e34 (2009).
23. Schwalbe, M., Ozenne, V., Bibow, S., Jaremko, M., Jaremko, L., Gajda, M., *et al.* Predictive atomic resolution descriptions of intrinsically disordered hTau40 and alpha-synuclein in solution from NMR and small angle scattering. *Structure.* **22**(2), 238–249 (2014).
24. Schweers, O., Schonbrunn-Hanebeck, E., Marx, A., Mandelkow, E. Structural studies of tau protein and Alzheimer paired helical filaments show no evidence for beta-structure. *J. Bio. Chem.* **269**(39), 24290–24297 (1994).
25. Mylonas, E., Hascher, A., Bernado, P., Blackledge, M., Mandelkow, E., Svergun, D. I. Domain conformation of tau protein studied by solution small-angle X-ray scattering. *Biochemistry.* **47**(39), 10345–10353 (2008).
26. Eliezer, D., Barre, P., Kobaslija, M., Chan, D., Li, X., Heend, L. Residual structure in the repeat domain of tau: echoes of microtubule binding and paired helical filament formation. *Biochemistry.* **44**(3), 1026–1036 (2005).
27. Luo, Y., Ma, B., Nussinov, R., Wei, G. Structural insight into tau protein's paradox of intrinsically disordered behavior, self-acetylation activity, and aggregation. *J. Phys. Chem. Lett.* **5**(17), 3026–3031 (2014).
28. Guo, J. P., Arai, T., Miklossy, J., McGeer, P. L. Aβ and tau form soluble complexes that may promote self aggregation of both into the insoluble forms observed in Alzheimer's disease. *Proc Natl Acad Sci U. S. A.* **103**(6), 1953–1958 (2006).
29. Ittner, L. M., Götz, J. Amyloid-β and tau — a toxic pas de deux in Alzheimer's disease. *Nat. Rev. Neurosci.* **12**(2), 67–72 (2011).
30. King, M. E., Kan, H. M., Baas, P. W., Erisir, A., Glabe, C. G., and Bloom, G. S. Tau-dependent microtubule disassembly initiated by prefibrillar beta-amyloid. *J. Cell. Biol.* **175**(4), 541–546 (2006).
31. Clavaguera, F., Bolmont, T., Crowther, R. A., Abramowski, D., Frank, S., Probst, A., *et al.* Transmission and spreading of tauopathy in transgenic mouse brain. *Nat. Cell. Biol.* **11**(7), 909–913 (2009).
32. Jucker, M., Walker, L. C. Self-propagation of pathogenic protein aggregates in neurodegenerative diseases. *Nature.* **501**(7465), 45–51 (2013).
33. Manczak, M., Reddy, P. H. Abnormal interaction of oligomeric amyloid-beta with phosphorylated tau: implications to synaptic dysfunction and neuronal damage. *J. Alzheimers Dis.* **36**(2), 285–295 (2013).
34. Bolmont, T., Clavaguera, F., Meyer-Luehmann, M., Herzig, M. C., Radde, R., Staufenbiel, M., *et al.* Induction of tau pathology by intracerebral infusion of amyloid-β-containing brain extract and by amyloid-β deposition in APP $\times$ tau transgenic mice. *Am. J. Pathol.* **171**(6), 2012–2020 (2007).
35. Walker, L. C., Diamond, M. I., Duff, K. E., Hyman, B. T. Mechanisms of protein seeding in neurodegenerative diseases. *JAMA Neurol.* **70**(3), 304–310 (2013).
36. Prusiner, S. B. A unifying role for prions in neurodegenerative diseases. *Science.* **336**(6088), 1511–1513 (2012).
37. Miller, Y., Ma. B., Nussinov, R. Synergistic interactions between repeats in tau protein and Aβ amyloids may be responsible for accelerated aggregation via polymorphic states. *Biochemistry.* **50**(23), 5172–5181 (2011).

38. Luhrs, T., Ritter, C., Adrian, M., Riek-Loher, D., Bohrmann, B., Dobeli H, *et al.* 3D structure of Alzheimer's amyloid-beta(1-42) fibrils. *Proc. Natl. Acad. Sci. U. S. A.* **102**(48), 17342–17347 (2005).
39. Do, T. D., Economou, N. J., Chamas, A., Buratto, S. K., Shea, J-E., Bowers, M. T. Interactions between amyloid-β and tau fragments promote aberrant aggregates: implications for amyloid toxicity. *J. Phys. Chem. B.* **118**(38), 11220–11230 (2011).
40. Clementi, M. E., Marini, S., Coletta, M., Orsini, F., Giardina, B., and Misiti, F. Abeta(31–35) and Abeta(25–35) fragments of amyloid beta-protein induce cellular death through apoptotic signals: role of the redox state of methionine-35. *FEBS Lett.* **579**(13), 2913–2918 (2005).
41. Pike, C. J., Walencewicz-Wasserman, A. J., Kosmoski, J., Cribbs, D. H., Glabe, C. G., and Cotman, C. W. Structure-activity analyses of beta-amyloid peptides: contributions of the beta 25–35 region to aggregation and neurotoxicity. *J. Neurochem.* **64**(1), 253–265 (1995).
42. Moore, C. L., Huang, M. H., Robbennolt, S. A., Voss, K. R., Combs, B., Gamblin, T. C., *et al.* Secondary nucleating sequences affect kinetics and thermodynamics of tau aggregation. *Biochemistry.* **50**(50), 10876–10886 (2011).
43. Li, W., Lee, V. M. Characterization of two VQIXXK motifs for tau fibrillization in vitro. *Biochemistry.* **45**(51), 15692–156701(2006).
44. Goux, W. J., Kopplin, L., Nguyen, A. D., Leak, K., Rutkofsky, M., Shanmuganandam, V. D., *et al.* The formation of straight and twisted filaments from short tau peptides. *J. Biol. Chem.* **279**(26), 26868–26875 (2004).
45. von Bergen, M., Barghorn, S., Li, L., Marx, A., Biernat, J., Mandelkow, E. M., *et al.* Mutations of tau protein in frontotemporal dementia promote aggregation of paired helical filaments by enhancing local beta-structure. *J. Biol. Chem.* **276**(51), 48165–48174 (2001).
46. Stancu, I.-C., Vasconcelos, B., Terwel, D., Dewachter, I. Models of β-amyloid induced tau-pathology: the long and "folded" road to understand the mechanism. *Mol. Neurodegener.* **9**(1) 1–14 (2014).
47. Lasagna-Reeves, C. A., Castillo-Carranza, D. L., Guerrero-Muoz, M. J., Jackson, G. R., Kayed, R. Preparation and characterization of neurotoxic tau oligomers. *Biochemistry.* **49**(47), 10039–10041 (2010).
48. Qi, R., Luo, Y., Wei, G., Nussinov, R., Ma, B. Aβ "stretching-and-packing" cross-seeding mechanism can trigger tau protein aggregation. *J. Phys. Chem. Lett.* **6**(16), 3276–3282 (2015).
49. Dinkel, P. D., Siddiqua, A., Huynh, H., Shah, M., and Margittai, M. Variations in filament conformation dictate seeding barrier between three- and four-repeat tau. *Biochemistry.* **50**(20), 4330–6 (2011).
50. Csermely, P., Palotai, R., and Nussinov, R. Induced fit, conformational selection and independent dynamic segments: an extended view of binding events. *Trends Biochem. Sci.* **35**(10), 539–546 (2010).
51. Ma, B., Kumar, S., Tsai, C.-J., Nussinov, R. Folding funnels and binding mechanisms. *Protein Eng.* **12**(9), 713–720, (1999).
52. Ma, B., Nussinov, R. Enzyme dynamics point to stepwise conformational selection in catalysis. *Curr. Opin. Chem. Biol.* **14**(5), 652–659 (2010).

53. Ma, B., Nussinov, R. Selective molecular recognition in amyloid growth and transmission and cross-species barriers. *J. Mol. Biol.* **421**(2–3), 172–184 (2012).
54. Yu, X., Luo, Y., Dinkel, P., Zheng, J., Wei, G., Margittai, M., *et al.* Cross-seeding and conformational selection between three- and four-repeat human tau proteins. *J. Biol. Chem.* **287**(18), 14950–14959 (2012).
55. Siddiqua, A., Luo, Y., Meyer, V., Swanson, M. A., Yu, X., Wei, G., *et al.* Conformational basis for asymmetric seeding barrier in filaments of three- and four-repeat tau. *J. Am. Chem. Soc.* **134**(24), 10271–10278 (2012).
56. Clavaguera, F., Akatsu, H., Fraser, G., Crowther, R. A., Frank, S., Hench, J., *et al.* Brain homogenates from human tauopathies induce tau inclusions in mouse brain. *Proc. Natl. Acad. Sci. U. S. A.* **110**(23), 9535–9340 (2013).
57. Goedert, M., Jakes, R. Mutations causing neurodegenerative tauopathies. *Biochim. Biophys. Acta.* **1739**(2–3), 240–250 (2005).
58. Larini, L., Gessel, M. M., LaPointe, N. E., Do, T. D., Bowers, M. T., Feinstein, S. C., *et al.* Initiation of assembly of tau(273–284) and its ΔK280 mutant: an experimental and computational study. *Phys. Chem. Chem. Phys.* **15**(23), 8916–8928 (2013).
59. Ganguly, P., Do, T. D., Larini, L., LaPointe, N. E., Sercel, A. J., Shade, M. F., *et al.* Tau assembly: the dominant role of PHF6 (VQIVYK) in microtubule binding region repeat R3. *J. Phys. Chem. B.* **119**(13), 4582–4593 2015.
60. Meyer, V., Dinkel, P. D., Luo, Y., Yu, X., Wei, G., Zheng, J., *et al.* Single mutations in tau modulate the populations of fibril conformers through seed selection. *Angew. Chem. Int. Ed. Engl.* **53**(6), 1590–1593 (2014).
61. Martin, L., Latypova, X., Terro, F. Post-translational modifications of tau protein: implications for Alzheimer's disease. *Neurochem. Int.* **58**(4), 458–471 (2011).
62. Cohen, T. J., Guo, J. L., Hurtado, D. E., Kwong, L. K., Mills, I. P., Trojanowski, J. Q., *et al.* The acetylation of tau inhibits its function and promotes pathological tau aggregation. *Nat. Commun.* **2**, 252 (2011).
63. Funk, K. E., Thomas, S. N., Schafer, K. N., Cooper, G. L., Liao, Z., Clark, D. J., *et al.* Lysine methylation is an endogenous post-translational modification of tau protein in human brain and a modulator of aggregation propensity. *Biochem. J.* **462**(1), 77–88 (2014).
64. Reynolds, M. R., Berry, R. W., Binder, L. I. Site-specific nitration differentially influences tau assembly in vitro. *Biochemistry.* **44**(42), 13997–14009 (2005).
65. Johnson, G. V. and Stoothoff, W. H. Tau phosphorylation in neuronal cell function and dysfunction. *J. Cell. Sci.* **117**(Pt 24):5721–5729 (2004).
66. Shih, H. H., Tu, C., Cao, W., Klein, A., Ramsey, R., Fennell, B. J., *et al.* An ultra-specific avian antibody to phosphorylated tau protein reveals a unique mechanism for phosphoepitope recognition. *J. Biol. Chem.* **287**(53), 44425–44434 (2012).
67. Chin Alexander, F., Toptygin, D., Elam, W. A., Schrank Travis, P., Hilser Vincent, J. Phosphorylation increases persistence length and end-to-end distance of a segment of tau protein. *Biophys. J.* **110**(2), 362–371 (2016).
68. Xu, L., Zheng, J., Margittai, M., Nussinov, R., Ma, B. How does hyperphopsphorylation promote tau aggregation and modulate filament structure and stability?. *ACS Chem. Neurosci.* **7**(5), 565–575 (2016).

69. Cohen, T. J., Friedmann, D., Hwang, A. W., Marmorstein, R., Lee, V. M. The microtubule-associated tau protein has intrinsic acetyltransferase activity. *Nat. Struct. Mol. Biol.* **20**(6), 756–762 (2013).
70. Min, S. W., Cho, S. H., Zhou, Y., Schroeder, S., Haroutunian, V., Seeley, W. W., *et al.* Acetylation of tau inhibits its degradation and contributes to tauopathy. *Neuron.* **67**(6), 953–966 (2010).
71. Friedhoff, P., von Bergen, M., Mandelkow, E. M., Mandelkow, E. Structure of tau protein and assembly into paired helical filaments. *Biochim. Biophys. Acta.* **1502**(1), 122–132 (2000).
72. Goedert, M., Jakes, R., Spillantini, M. G., Hasegawa, M., Smith, M. J., and Crowther, R. A. Assembly of microtubule-associated protein tau into Alzheimer-like filaments induced by sulphated glycosaminoglycans. *Nature.* **383**(6600), 550–553 (1996).
73. Paudel, H. K., Li, W. Heparin-induced conformational change in microtubule-associated protein tau as detected by chemical cross-linking and phosphopeptide mapping. *J. Biol. Chem.* **274**(12), 8029–8038 (1999).
74. Sibille, N., Sillen, A., Leroy, A., Wieruszeski, J. M., Mulloy, B., Landrieu, I., *et al.* Structural impact of heparin binding to full-length tau as studied by NMR spectroscopy. *Biochemistry.* **45**(41), 12560–12572 (2006).
75. Minoura, K., Yao, T-M., Tomoo, K., Sumida, M., Sasaki, M., Taniguchi, T., *et al.* Different associational and conformational behaviors between the second and third repeat fragments in the tau microtubule-binding domain. *Eur. J. Biochem.* **271**(3), 545–552 (2004).
76. Adamcik, J., Sánchez-Ferrer, A., Ait-Bouziad, N., Reynolds, N. P., Lashuel, H. A., Mezzenga, R. Microtubule-binding R3 fragment from tau self-assembles into giant multistranded amyloid ribbons. *Angew. Chem.* **128**(2), 628–632 (2016).
77. Akoury, E., Mukrasch, M. D., Biernat, J., Tepper, K., Ozenne, V., Mandelkow, E., *et al.* Remodeling of the conformational ensemble of the repeat domain of tau by an aggregation enhancer. *Protein Sci.* **25**(5), 1010–1020 (2016).
78. Luo, Y., Dinkel, P., Yu, X., Margittai, M., Zheng, J., Nussinov, R., *et al.* Molecular insights into the reversible formation of tau protein fibrils. *Chem. Commun. (Camb).* **49**(34), 3582–3584 (2013).
79. Levine, Z. A., Larini, L., LaPointe, N. E., Feinstein, S. C., and Shea, J.-E. Regulation and aggregation of intrinsically disordered peptides. *Proc. Natl. Acad. Sci. U. S. A.* **112**(9), 2758–2763 (2015).
80. Eschmann, N. A., Do, T. D., LaPointe, N. E., Shea, J.-E., Feinstein, S. C., Bowers, M. T., *et al.* Tau aggregation propensity engrained in its solution state. *J. Phys. Chem. B.* **119**(45), 14421–14432 (2015).
81. Dobson, C. M. Protein misfolding, evolution and disease. *Trends Biochem. Sci.* **24**(9), 329–332 (1999).
82. Klein, W. L., Krafft, G. A., and Finch, C. E. Targeting small Abeta oligomers: the solution to an Alzheimer's disease conundrum?. *Trends Neurosci.* **24**(4), 219–224 (2001).

Design and Application of Anti-Amyloid Peptide-Based Inhibitors as a Therapeutic Strategy of Alzheimer's Disease

3

Rita P.-Y. Chen

Institute of Biological Chemistry, Academia Sinica, Taiwan

The success of anti-amyloid β-protein passive immunotherapy on patients with mild cognitive declines supports the amyloid cascade hypothesis and indicates the potential of anti-amyloid strategy in treating Alzheimer's disease (AD). Synthetic peptide inhibitors have the advantages of target specificity, low cost, and low side effect. Many peptides have been designed against amyloid β-protein (Aβ) association, and some of them have been tested in animal models. This review summarizes the principles of peptide design, the modifications that can be used to increase the inhibition effect and bioavailability of these peptides, and the *in vitro* methods commonly used for efficacy evaluation. Several animal models used to validate the efficacy of anti-AD drugs and the peptide delivery strategies in the animal experiments are also introduced. At the end, the designed anti-amyloid peptides that have been tested *in vivo* are summarized and compared. There are eight peptides effective in reducing the Aβ burden in the brain, and five of them are able to retard the Aβ-induced cognitive decline. Considering the high cost of anti-Aβ passive immunotherapy, anti-amyloid peptide drugs could be a less expensive alternative in AD therapy.

1. Introduction

Alzheimer's disease (AD) is the most common form of dementia (~62% in 37 million patients with dementia). There is a high rate of AD occurrence in people over the age of 65; thus, its patient number keeps increasing with the extended human lifespan. AD is an incurable disease and only symptomatic treatments are currently available on the market.[1] The pathological characteristics of AD include intracellular neurofibrillary tangles formed from hyperphosphorylated tau protein and extracellular amyloid plaque deposition formed from Aβ peptides. The amyloid cascade hypothesis emphasizes the importance of amyloid plaque deposition in the onset of AD development.[2,3] Recently, in the clinical trials of passive anti-amyloid immunotherapy, aducanumab and solanezumab showed positive responses in retarding cognitive decline in people with prodromal or mild AD,[4–6] suggesting that anti-amyloid is a promising strategy in delaying the onset of AD.[7,8]

Alzheimer's amyloid plaques are formed via intermolecular hydrogen bonds among the Aβ peptides, which are the catabolic products from a type I membrane protein called amyloid precursor protein (APP). The Aβ peptides self-associate to form a cross-β structure. Three integral membrane proteases, α-secretase, β-secretase, and γ-secretase, process APP (Fig. 1).[9,10] In the nonamyloidogenic pathway, α-secretase cleaves APP at a position about 12 residues above the membrane to yield sAPPα and C83, and the latter is further cleaved by γ-secretase to yield p3 and a C-terminal fragment named APP intracellular domain (AICD). In the amyloidogenic pathway, β-secretase cleaves APP at a higher position to yield

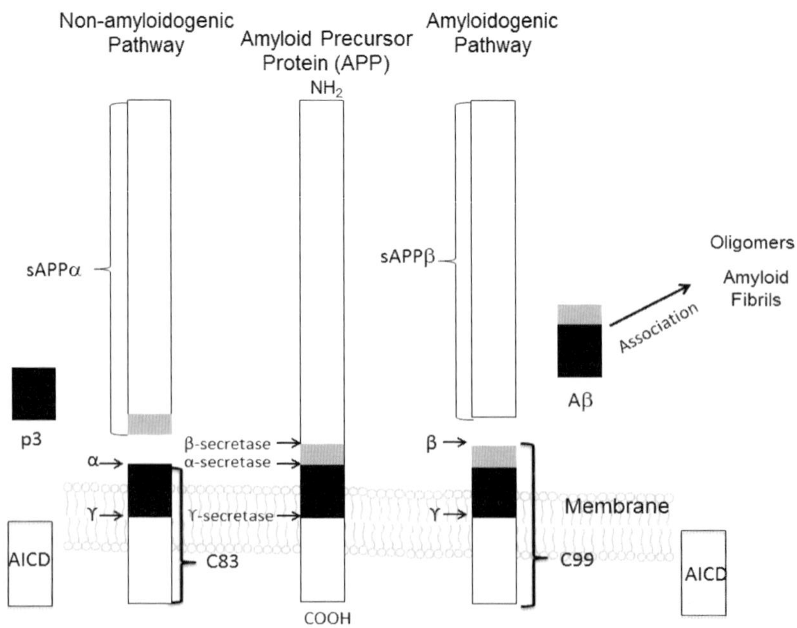

Figure 1: APP processing pathways. Depending on the cut site of γ-secretase, different lengths of Aβ peptides are produced, such as Aβ39, Aβ40, and Aβ42.

sAPPβ and C99, which is further cleaved by γ-secretase to yield the Aβ peptides and AICD. Depending on the cleavage site of γ-secretase, Aβ peptides could be 39–43 amino acid long. Aβ40 (containing 40 amino acid residues) accounts of ~90% of all Aβ peptides.[11] On the other hand, the predominant form found in amyloid plaques is Aβ42, with two more C-terminal residues, Ile and Ala.

It is widely accepted that amyloid propagation follows a nucleation-dependent polymerization mechanism (Fig. 2). In this model, the formation of nuclei (seed) is the rate-determination step, and the time required for nuclei formation is the lag time observed in the kinetics of amyloidogenesis. This is why all the amyloid-related diseases have a long latent period. The nucleation rate is proportional to x order of the concentration of the amyloid precursor state. When the amount of nuclei reaches a critical concentration, the elongation phase starts. The elongation rate is proportional to the concentrations of seed and amyloid precursor state. Because Aβ amyloid plaque deposition occurs earlier than cognitive decline, inhibiting the

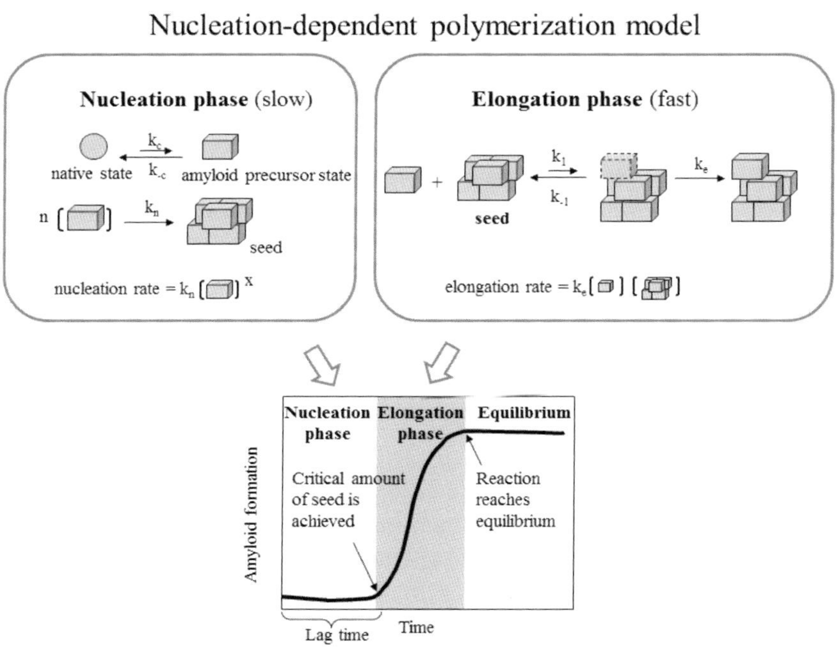

Figure 2: The nucleation-dependent polymerization model for amyloid formation. The native state of the target protein or peptide prone to amyloid formation is represented by ○. Under certain conditions, the target protein or peptide might favor populating an amyloid precursor state (represented by □), which tends to associate into nuclei. The conversion rate constants between these two states are k_c and k_{-c}. k_n and k_e are the nucleation rate constant and elongation rate constant, respectively.

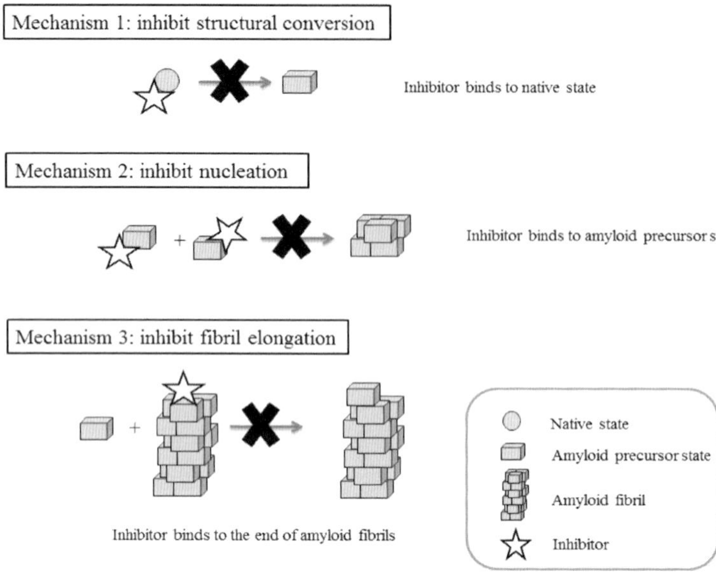

Figure 3: Schematic presentation of three anti-amyloid inhibition mechanisms. In mechanism (1), the inhibitor binds to the native state of the target in order to stabilize it from converting into the amyloid precursor state. In mechanism (2), the inhibitor binds to the amyloid precursor state, which is prone to association, thereby inhibiting the nucleation step. In mechanism (3), the inhibitor binds to the growing end of the amyloid fibril. This fibril can no longer interact with the amyloid precursor state, so fibril elongation is blocked.

formation of such aggregates is considered a potential approach in the prevention or treatment of AD. According to the nucleation-dependent polymerization model, inhibitor design can be based on the following three mechanisms: (1) interacting with the native state and stabilizing it from converting into the amyloid precursor state, (2) interacting with the amyloid precursor state to prevent it from associating into nuclei, and (3) interacting with nuclei or fibrils to block the surface for further elongation (Fig. 3). Because the amyloid precursor state is not a stable and major species, most peptide inhibitors are designed to work on mechanisms (1) and (3).

2. Design of Peptide Inhibitors

2.1. *Selection of peptide sequence based on amyloidopathy*

Because amyloid propagation is a self-association process, designing a peptide inhibitor to inhibit $A\beta$ formation is a technically sound strategy. There are many peptides reported to have anti-amyloid efficacy.[12–14] The peptide sequence can be directly adopted from the Aβ sequence, obtained from screening, or determined by rational design.

2.1.1. *Adopted from* Aβ *sequence*

(1) LVFFA (Aβ^{17-21}) and KLVFF (Aβ^{16-20})

The C terminus of the Aβ peptides contains the upper half of the membrane-spanning region of APP. As shown in Fig. 4, the C terminus is extremely hydrophobic, whereas the N terminus contains many hydrophilic and charged residues, except the LVFFA sequence (residues 17–21). There are quite a few reports discussing the relationship between the Aβ sequence and its toxicity or aggregation propensity.[15,16] Williams *et al.* substituted Ala or Pro at different positions of the Aβ40 sequence and examined the effect of each substitution in the fibril-elongation thermodynamics by measuring the peptide critical concentration for fibril growth.[17–19] They reported three sequence elements, 18–21, 25–26, and 32–33, that were particularly sensitive to substitutions. In addition,

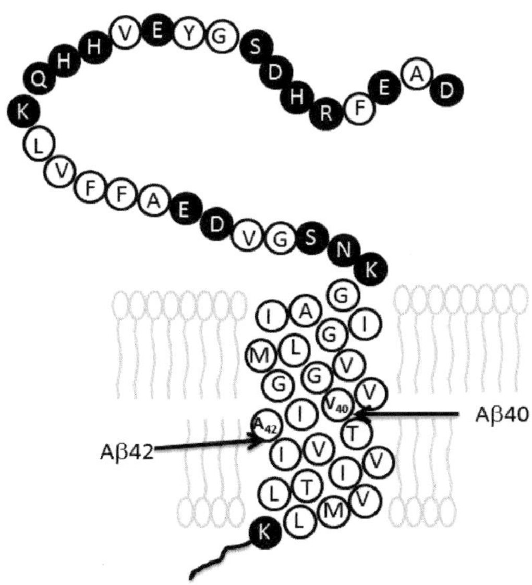

Figure 4: Schematic presentation of the amino acid sequence of Aβ and the other residues in the membrane-spanning region of APP. Each circle represents an amino acid residue, and the name of each residue is shown. The hydrophilic residues are marked as black circles. The cleavage sites for obtaining Aβ40 and Aβ42 are indicated with arrows.

the residues in the C terminus (about two-thirds of the entire Aβ40 sequence) are involved in the formation of a hydrogen-bonded β-sheet. Pro substitution at any residue of sequence 17–23 (LVFFAED) lead to the loss of fibril formation of $A\beta^{12-26}$.[20] Kim and Hecht replaced 12 hydrophobic residues in the 17–20, 31–32, 34–36, and 39–41 sequences of Aβ42 with random nonpolar residues and reported that (1) particular hydrophobic side-chains are not required for Aβ42 aggregation[21] and (2) hydrophobic residues Ile41 and Ala42 are important for the superior aggregation propensity of Aβ42 compared to Aβ40.[22] These two residues also cause the solid state peptide synthesis of Aβ42 to be much more difficult than Aβ40.

The toxicity and the sequence of various Aβ peptides are compared and summarized in Fig. 5. The most dramatic toxicity decrease is shown in the peptide without the 17–21 sequence, followed by the peptide without the 25–35 sequence. Most peptide inhibitor designs are centered on these two segments. Lin and Chu reported that the addition of the peptide $A\beta^{15-20}$ (sequence QKLVFF) can inhibit β-sheet formation of Aβ40.[23] Austen *et al.* designed a peptide named OR1 with the sequence RGKLVFFGR by adding Arg and Gly at both ends of the KLVFF sequence to increase peptide solubility.[24]

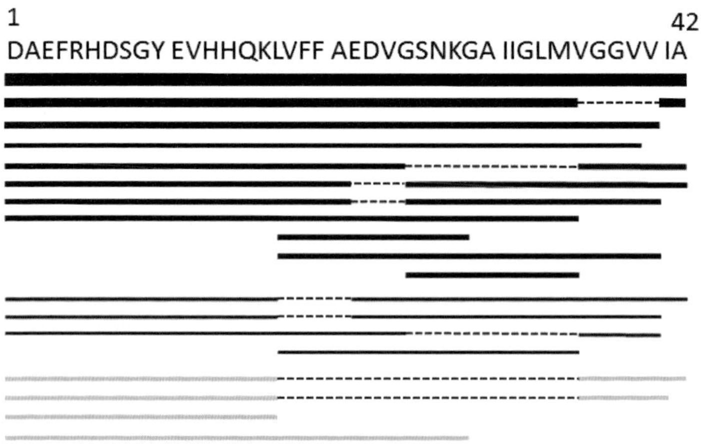

Figure 5: Summary of Aβ peptide studies with regard to the toxicity of the peptides. The toxic peptides are represented as black lines. The dotted line represents the segment not included in the peptide. Line thickness is proportional to the peptide toxicity. The peptides without significant toxicity are represented as gray lines. The toxicity data are referred to in the literature.[15,16]

In the Thioflavin T (ThT) binding assay, OR1 reduces Aβ40 and Aβ42 aggregation to less than 20% when coincubated in a molar ratio of 2:1 (OR1:Aβ40). The inhibition of aggregation was enhanced in C-end amidated OR1, denoted OR2. The addition of Arg and Gly in the design is crucial, because the C-end amidated KLVFF peptide was not effective.

In 1996, Tjernberg *et al.* systematically synthesized 31 10-mer peptides corresponding to different Aβ segments from $A\beta^{1-10}$ to $A\beta^{31-40}$ on a cellulose membrane matrix and used ^{125}I-labeled Aβ40 to screen the peptides with the capability to bind with Aβ by radioligand binding.[25] They reported that the peptide KLVFF ($A\beta^{16-20}$) can prevent Aβ40 from assembling into amyloid fibrils.

At the same time, Ghanta *et al.* tested the effects of $A\beta^{18-25}$, $A\beta^{16-25}$, and a hybrid sequence comprising $A\beta^{15-25}$ plus an oligolysine at its C-end on inhibiting Aβ39 aggregation.[26] Unexpectedly, the aggregation of Aβ39 observed by light scattering was actually promoted by the addition of these peptides. No fluorescence was emitted when these peptides were mixed with ThT. However, in the cell viability test, the hybrid peptide (2-fold and 4-fold molar ratio) was effective in decreasing the Aβ39 toxicity.

Pallitto *et al.* added six Lys residues as a disrupting element to the C-terminal end of three peptides: $A\beta^{16-20}$ (KLVFF), $A\beta^{20-25}$ (FAEDVG), and $A\beta^{15-25}$ (QKLVFFAEDVG).[27] By mixing the hybrid peptides with Aβ39 at a molar ratio of 1:1 or 2:1 (peptide: Aβ), all of these peptides are effective in decreasing Aβ39 toxicity. The efficacy of these hybrid peptides is ranked in the order $A\beta^{16-20}$–K6 > $A\beta^{20-25}$–K6 > $A\beta^{15-25}$–K6. Surprisingly, a hybrid peptide with the LVFFA sequence scrambled to VLFKF can still effectively decrease the cytotoxicity of Aβ39 to PC-12 cells, challenging the idea that the exact Aβ sequence is necessary in the design of peptide inhibitor. Also interesting is the finding that these peptides did not inhibit Aβ aggregation and, on the contrary, an increase (not decrease) in Aβ aggregate size was observed by dynamic light scattering. This observation is very similar to that published by Ghanta *et al.*[26] However, there has been no further structural characterization to identify whether the aggregates are oligomers, fibrils, or amorphous aggregates. This finding disrupts the relationship between toxicity and aggregation. "Anti-amyloid" is not equal to "anti-aggregation". It is very likely that these hybrids divert the Aβ aggregation pathway from forming toxic oligomers or fibrils to associating as nontoxic amorphous aggregates.[28]

Soto *et al.* proposed that a peptide with the ability to bind Aβ but has low propensity to adopt a β-sheet conformation can inhibit Aβ fibrillogenesis. Based on the LVFFA sequence ($A\beta^{17-21}$), they designed an 11-amino-acid peptide with the sequence RDLPFFAVPID, which they called inhibitor of Aβ fibrillogenesis peptide 1 (iAβ1).[29] In the *in vitro* Aβ fibrillogenesis experiment using the ThT fluorescence assay, iAβ1 can inhibit Aβ40 and Aβ42 fibrillization. The same inhibition can also be observed when the L-amino acids in iAβ1 are replaced by D-amino acids. Interestingly, iAβ1 (40-fold molar excess) can disassemble half of the preformed Aβ42 fibrils through an unknown mechanism. Considering the transportation of peptides across the blood–brain barrier and the possibility of generating immune response, Soto *et al.* used iAβ1 as a template to design another six shorter peptides. They found that the shorter peptide with the sequence LPFFD, denoted as iAβ5, had the strongest inhibition effect, and they named it "β-sheet breaker" peptide.[29] The rats injected with the mixture of iAβ5 and Aβ42 (molar ratio iAβ5 : Aβ42 = 20:1) into the amygdala had fewer Aβ deposits in the brain than the rats injected with Aβ42 alone, indicating that iAβ5 can inhibit Aβ42 fibril formation *in vivo*.[30] To further examine whether iAβ5, the "β-sheet breaker", could dissolve preexisting fibrils, Aβ42 was injected into the rat brain 8 days before the injection of iAβ5 (20 or 40 molar excess). The 8-day period was designed for Aβ42 to form amyloid fibrils in the brain. The results showed that, when more iAβ5 was injected, the Congo red–positive deposits observed in the rat brain were smaller.[31] The data suggested that iAβ5 has the potential to reverse Aβ deposition in AD pathology. To increase the *in vivo* biostability of iAβ5, N-acetylated and C-amidated iAβ5 (denoted iAβ5p) was designed and tested on the AD transgenic mice.[32] iAβ5p was given to the $APP_{lon}/PS1_{A246E}$ transgenic mice by intracerebroventricular (i.c.v.) infusion via microosmotic Alzet pump for 8 weeks from the age of 6–7 months (total dosage 2.5 mg). This resulted in a 67% reduction of Aβ amyloid load in the hippocampus.[32]

Szegedi *et al.* compared five pentapeptides based on Aβ sequences 4–8, 17–21, 30–34, or 38–42, and reported that the peptide LPYFDa (a represents amidation) had the most neuroprotective effect.[33] This peptide sequence is very similar to iAβ5, except one Phe was replaced by Tyr and the C-terminal carboxyl group was amidated. LPYFDa not only protects SH-SY5Y cells against the toxic effect of Aβ42 (molar ratio peptide:Aβ42 = 5:1), but also decreases Aβ42-induced neurite degeneration and tau aggregation in the differentiated SH-SY5Y cells.[34] Using tritium-labeled LPYFDa and intravenous injection in rats, radioactivity can be detected in the brain (0.29%), suggesting that LPYFDa or its metabolites can cross the blood–brain barrier.[35] Intraperitoneally administered LPYFDa in rats can protect the synapses against fibrillary Aβ42-induced neuronal excitation.[35]

(2) GxxxG segment in the C terminus of Aβ

In addition to the LVFFA sequence, studies also focused on another specific feature in the Aβ sequence, the GxxxG motif (also called glycine zipper). This motif also appears in the amyloidogenic region of mammalian prion protein[36,37] There are three GxxxG segments in the Aβ sequence: GSNKG ($A\beta^{25-29}$), GAIIG ($A\beta^{29-33}$), and GLMVG ($A\beta^{33-37}$). The latter two segments reside in the transmembrane region of APP. The two glycine residues in this motif are at i and $i+4$ positions, which are located on the same face of the helix in membrane. It has been reported that this GxxxG motif is involved in the helix-helix dimerization in membrane[38–41] Liu *et al.* proposed that these glycines in the GxxxG motif mediate helix interactions when this segment resides in the membrane environment and can function as molecular notches or grooves on the surface of β-sheets to mediate sheet-to-sheet packing when this segment occurs in the amyloid cross-β structure.[42] Sato *et al.* designed a series of peptide inhibitors that have alternating glycine and aromatic residues on one face of a β-strand. They reported that the peptide I10, with the sequence RGTWEGKW and C-end amidation, reduces Aβ40 fibrillization significantly in a peptide:Aβ molar ratio of 5:1 by ThT fluorescence assay and electron microscopy (EM). It also increases cell survival of cultured rat cortical neurons incubated with Aβ42.[43] They also showed that the efficacy of I10 is better than the C-end amidated iAβ5 in reducing the ThT fluorescence of Aβ40, but I10 cannot depolymerize preformed Aβ40 fibrils. However, the efficacy of I10 has not been tested *in vivo*.

2.1.2. *Random screening*

Wiesehan *et al.* used a method named "mirror image phage display" to screen for the D-peptide sequence capable of binding to *Aβ*42.[44] D-Peptide describes a peptide composed of D-form amino acids. This method uses D-amino acids to replace its L-enantiomer in the target protein and screens for peptide sequences (composed of L-amino acids) that can bind to this D-form target by phage display technique. Then, the screened peptide sequence is transformed into its D-form analog, which should be able to bind to the original target.[45] The screened D-peptide, denoted ^{D}pep or D1, has the sequence qshyrhispaqv (D-amino acids are represented by lowercase single letter code of L-amino acids) and has no sequence correlation with Aβ at all. D1 can decrease Aβ42 cytotoxicity on PC12 cells, but at a high concentration (100-fold molar excess of D1 to Aβ42).[46] After reducing the concentration of the screened target D-Aβ42 from 250 to 2 nM, in which monomeric Aβ and oligomeric Aβ are the dominant species, the same group screened for another peptide, D3.[47] The D3 sequence (rprtrlhthrnr) is completely different from both D1 and Aβ and contains many basic residues. Comparing D1 and D3, K_D of D3 in Aβ42 binding is 200 times lower than D1, and so 10-fold less D3 was required to effectively reduce Aβ42 cytotoxicity.[44,47] After infusing D1 or D3 to the hippocampus of the transgenic mice ($APP_{swe}/PS1\Delta E9$) for 30 days from the age of 8 months (9μg/day/mouse), they reported that D3 (but not D1) reduces Aβ load in the mouse brain by 33%. However, neither D3 nor D1 improves cognitive deficits in the water maze test. On contrary, the spatial learning performance of the D1-treated mice is even worse than the control mice.[48] Taking the advantage of its protease resistance property, D3 can be given orally. Orally feeding $APP_{swe}/PS1\Delta E9$ mice with D3 (0.5–1 mg/day in drinking water) for a longer period (8 weeks) improves learning ability in the young mice (4-month-old).[49]

Frydman-Marom *et al.* investigated a β-breaker peptide screened from a small peptide library.[50] Each peptide contains at least one aromatic residue (for target recognition) and at least one residue that does not favor β-sheet structure, e.g., Pro, ^{D}Pro, or α-aminoisobutyric acid (Aib). Pro can be a β-breaker due to its particular ϕ angle ($-60°$). Aib, an α-methylated alanine, is an even better β-breaker than Pro because there are two methyl groups attached to its α-carbon, which favors helical conformation.[51,52] Unlike other studies, they screened the library by the ability to inhibit Aβ42 oligomer formation. The screened di-peptide, named D-Trp-Aib, contains only one D-form Trp and one Aib and has a molecular mass of only 289 Da. D-Trp-Aib partly reduces the cytotoxicity of Aβ42 oligomers toward PC12 cells in a peptide:Aβ molar ratio of 10:1 (cell viability raised from ~33 to ~50%). Although its protection effect is not dramatic compared with other reported peptide inhibitors, the authors of the work claimed that D-Trp-Aib has the following advantages: (1) it is nontoxic *in vivo*, (2) it is stable in serum, (3) it has oral bioavailability, and (4) it can cross the blood–brain barrier by intravenous, oral, or intranasal (i.n.) administration. The $APP_{swe/lon}$ transgenic mice treated with intraperitoneal (i.p.) injection of D-Trp-Aib three times daily for 120 consecutive days since the age of 4.5 months (1 mg/kg · BW/day) have reduced Aβ plaque in the brain and improved cognitive ability.[50]

2.2. *Selection of peptide sequence based on tauopathy*

Sievers *et al.* designed a peptide inhibitor against tau aggregation based on the aforementioned mechanism 3: capping the fibril ends to stop fibril elongation (Fig. 3).

They used the crystal structure of segment VQIVYK from human tau sequence 306–311, which has been reported to form amyloid fibrils, as the template for inhibitor design. Following the steric zipper model of amyloid fibrils they used computer modeling to uncover which side-chain conformation of nonnatural D-amino acids in a β-strand backbone can bind to the end of the steric zipper template. From four potential candidates, they identified the D-peptide with the sequence tlkivw as the most effective sequence by its ability to inhibit fibril formation of tau K12 (tau sequence 244–394) observed by ThT fluorescence assay. Because the peptide design is based on fibril structure but not based on association property like previously mentioned anti-Aβ peptides, the inhibition effect is sequence specific. Scrambling the tlkivw sequence or changing the residues to L-amino acids decreases the inhibition efficacy.[53]

2.3. *Other peptide inhibitors targeting the signal transduction pathway*

Along with extracellular amyloid plaques, another pathological hallmark of AD is the intracellular neurofibrillary tangles formed from hyperphosphorylated tau protein and phospho-neurofilament proteins. Cyclin-dependent kinase (cdk5) is a kinase hyperactivated in AD brains.[54] When cdk5 forms a complex with p25, which is a truncated form of p35 (the cdk5 activator), the cdk5/p25 complex is hyperactive and produces hyperphosphorylation of tau and filament.[55,56] Shukla *et al.* designed a peptide TFP5 containing a 24-residue sequence from p35, a Tat (transactivator of transcription) sequence for cell penetration, and one fluorescein isothiocyanate (FITC) fluorescence tag for visualization.[57] TFP5 can inhibit hyperactive cdk5/p25 activity without affecting the endogenous cdk5/p35 activity. When they injected TFP5 (40 mg/kg · BW/day) into the 5XFAD double-transgenic AD mice via i.p. administration for three consecutive days, they observed a reduction of hyperphosphorylation in tau and neurofilaments. Interestingly, amyloid plaque deposition was also reduced because TFP5 reduces APP phosphorylation and thus further affect Aβ generation.[57]

Another peptide, denoted wtNBD, was designed to inhibit NF–κB activation, which is negatively correlated with cognitive function in AD patients.[58] The peptide contains a segment with cell internalization ability followed by a segment that can inhibit the induction of NF–κB activation. The peptide was given to 5-month-old 5XFAD mice via i.n. administration (0.1 mg/kg · BW/2 days for 30 days). Not only is NF–κB activation reduced, but the treatment also suppresses microglial activation, reduces plaque deposition, and improves cognitive ability.[57]

3. Modification of Peptide Inhibitors

Peptide inhibitors are commonly modified to increase efficacy. The modifications can be classified as backbone modification, end modification, and polymerization.

3.1. *Backbone modification*

Backbone modification enhances the inhibition effect by introducing steric hindrance to increase lipophilicity and membrane permeability, and to increase peptide biostability against peptidase attack.[59]

3.1.1. *Pro substitution*

Pro is the only cyclic amino acid within the 20 natural amino acids. Because amide nitrogen and α-carbon are in the ring structure, the torsion angle of $N-C^{\alpha}$ bond (ϕ angle) is restricted. Pro disfavors both α-helical conformation and β-strand conformation. It appears in the turn of β-sheets or the C-terminal end of α-helices. Therefore, Pro substitution can be used to avoid peptides from adopting cross-β structure.[29]

3.1.2. *D-Amino-acid-containing peptide*

One major concern with peptide drugs is stability. Peptides are susceptible to proteolytic degradation by various proteases and peptidases in body. Replacing naturally occurring L-amino acids with their stereoisomer D-amino acids in peptide drugs prolongs the half-life of the peptides.[29,60] Taylor *et al.* designed the retro-inverso peptide RI-OR2 (Ac-rGffvlkGr-NH_2) in which the sequence was adopted from the reverse sequence of the OR2 peptide (RGKLVFFGR-NH_2), all L-amino acids were replaced by D-amino acids, and both ends of the peptide were protected.[61] RI-OR2 is stable in either human plasma or brain extract and is a more effective aggregation inhibitor for Aβ40 and Aβ42 than OR2.

Substitution with ^{D}Pro can change Aβ aggregation pathway dramatically. Chang *et al.* found that, when the residues V24, I32, or V36 of Aβ40 are replaced by ^{D}Pro, the mutated peptides cannot form amyloid fibrils but form nontoxic amorphous aggregates instead.[28] It is because the ^{D}Pro plus the following Gly in the Aβ sequence can form a type-II' turn which is a favored turn type for β-hairpin. The backbone orientation of the β-strands in a β-hairpin is different from that in the cross-β structure of amyloid fibrils. Therefore, the ^{D}Pro-containing peptides can no longer form amyloid fibrils. Based on this observation, Lin *et al.* designed a peptide inhibitor V24P(10–40)-PEI that can associate with Aβ40 or Aβ42 to form amorphous aggregates and thus decrease their toxicity.[62] This peptide has the sequence of Aβ^{10-40}, except the Val24 residue is replaced by a ^{D}Pro and is cationized by conjugating with polyethylimine (PEI) for membrane penetration. In the *in vivo* animal studies this peptide can reduce Aβ accumulation in the brain of the APP$_{swe}$/PS1ΔE9 mice, indicating that diverting the Aβ association pathway is a feasible direction in designing anti-amyloid peptide inhibitors.

3.1.3. *Amide nitrogen methylated peptide*

When the amide proton in the peptide backbone is replaced by an N-methyl group, the backbone can no longer be cut by peptidases. Moreover, this modification creates a "blocking surface" to prevent fibril elongation. β-sheet structure can be thought of as a helix with two residues per turn. The side chain of the i residue faces the same side as those of the $i - 2$ and $i + 2$ residues, while its amide proton and carbonyl group face another side for hydrogen bond formation. Gordon *et al.* designed a pentapeptide A$\beta^{16-20\,m}$ (Ac–K^{me}LV^{me}FF–NH$_2$), in which *N*-methyl Leu and *N*-methyl Phe were used in residues 17 and 19.[59] While ^{me}Leu17 and ^{me}Phe19 can no longer act as hydrogen bond donors, the amide proton and carbonyl group of Lys16, Val18, and Phe20 of this pentapeptide can still form hydrogen bonds with another β-strand. However, the inhibition effect of this peptide against Aβ40 fibrillization is only slightly better than the nonmethylated congener Ac-KLVFF-NH$_2$. Moreover, scrambling the peptide sequence with the same N-methylated residues (peptide Ac–V^{me}FK^{me}LF–NH$_2$)does not affect the inhibition effect. Another N-methyl peptide A$\beta^{16-20m2}$ (Ac–KL^{me}VF^{me}F–NH$_2$), which has N-methyl amino acids at another side of the peptide (^{me}Val18 and ^{me}Phe20), is slightly more effective than A$\beta^{16-20\,m}$ and the nonmethylated congener. These data suggest that these N-methylated peptides do not work on blocking the fibril-elongation surface, but contribute to preventing self-aggregation of the peptide.

Hughes *et al.* reported that Aβ^{25-35} derivatives with single N-methylation at different residues (residues 25, 29, 30, 31, 33, and 34) produce different inhibition effects. The peptide with ^{me}Gly25 is toxic. The peptide with ^{me}Gly33 prevents Aβ^{25-35} fibrillization and decreases its toxicity to PC12 cells when mixing in a 1:1 molar ratio. The other modified peptides are less effective.[63] Adessi *et al.* reported that using N-methylated Phe at the third residue of iAβ5p (sequence Ac-LPFFD-NH$_2$) largely increases the stability of the modified peptides in rat brain homogenate. However, methylation at the amide nitrogen of another Phe reduced the inhibition of the modified peptides.[64] These data indicate the importance of selecting the N-methylation site in the peptide design.

The N-methyl peptides have target specificity. An unrelated N-methyl prion peptide (PrPm) cannot inhibit Aβ40 fibrillization.[65] In addition, it was found that these N-methyl peptides have higher solubility and can enter into monkey kidney fibroblast-like COS cells.[59] Moreover, the N-methyl peptide Ac–$K^{me}LV^{me}FF^{me}$AE–NH_2 can resist chymotrypsin digestion (chymotrypsin can cut peptides containing aromatic residues Trp, Tyr, and Phe) while more than half amount of its nonmethylated congener peptide has been cut.[65] Membrane permeability and biostability are two important advantages of backbone N-methylation.

3.2. *End modification*

The purposes of end modification include increasing peptide solubility and membrane permeability, enhancing inhibition effects by charge repulsion, and protecting peptides from exopeptidase digestion.

3.2.1. *N- and C-end protection*

Acetylation or succinylation are commonly used to protect free N-terminal amino groups. These modifications can be simple, using acetic anhydride or succinic anhydride in the final step of peptide synthesis. C-terminal amidation can be easily achieved by choosing a suitable resin (e.g., Rink amide AM resin) in solid phase peptide synthesis. Adessi *et al.* proved that N-acetylated and C-amidated iAβ5, named iAβ5p, remains intact in human plasma at 37° for more than 24 hours by high-performance liquid chromatography analysis, while the half-life of unmodified iAβ5 is only 7 ± 3 min.[32]

3.2.2. *Addition of a charge tail*

Pallitto *et al.* reported that the peptide KLVFF plus 6 Lys at its C terminus decreases Aβ39 toxicity.[27] Lowe *et al.* reported that the efficacy of this peptide inhibitor is proportional to the number of Lys residues. At least 3 Lys are required to decrease Aβ40 toxicity, and KLVFF-K3, KLVFF-K4, and KLVFF-K6 can protect PC-12 against Aβ40-induced cytotoxicity at lower-than-equimolar ratios.[66] Aβ40 contains 6 negatively charged residues and three positively charged residues at pH 7 (net charge is -3). Positively charged residues in the peptide inhibitor are expected to interact with negatively charged residues in Aβ40. However, in using Glu or Ser instead of Lys, KLVFF-E4 is more effective than KLVFF-K4, whereas KLVFF-S4 is ineffective. Although it is not clear why the Glu tail works better than the Lys tail, it can be concluded that both positively charged and negatively charged tails can enhance the inhibition effect of peptide inhibitors by charge repulsion.

Poduslo *et al.* reported that polyamine putrescine (PUT) modified D-peptide D-YiAβ11, derived from the sequence of iAβ1, has higher permeability at the blood–brain barrier than the peptide without putrescine modification.[67] This modification also enhances the ability of fibril disassembly and fibrillogenesis inhibition.

3.2.3. *Addition of a hydrophobic tail*

Findeis *et al.* tested more than 80 peptides containing different Aβ segments and different modifications such as cholyl, *N*-acetylneuraminyl, 2-iminobiotinyl, fluoresceincarbonyl modifications at N-terminal amino group, or containing D-amino acids. The inhibition efficacy was measured by the effect of these peptides on extending the lag time and on decreasing the final ThT fluorescence of Aβ40 fibrillization. Interestingly, the results showed that many cholyl-modified peptides are particularly effective. Cholyl-modified peptide sequences dramatically affect the inhibition efficacy. Cholyl-Aβ^{1-15}, cholyl-Aβ^{6-20}, cholyl-Aβ^{7-20}, cholyl-Aβ^{15-20}, cholyl-Aβ^{16-20}, cholyl-Aβ^{17-21}, and cholyl-Aβ^{16-30} can achieve 100% inhibition of Aβ40 fibrillization whereas cholyl-Aβ^{8-20}, cholyl-Aβ^{9-20}, cholyl-Aβ^{10-20}, and cholyl-Aβ^{11-20}, cholyl-Aβ^{12-20} have 0% inhibition. It is interesting that the inhibition effect is so sensitive to the Aβ sequence adopted in the peptide design. For example, cholyl-Aβ^{7-20} and cholyl-Aβ^{15-20} are effective, but the peptides having the Aβ sequence between these two peptides are not effective. The data imply that the inhibition effect is related to the peptide solubility. ^{14}C-labeled cholyl-Aβ^{17-21} (cholyl-LVFFA, also named PPI-368) can be incorporated into Aβ40 fibrils to inhibit further fibril elongation, and that inhibition effect is specific to Aβ40 but not to other amyloidogenic proteins such as transthyretin (TTR) or islet amyloid polypeptide (IAPP).[68] Using D-amino acids plus C-terminal amidation, the redesigned D-peptide (cholyl-lvffa-NH_2, denoted PPI-457) is better than PPI-368, based not only on its fibrillization inhibition effect, but also its stability in cerebral spinal fluid.[60] PPI-457 was further redesigned into PPI-1019 (*N*-methyl-lvffl-NH_2).[12] The safety of PPI-1019 (Apan^{TM}) was tested in human Phase 1 clinical trials sponsored by PRAECIS Pharmaceuticals Inc. The trials were completed in 2005, but the results have not been disclosed and no further action was taken afterward.

Kokkoni *et al.* created a peptide library based on the LVFFA sequence with the various modifications mentioned above and screened for the best peptide inhibitor against aggregation and toxicity of Aβ40 and Aβ42.[69] SEN304 ([^{ch}G–y–^{ch}G–^{ch}G–^{me}l]–NH_2) (^{ch} stands for cyclohexyl; ^{me} stands for N-methylation) was selected as the most effective inhibitor.[70] Its K_D value for binding to Aβ42 and Aβ40 monomers is 2.5μM and 14μM, respectively.[71]

3.2.4. *PEGylation*

The covalent attachment of polyethylene glycol (PEG) to a peptide drug can "mask" the peptide to reduce its immunogenicity and prolong its circulatory time *in vivo*.[72–74] Rocha *et al.* added PEG to the N terminus or C terminus of iAβ5 and reported that the PEGylation did not affect the inhibition effect of iAβ5 in Aβ42 fibrillization and toxicity.[75]

3.3. *Polymeric peptide*

To improve the binding affinity and specificity of the KLVFF sequence, Chafekar *et al.* synthesized a KLVFF dendrimer as the amyloid inhibitor, taking advantage of multivalent interaction. Four KLVFFGG peptides were ligated to a Cys dendrimer to form $[\text{KLVFFGG}]_4$ dendrimer.[76] The dendrimeric peptide is more effective than the KLVFFGG peptide in inhibiting Aβ fibrillization observed by ThT fluorescence assay. No toxicity assay was done for this dendrimer. Zhang *et al.* used eight-arm-branched amino-PEG as a scaffold to conjugate two biotin with six ckffvlk sequences (the reversed KLVFF sequence synthesized by using D-amino acids instead of L-amino acids plus one extra ^{D}Lys and ^{D}Cys). The efficacy of the conjugated polymeric peptide was evaluated according to its ability to bind to the preformed Aβ fibrils (inhibition mechanism [3] in Fig. 3).

4. Efficacy Comparison of Peptide Inhibitors

4.1. *Anti-Aβ aggregation assay*

Each anti-amyloid peptide inhibitors was designed based on its ability to interact or bind with Aβ to inhibit Aβ association. The most simple efficacy evaluation is the *in vitro* anti-Aβ aggregation assay. Aβ peptides are prone to aggregate, so Aβ stock preparation is an important issue, because peptide concentration in the stock solution is usually very high, and aggregation kinetics is extremely concentration dependent. There are many different solvents, acids, and chemicals used to prepare Aβ stock, including dimethyl sulfoxide (DMSO), hexafluoro-2-propanol (HFIP), trifluoroethanol (TFE), trifluoroacetic acid (TFA), formic acid, acetonitrile, and NaOH.[77] Different experiments require different solvent selection. DMSO is the most common solvent used to prepare stock solution for Aβ40 or other shorter Aβ peptides.[78] Aβ42 is more aggregation-prone than Aβ40 and can aggregate in DMSO. Moreover, DMSO has absorption at wavelengths lower than 220 nm, so it is not suitable for Far-ultraviolet (UV) circular dichroism (CD) measurement.

An organic solvent such as TFE is toxic and not suitable for cell viability assay. NaOH (0.1%) and TFA (0.1%) have been used to prepare Aβ stock but peptide stability during storage should be measured.[66,79]

According to the properties of Aβ aggregates, there are several spectroscopic methods commonly used to examine Aβ aggregation:

(1) ThT fluorescence assay

The name "amyloid" comes from the amylose-like spectrum shifting property of amyloid fibrils upon dye binding. ThT fluorescence assay is commonly used to monitor the kinetics of amyloidogenesis.[78] Amyloid fibrils have cross-β structure, which has intermolecular hydrogen bonds parallel to the fibril extension direction and residue side-chains perpendicular to the fibril direction. The distance between C^{α} of residues i and i + 2 on a β-strand is about 6.5 Å. This distance allows the ThT molecule (about 15 Å long, 6.1 Å wide, 4.3 Å thick) to insert into amyloid fibrils with its long axis parallel to the elongation axis of fibrils.[80] ThT has a benzothiazole ring and a dimethylaminobenzene ring. The rotation of the bond connecting these two rings, which is favored in a polar solvent, allows the formation of a nonfluorescent twisted internal charge transfer (TICT) state.[81,82] Therefore, ThT in water has a strong absorption peak centered at 412 nm but low fluorescence emission upon excitation. When ThT binds to amyloid fibrils, the rigidity of its microenvironment (β-sheet groove) restricts the rotation of the bond between these two rings and traps the excited ThT in an emissive locally excited state. The absorption maximum of ThT/amyloid complex is red-shifted to 442 nm, and the complex emits strong fluorescence at 478 nm when excited. The fluorescence enhancement effect is very strong; thus, the ThT binding assay is widely employed in detecting amyloid fibrils. However, there are a few considerations to bear in mind: (1) the pH value of the ThT/amyloid complex affects the fluorescence intensity. ThT-emitted fluorescence decreases when pH value of the solution decreases and is maximal at pH around 9.[83,84] (2) ThT fluorescence is not an absolute indicator for amyloid formation. Some amorphous aggregates can bind ThT and emit strong fluorescence.[28] ThT can even bind to non-β-sheet molecules, such as γ-cyclodextrin, and emit fluorescence as long as the molecule has a cavity with a suitable diameter and length to accommodate it.[85] Some amyloid fibrils formed from the Gly-rich peptides do not show fluorescence emission upon ThT binding.[86] (3) ThT fluorescence cannot be used to compare the amount of amyloid fibrils formed from different proteins/peptides.[43] For example, the ThT fluorescence of the fibrils formed from human prion peptide (sequence 108–144) with Val at the residue-129 is five times higher than that with Met at the same position.[87] (4) ThT can be demethylated or oxidized during storage or after

UV irradiation. These modifications greatly decrease the fluorescence enhancement effect of ThT upon amyloid-binding.[88]

(2) Light scattering

Light scattering can sensitively detect protein and peptide aggregation. Right-angle light scattering can be obtained simply by measuring the scattered excitation light at the 90 degree direction at a wavelength where no absorption occurs on a spectrofluorometer.[59] The aggregate size can be obtained and compared using dynamic light scattering instrument. However, light scattering data analysis requires a measure of care. Although the designs of many anti-amyloid peptide inhibitors aimed to inhibit Aβ association into amyloid, it was commonly reported that the aggregate size is bigger in the peptide/Aβ mixture than the sample containing Aβ alone (i.e., aggregation is promoted rather than inhibited in the presence of the peptide inhibitor). For example, Lowe *et al.* analyzed the relationship between cell viability and the hydrodynamic diameter of aggregates in the mixture of Aβ40 and the peptides composed of the KLVFF sequence plus different numbers of Lys. They found that the most effective peptide had the biggest aggregate diameter and the fastest aggregate growth rate.[66] Therefore, anti-fibrillization should be uncoupled from anti-aggregation. Aβ peptides have three distinct association pathways: fibrillization, oligomerization, and formation of amorphous aggregate. The last one is less toxic than the other two. Diverting the association pathway to promote amorphous aggregate formation, i.e., the inhibition mechanism (2): interacting with the amyloid precursor state to inhibit its association into nucleus, is an effective strategy in the design of anti-amyloid peptide (Fig. 3). The peptide V24P(10–40)-PEI was designed based on this mechanism.[62]

(3) Other spectroscopies and microscopies

Cross-β structure and fibril morphology are two major structural features of amyloid. CD spectroscopy (appearance of a negative peak at 218 nm) and Fourier transform infrared (FTIR) spectroscopy (appearance of an absorption peak at $\sim$1624 cm^{-1}) are commonly utilized to detect β-structure. Transmission electron microscopy (TEM) and atomic force microscopy (AFM) are used to observe fibril morphology. These techniques can provide structural information but cannot be used for quantitative comparison.

4.2. *Anti-Aβ toxicity assay*

Cell viability assay is commonly used to examine whether the inhibitors can decrease Aβ-induced cytotoxicity. Human neuroblastoma cells (SH-SY5Y),[16,24] human

HeLa cells,[89] human testis embryonal carcinoma cells (NT2),[68] mouse neuroblastoma cells (N2a),[28,62] and rat pheochromocytoma cells (PC-12)[26,27,46,66,89] are commonly used cell cultures. Primary hippocampal and cortical neuronal cultures have also been used. Shearman *et al.* compared the toxicity of $A\beta^{25-35}$ using different cell lines and reported that PC-12 cells and primary cortical culture were more sensitive than SH-SY5Y cells and HeLa cells and that several cell lines such as human neuroepithelioma SK-N-MC cells and human glioblastoma astrocytoma cell (U-373MG) were insensitive to the peptide.[89,90]

The (3-(4,5-cimethylthiazol-2-yl)-2,5-diphenyltetrazolium bromide (MTT) reduction assay is the most prevalent technique for measuring cell survival.[16,66,79,89] MTT is a colorimetric agent. Mitochondrial reductase in live cells can convert MTT into an insoluble purple formazan. The formazan can be dissolved in the acid cell lysis buffer containing sodium dodecyl sulfate (SDS) and organic solvent such as isopropanol or N,N-dimethylformamide. The purple color can be measured by absorption at 570 nm.

$A\beta40$[28,66,68] and $A\beta42$[34,43,47,64,68] are most commonly used to evaluate $A\beta$-induced cytotoxicity. Several reports also use $A\beta25-35$[63] and $A\beta39$.[26,27] Because $A\beta42$ is the most toxic $A\beta$ peptide, usually 1 to 10 μM $A\beta42$ is enough to kill cells. Higher concentrations (25 or 30 μM) are used for the other peptides.

To compare the ability of peptide inhibitors to reduce $A\beta$-induced cell death, the peptide inhibitor is mixed with $A\beta$ and preincubated before being added to the culture medium. For some peptides that claim to break preexisting fibrils, the peptide inhibitor is mixed with the $A\beta$ fibrils formed in advance. The molar ratio of inhibitor:$A\beta$ usually ranges from 1:1 to 10:1.

The peptide-induced cell death varies depending on the assay conditions. For example, Sato *et al.* reported that different preincubation times affect the cell viability of rat cortical neurons.[43] Other conditions, such as $A\beta$ peptide quality, the solvent for preparing $A\beta$ stock solution, and the $A\beta$ incubation condition (buffer composition, temperature, time, etc.), can affect cell viability as well. This is because $A\beta$ can proceed along several different association pathways to form oligomer, fibrils, or amorphous aggregates, which show different toxicities. Aged samples might preaggregate, even in the stock solution. In addition, different cell lines and cell treatment times might also affect the toxicity results. Different groups obtain differing results. For example, Amijee *et al.* reported that the "β-sheet breaker", i$A\beta$5p, does not have any effect on inhibiting 10 μM $A\beta42$ aggregation at concentrations of up to 50 μM by ThT binding assay and MTT cytotoxicity assay, while the other peptides, such as SEN304, SEN606, and PPI-1019, had IC_{50} of several micromolar under the same testing conditions.[69,70]

5. Validation of Peptide Inhibitors

The anti-amyloid peptide inhibitors are designed against amyloid deposition. Because Aβ amyloid plaques are formed in the brain, it is crucial for the designed peptides to target the right place *in vivo*.[91] The peptides effective in the *in vitro* assay do not necessarily function *in vivo*. There are quite a few animal models used to validate the efficacy of anti-AD drugs, including transgenic mouse models and Aβ-injection rat or mouse models. Moreover, due to the poor biostability of peptides, i.c.v. infusion, i.p. injection, i.n. administration, and oral feeding have been employed to deliver the peptides. The *in vivo* efficacy of the anti-amyloid peptides can be evaluated by:

(1) Total Aβ amount in the brain. The Aβ amount can be assayed from brain homogenate by the enzyme-linked immunosorbent assay (ELISA). Aβ in the brain should be extracted by a detergent such as SDS, a denaturant such as GdnHCl, or formic acid. Some senile plaques that cannot be extracted by 2% SDS can be solubilized by 70% formic acid.[92]
(2) Aβ amyloid plaque load in the brain. Amyloid plaque accumulation can be quantified by stained pathological images. Plaque number, plaque area, and plaque size can be visualized and quantified by ThS staining, Congo red staining, or immunohistostaining using various anti-Aβ antibodies.
(3) Aβ accumulation will activate astrocyte and microglial cells and induce neuroinflammation. Astrocyte activation is usually monitored by anti-GFAP (glial fibrillary acidic protein) antibody. Microglia activation is usually monitored by anti-CD11b antibody (OX42) or anti-Iba1 (ionized calcium-binding adapter molecule 1) antibodies. Biochemical markers of neuroinflammation include proinflammatory cytokine TNF-α (tumor necrosis factor-α), IL-1β (interleukin-1β), and total nitrite.[93]
(4) Neuroinflammation leads to neuronal damage and cognitive impairments. Many tests, such as the Morris water maze, Y-maze, T-maze, Barnes maze, and novel object recognition test, etc., have been employed to evaluate cognitive declines in the AD mouse models. The Morris water maze is the most popular behavioral test for spatial memory. Different mouse ages and different mouse models affect the applicability of these behavioral tests.[94]

5.1. *In vivo animal models*

5.1.1. *Transgenic mouse model*

One strong piece of evidence supporting the amyloid hypothesis is familial AD (FAD). Duplication or mutations in the APP gene and mutations in presenilin (PS)

1 or 2, the catalytic component of γ-secretase complex, give rise to early onset AD.[9,95,96] The Swedish double mutation (K670 $\rightarrow$ N, M671 $\rightarrow$ L) in the APP gene results in enhanced β-secretase cleavage (cutting site is between residues 671 and 672). The γ-secretase cleavage sites are between residues 711 and 712 (for Aβ40) and between residues 713 and 714 (for Aβ42). The mutations in residues 714–717 of the APP gene or the mutations at PS1 or PS2 affect γ-secretase cleavage and cause an increased Aβ42/Aβ40 ratio[97,98]

Quite a few AD transgenic mouse models were developed to mimic AD pathology[99–103] The following models are commonly used mouse models or the models which have been used to test the peptides mentioned in section 2:

(1) The transgenic mice Tg2576 express 695 amino-acid human APP with the Swedish mutation under the control of the hamster prion promoter. Aβ level in the mouse brain increases with age. Amyloid plaques are shown by 8 months.[92] The Aβ accumulation is age dependent. Between young (2- to 8-month-old) and aged (11- to 13-month-old) mice, the Aβ levels show 5-fold (Aβ40) and 14-fold (Aβ42) difference. Cognition impairment can be observed in the 9 to 10 month-old mice.[104]
(2) The transgenic mice APP23 express 751 amino-acid human APP under the control of the mouse Thy-1 promoter and show Aβ plaques earlier (about 6 months) than Tg2576. The APP23 mice have been used to examine the efficacy of valproic acid in Aβ production and behavioral deficit.[105,106]
(3) The J20 mice have the Swedish double mutation and the Indiana mutation (V717L). The expression of $APP_{swe/ind}$ is controlled by the PDGF-β promoter. Amyloid plaque formation occurs at the same time as APP23.
(4) The $APP_{lon}/PS1_{A246E}$ mice were generated by crossbreeding the single transgenic line expressing human APP with the V717I mutation (London mutation) and another single transgenic line expressing human PS1 with the A246E mutation. Both genes are controlled by the mouse Thy-1 promoter. Because of the A246E mutation at PS1, the Aβ42/Aβ40 ratio in the $APP_{lon}/PS1_{A246E}$ mice is 3.11, much higher than the ratio in the parental single transgenic line APP_{lon} (0.43). The $APP_{lon}/PS1_{A246E}$ mice show more aggressive amyloidosis (by 6–9 months) than the APP_{lon} mice (by 12–15 months).[107] Spatial learning deficit can be observed in the $APP_{lon}/PS1_{A246E}$ at the age of 5 months. This mouse model was used to test the *in vivo* efficacy of the peptide iAβ5p.[32]
(5) The $APP_{swe/lon}$ mice express human APP751 with the Swedish and London mutations under the control of mouse Thy-1 promoter. Amyloid plaques are detected by 6 months.[108] This mouse model was chosen to test the efficacy of the peptide D-Trp-Aib.[50]

(6) One very popular mouse model, APP_{swe}/PS1ΔE9, coexpresses human APP with the Swedish double mutation and exon-9-deleted PS1 under the control of the mouse prion promoter.[109] These two transgenes are cointegrated to the same chromosome location, so breeding this mouse line is easy. PS1 is predominantly endoproteolytically cleaved at the loop region of exon 9 into two domains.[110,111] These two domains form an active heterodimer that is more stable than full-length PS1. The exon-9-deleted PS1, which cannot be endoproteolytically cleaved, exhibits a longer half-life[112] and produced a higher Aβ42/Aβ40 ratio than other PS1 variants M146L and A246E.[98] The APP_{swe}/PS1ΔE9 mice show Aβ deposition in the brain by 6 months and spatial learning deficits by 7 months.[113,114] This mouse model was used to test the *in vivo* efficacy of the peptides D3 and V24P(10–40)-PEI.[47–49,62,115]

All of the aforementioned mice have age-dependent accumulation of Aβ plaques in the brain and are commonly used to assess the validity of drugs on Aβ accumulation and cognitive performance. They cannot be used to evaluate the efficacy of drugs targeting on tauopathy, as no neurofibrillary tangles can be found in these mouse models.

(7) The 5XFAD mouse is the model with the earliest onset of amyloid plaque deposition because it has five FAD mutations. Human APP695 with the Swedish, London (V717I), and Florida (I716V) mutations and human PS1 with M146L and L286V mutations are coexpressed under the control of the mouse Thy-1 promoter. The mice produce an extremely large amount of Aβ42 (~140 ng/mg total brain protein) and Aβ deposition starts as early as 2 months.[116] Alongside extracellular Aβ amyloid deposition, these mice show intraneuronal Aβ aggregates by 1.5 months, which lead to neurodegeneration (synapse degeneration is represented by the decrease of the synaptophysin level) and neuron loss. Spatial memory deficits can be observed at the age of 4–5 months. Although the 5XFAD mice do not have neurofibrillary tangles either, this mouse model has abnormal tau phosphorylation. Tau phosphorylation might result from the increased p25 level. P25 forms a hyperactive complex with cdk5 and enhances tau phosphorylation. Therefore, 5XFAD is used to test the peptides TFP5[57] and wtNBD[58] targeting in the tau phosphorylation pathway.

(8) The 3XTg mouse is a triple transgenic mouse model with neurofibrillary tangle formation.[117] The 3XTg mice coexpress APP_{swe}, tau_{P301L}, and $PS1_{M146V}$, controlled by the mouse Thy-1.2 promoter. The P301L mutation decreases the microtubule-binding property of tau, thus promoting tau aggregation. Amyloid deposition begins at around 6 months, earlier than tangle formation

(12 months). Cognitive deficits occur at 4–6 months. Unlike 5XFAD, no neuron loss can be observed.

It is not clear yet why these AD transgenic mice develop so many amyloid plaques but not neurofibrillary tangle formation unless tau mutation is included, despite the fact that there is no tau mutation in AD patients. The htau mice, which express human tau but not mouse tau, have phosphorylated neurofibrillary tangles in the brain in the pattern very similar to that in the AD brains and develop age-dependent cognitive impairments,[118,119] while the 8c mice which express both human tau and mouse tau do not.[120] The data imply that mouse tau protein might inhibit neurofibrillary tangle formation. The transgenic mice with APP and PS1 mutations and human tau, but without mouse tau, might serve as a better AD model reflecting the real AD pathology.

5.1.2. *Aβ-injected mouse or rat model*

In addition to cost considerations, one drawback of current AD transgenic mouse models is that their pathology does not really replicate the symptoms and lesions of AD patients. The APP transgenic mice have amyloid deposits near where astrogliosis and microglial activation can be observed. Neurite dystrophy and cognitive deficits are shown in several models, but neuron loss and neurofibrillary tangles are scarcely reported.[103] Another disadvantage is that using transgenic mice is time consuming. It takes several months for amyloid to accumulate in the mouse brain. Therefore, Aβ-injected mice and rats provide alternative models to validate drug efficacy on degrading Aβ, inhibiting Aβ association, dissolving existing Aβ fibrils, protecting neurons, or reducing Aβ-induced cognitive declines by intracerebrally (i.c.) injecting soluble Aβ or preformed Aβ fibrils into the brain of rats or mice.[30,31,121–124] After intrahippocampal (i.h.) or i.c.v. injection of Aβ40, Aβ42, or even $A\beta^{25-35}$, the mice or rats show behavioral deficits.[122,125–128] iAβ5p efficacy was proved with the ability of inhibiting Aβ fibrillogenesis, disassembling fibrillary amyloid deposits, and preventing Aβ-induced memory impairments by using the Aβ-injection rat model.[30,31,128]

5.2. *Delivery of anti-AD peptide inhibitors*

5.2.1. *Intracerebroventricular or intrahippocampal infusion*

Brain infusion is the most direct way to ensure the peptide enters into the brain. van Groen *et al.* infused the FITC-labeled D3 peptide (FITC-D3) into the hippocampus of the 21-month-old $APP_{swe}/PS1\Delta E9$ mice, FITC labeling can be seen in the dense core of Aβ plaques, the microglial cells surrounding the plaques, and the dystrophic axons surrounding the plaques.[115] Permanne *et al.* examined the tissue distribution of ^{3}H-iAβ5p in rat via intravenous injection. Most of the radioactivity appeared in organs or tissues for excretion and degradation, but only trace amounts of radioactivity can be found in the brain.[32] They used a microosmotic Alzet pump to deliver a total of 2.5 mg iAβ5p into the 6- to 7-month-old $APP_{lon}/PS1_{A246E}$ mice for 8 weeks. Immunohistochemical analysis of hippocampus showed 67% reduction in Aβ amyloid burden. When i.p. injection was used, even 10 times the amount of iAβ5p cannot yield the same reduction effect as i.c.v. infusion.

5.2.2. *Intraperitoneal injection*

Intraperitoneal (IP) injection was used to test the efficacy of the peptides iAβ5p,[32,128] iAβ5,[31] D-Trp-Aib,[50] and TFP5.[57] iAβ5 and iAβ5p are pentapeptides (LPFFD), and end protection was used in iAβ5p to improve its biostability. D-Trp-Aib is a dipeptide with excellent bioavailability because it contains only one D-form Trp and one unnatural amino acid Aib. TFP5 is a long peptide (38 residues) constituted solely of natural amino acids, so an 11-aa Tat sequence is included in its design to improve its cell permeability. Shukla *et al.* reported that FITC-labeled TFP5 can be detected in the cortex, hippocampus, and cerebellum of mice on the fourth day after i.p. injection.[57]

5.2.3. *Oral feeding*

Oral feeding is the most convenient route of drug administration. However, oral bioavailability of peptides is limited by the degradation in the gastrointestinal tract and the inability to cross the epithelial barrier. In all the designed anti-AD peptides, only D3 was tested in mice via oral administration because it contains only D-amino acids to avoid protease digestion and is Arg-rich, which might assist adsorptive-mediated transcytosis. Funke *et al.* compared the efficacy of D3 administered by oral feeding or i.h. infusion to the 4-month-old $APP_{swe}/PS1\Delta E9$ for eight weeks.[49] The dosage of oral feeding was 0.5–1 mg/mouse/day, which is 56- to 112-folds higher than that of i.h. infusion (total 0.5 mg per mouse). Similar amyloid plaque reduction ratios in the hippocampus were found in these two treatment groups compared with the untreated mice, but only the oral treatment group showed significant reduction in the number of Congo red positive plaques, and signal intensities for astrocyte and microglia. Moreover, the mice receiving oral treatment showed more improved cognitive performance. Although oral feeding requires much higher peptide levels, the data showed that anti-AD peptide can function in the brain via oral delivery.

5.2.4. *Intranasal delivery*

For the drugs with poor bioavailability, i.n. administration is a good noninvasive route for drug delivery. Moreover, because the blood–brain barrier is the major obstacle for drugs targeting the central nervous system, nasal drug delivery through olfactory epithelium in the nasal cavity provides a possibility to bypass the blood–brain barrier.[129–131] Hydrophilic drugs typically have poor nasal absorption ratio. Protein cationization, either by conjugating with compounds carrying multiple positive charges or by linking with cell-penetrating peptides, enhances the protein transcytosis ability by binding with negatively charged glycolipids and glycoproteins via absorptive-mediated endocytic mechanism[132–134] Loftus *et al.* reported that PEI-labeled green fluorescent protein can enter the brain by i.n. administration.[135] Lin *et al.* found that using PEI-conjugated peptide V24P(10–40)-PEI as nose drops to treat the 4-month-old $APP_{swe}/PS1\Delta E9$ mice for 4 months can reduce Aβ accumulation in the mouse brain.[62] Rangasamy *et al.* used the antennapedia homeodomain (containing many basic residues) in designing the peptide wtNBD. Giving wtNBD to the 5-month-old 5XFAD mice via i.n. administration for 30 days suppresses microglial activation, reduces Aβ plaque deposition, and improves cognitive performance of the mice.[58]

6. Summary of Peptide Inhibitors Examined in Animal Experiments

There are many peptides claimed to be effective in inhibiting Aβ aggregation and reducing Aβ cytotoxicity but only a small number were tested *in vivo*. The *in vivo* results of these anti-amyloid peptides are summarized and compared in Table 1. Here are several preliminary conclusions from the limited data:

(1) A peptide with the ability to bind Aβ is not necessarily effective in inhibiting Aβ accumulation in the brain. For example, D1 binds Aβ42 with a K_D value of 0.4 μM,[44] but it cannot reduce Aβ plaque *in vivo*. Moreover, the cognitive performance of the transgenic mice treated with D1 is even worse than the untreated mice.

(2) Amyloid reduction and decrease of astrocyte and microglia signals are not proportional to the cognitive performance. D3 has been tested using oral or i.h. infusion (D3 dosages are different). Similar Aβ plaque reduction in the hippocampus is observed in both treatments by IHC, but only the oral feeding group has significantly improved cognitive performance.

Table 1: Summary of the anti-AD peptide inhibitors tested in animal experiments.

Peptide	Animal	Age of starting treatment	Peptide delivery[a]	Dosage	Aβ reduction by ELISA[b]	Aβ plaque reduction[b]	Cognitive ability[b]	Others[c]
iAβ5p[128]	Male rat injected with 40 μg Aβ42 fibrils in HIP; one day before peptide treatment	3 m	i.p.	4.5 mg/injection 3 injection/week 1 month	—	45%↓(HIP) by ThS	Morris water maze↑	GFAP↓
iAβ5p[32]	APP_{lon}/ $PS1_{A246E}$	6–7 m	i.c.v. infusion	Total 2.5 mg for 8 weeks	—	67%↓(HIP) by IHC	—	21%↑ in neuronal survival
iAβ5p[32]	APP_{lon}/ $PS1_{A246E}$	8–9 m	i.p.	1 mg/injection 3 injection/week 8 weeks	—	47%↓ (HIP)	—	17%↑ in neuronal survival
iAβ5p[32]	APP_{lon}	15 m	i.p.	1 mg/injection 4 injection/week 3 months	Aβ40 43% ↓ Aβ42 26% ↓	45%↓(CTX) 29%↓(HIP) by IHC	—	—

iAβ5[30]	Rat coinjected with 22.5μg soluble Aβ42 and iAβ5 in amygdala	—	i.c.v.	iAβ5(20× molar excess)	—	49%↓ 8 days after surgery by IHC	—	—
iAβ5[31]	Rat injected with iAβ5 8 days after 22.5μg soluble Aβ42 was injected in amygdala	—	i.c.v.	iAβ5 (20 or 40× molar excess)	—	29%↓ (20 × iAβ5) 67%↓ (40 × iAβ5) 8 days after last injection by IHC	—	IL-1β ↓
D1[47,48]	APP_{swe}/PS1ΔE9	8 m	i.h. infusion	9μg/day 30 days	—	no	Morris water maze ↓	GFAP n.s.d. Microglia n.s.d.
D3[47,48]	APP_{swe}/PS1ΔE9	8 m	i.h. infusion	9μg/day 30 days	—	32%↓(CTX) 33%↓(HIP) by IHC 36%↓(CTX) 29%↓(HIP) by CR	Morris water maze n.s.d.	GFAP↓ Microglia ↓
D3[49]	APP_{swe}/PS1ΔE9	4 m	Oral	0.5–1 mg/day 8 weeks	—	32%↓(HIP) by IHC 42%↓(HIP) by CR ↓	Morris water maze↑	GFAP↓ microglia↓
D3[49]	APP_{swe}/PS1ΔE9	4 m	i.h. infusion	Total 0.5 mg 8 weeks	—	32%↓(HIP) by IHC n.s.d. by CR	Morris water maze n.s.d.[d]	GFAP n.s.d. microglia n.s.d.

(Continued)

Table 1: (*Continued*)

Peptide	Animal	Age of starting treatment	Peptide delivery[a]	Dosage	Aβ reduction by ELISA[b]	Aβ plaque reduction	Cognitive ability	Others
FITC-D3[115]	APP_{swe}/ PS1ΔE9	21 m	i.h. infusion	Total 0.5 mg 8 weeks	Aβ40 66%↓ Aβ42 28%↓	n.s.d. in CTX In HIP 64%↓ by IHC 53%↓ by CR	Morris water maze n.s.d. Barnes maze n.s.d.[d]	GFAP↓ microglia↓
V24P(10-40)-PEI[62]	APP_{swe}/ PS1ΔE9	4 m	i.n.	8μg/mouse/day 6 days/week 4 months	Aβ40 72%↓ Aβ42 40%↓ (HIP) Aβ40 87%↓ Aβ42 32%↓ (CTX)	65%↓(CTX) 50%↓(HIP) by ThS	Morris water maze n.s.d.[d]	—
D-Trp-Aib[50]	$APP_{swe/lon}$	4.5 m	i.p.	1 mg/kg · BW per day 120 days	—	75%↓(CTX) 47%↓ (HIP) by ThS	Morris water maze ↑	—
TFP5[57]	5XFAD	5 m	i.p.	40 mg/kg · BW per injection 3 injections	—	38%↓ by IHC	Y-maze↑	p-tau↓ p-APP↓ inflammation↓ apoptosis↓
wtNBD[58]	5XFAD	5 m	i.n.	0.1 mg/kg · BW every 2 days 30 days	Aβ42 57%↓ (CTX) 64%↓ (HIP)	69%↓(CTX) 74%↓ (HIP) by IHC	Barnes maze↑	inflammation↓

[a] i.p., intraperitoneal; i.c.v., intracerebroventricular; i.h., intrahippocampal; i.n., intranasal.
[b] CTX, cortex; HIP, hippocampus; ThS, thioflavin S staining analysis; IHC, immunohistochemical analysis; CR, Congo red staining; n.s.d., no significant difference.
[c] GFAP, glial fibrillary acidic protein; IL-1β, interleukin-1β.
[d] No significant difference except 1 day.

(3) Peptide inhibitors can function in the brain via a noninvasive route such as oral delivery or i.n. delivery. D3, V24P(10–40)-PEI, and wtNBD are three examples. But, it should be noted that D3 contains only D-amino acids, so it can escape protease digestion in the gastrointestinal tract. Although D3 does not conjugate with a cell-penetration peptide, its sequence contains many Arg residues, which might contribute in its localization. wtNBD also has a cell-penetrating peptide. It can travel from the nose to function in the brain, though only L-amino acids are used. V24P(10–40)-PEI has N-terminal acetylation and PEI conjugation at its C terminus. The positively charged PEI moiety can help the peptide bypass the blood–brain barrier via i.n. route.

(4) Peptide therapy seems to work better in young mice. Aβ accumulation results from the imbalance of Aβ production and Aβ clearance. Because amyloid propagation is a nucleation-dependent process, Aβ accumulation rate increases with age. For example, the total Aβ level in the Tg2576 mice increases only 2% from the age of 5–6 months but increases 22% from the age of 6–7 months. At the age of 21 months, the Aβ amount is 377 times higher than at the age of 7 months.[92] Therefore, although amyloid reduction can still be seen in old mice, the amyloid burden might exceed a threshold and cause irreversible cognitive decline. Similar results were reported in the clinical trials of passive anti-amyloid immunotherapy. Solanezumab and aducanumab have a significant effect on cognition only in the patients with mild cognitive decline but not in the patients with moderate cognitive decline.[7]

In conclusion, several anti-amyloid peptides effectively reduce amyloid burden, astrogliosis, microglia activation, neuroinflammation, neuronal loss, and cognitive declines in various mouse models. Some of them can even function via a noninvasive route. These encouraging results suggest the feasibility of anti-amyloid peptide therapy in the treatment of AD or even in other amyloid-related neurodegenerative disorders. So far no FDA-approved drugs are available to stop or slow AD progression. There is a hope that the anti-amyloid peptide-based inhibitors can be tested in the human clinical trials.

References

1. Mount, C. and Downton, C. Alzheimer disease: progress or profit? *Nat. Med.* **12**, 780–784 (2006).
2. Hardy, J. and Selkoe, D. J. The amyloid hypothesis of Alzheimer's disease: progress and problems on the road to therapeutics. *Science*. **297**, 353–356 (2002).

3. Hardy, J. A. and Higgins, G. A. Alzheimer's-disease: the amyloid cascade hypothesis. *Science*. **256**, 184–185 (1992).
4. Patel, K. R. Biogen's aducanumab raises hope that Alzheimer's can be treated at its source. *Manag. Care.* **24**, 19 (2015).
5. Reardon, S. Antibody drugs for Alzheimer's show glimmers of promise. *Nature.* **523**, 509–510 (2015).
6. Underwood, E. Alzheimer's amyloid theory gets modest boost. *Science.* **349**, 464 (2015).
7. Moreth, J., Mavoungou, C., and Schindowski, K. Passive anti-amyloid immunotherapy in Alzheimer's disease: what are the most promising targets? *Immun. Ageing.* **10**, 18 (2013).
8. Toyn, J. What lessons can be learned from failed Alzheimer's disease trials? *Expert. Rev. Clin. Pharmacol.* **8**, 267–269 (2015).
9. Goedert, M. and Spillantini, M. G. A century of Alzheimer's disease. *Science.* **314**, 777–781 (2006).
10. Martins, I. J., Berger, T., Sharman, M. J., Verdile, G., Fuller, S. J., and Martins, R. N. Cholesterol metabolism and transport in the pathogenesis of Alzheimer's disease. *J. Neurochem.* **111**, 1275–1308 (2009).
11. Verdile, G., Fuller, S., Atwood, C. S., Laws, S. M., Gandy, S. E., and Martins, R. N. The role of beta amyloid in Alzheimer's disease: still a cause of everything or the only one who got caught? *Pharmacol. Res.* **50**, 397–409 (2004).
12. Findeis, M. A. Peptide inhibitors of beta amyloid aggregation, *Curr. Top. Med. Chem.* **2**, 417–423 (2002).
13. Sun, N., Funke, S. A., and Willbold, D A survey of peptides with effective therapeutic potential in Alzheimer's disease rodent models or in human clinical studies. *Mini Rev. Med. Chem.* **12**, 388–398 (2012).
14. Funke, S. A. and Willbold, D. Peptides for therapy and diagnosis of Alzheimer's disease. *Curr. Pharm. Des.* **18**, 755–767 (2012).
15. Liao, M. Q., Tzeng, Y. J., Chang, L. Y., Huang, H. B., Lin, T. H., Chyan, C. L., and Chen, Y. C. The correlation between neurotoxicity, aggregative ability and secondary structure studied by sequence truncated Abeta peptides. *FEBS Lett.* **581**, 1161–1165 (2007).
16. Liu, R., McAllister, C., Lyubchenko, Y., and Sierks, M. R. Residues 17-20 and 30-35 of β-amyloid play critical roles in aggregation, *J. Neurosci. Res.* **75**, 162–171 (2004).
17. Williams, A. D., Portelius, E., Kheterpal, I., Guo, J. T., Cook, K. D., Xu, Y., and Wetzel, R. Mapping Aβ amyloid fibril secondary structure using scanning proline mutagenesis. *J. Mol. Biol.* **335**, 833–842 (2004).
18. Williams, A. D., Shivaprasad, S., and Wetzel, R. Alanine scanning mutagenesis of Aβ(1-40) amyloid fibril stability. *J. Mol. Biol.* **357**, 1283–1294 (2006).
19. Wetzel, R., Shivaprasad, S., and Williams, A. D. Plasticity of amyloid fibrils. *Biochemistry.* **46**, 1–10 (2007).
20. Wood, S. J., Wetzel, R., Martin, J. D., and Hurle, M. R. Prolines and amyloidogenicity in fragments of the Alzheimer's peptide β/A4. *Biochemistry.* **34**, 724–730 (1995).

21. Kim, W. and Hecht, M. H. Generic hydrophobic residues are sufficient to promote aggregation of the Alzheimer's Aβ42 peptide. *Proc. Natl. Acad. Sci. U. S. A.* **103**, 15824–15829 (2006).
22. Kim, W. and Hecht, M. H. Sequence determinants of enhanced amyloidogenicity of Alzheimer Aβ42 peptide relative to Aβ40. *J. Biol. Chem.* **280**, 35069–35076 (2005).
23. Lin, S. Y. and Chu, H. L. Fourier transform infrared spectroscopy used to evidence the prevention of beta-sheet formation of amyloid $\beta(1-40)$ peptide by a short amyloid fragment. *Int. J. Biol. Macromolec.* **32**, 173–177 (2003).
24. Austen, B. M., Paleologou, K. E., Ali, S. A. E., Qureshi, M. M., Allsop, D., and El-Agnaf, O. M. A. Designing peptide inhibitors for oligomerization and toxicity of Alzheimer's β-amyloid peptide, *Biochemistry.* **47**, 1984–1992 (2008).
25. Tjernberg, L. O., Naslund, J., Lindqvist, F., Johansson, J., Karlstrom, A. R., Thyberg, J., Terenius, L., and Nordstedt, C. Arrest of beta-amyloid fibril formation by a pentapeptide ligand. *J. Biol. Chem.* **271**, 8545–8548 (1996).
26. Ghanta, J., Shen, C. L., Kiessling, L. L., and Murphy, R. M. A strategy for designing inhibitors of beta-amyloid toxicity. *J. Biol. Chem.* **271**, 29525–29528 (1996).
27. Pallitto, M. M., Ghanta, J., Heinzelman, P., Kiessling, L. L., and Murphy, R. M. Recognition sequence design for peptidyl modulators of beta-amyloid aggregation and toxicity, *Biochemistry.* **38**, 3570–3578 (1999).
28. Chang, E. S., Liao, T. Y., Lim, T. S., Fann, W., and Chen, R. P. A new amyloid-like β-aggregate with most of the amyloid characteristics except fibril morphology. *J. Mol. Biol.* **385**, 1257–1265 (2009).
29. Soto, C., Kindy, M. S., Baumann, M., and Frangione, B. Inhibition of Alzheimer's amyloidosis by peptides that prevent β-sheet conformation. *Biochem. Biophys. Res. Commun.* **226**, 672–680 (1996).
30. Soto, C., Sigurdsson, E. M., Morelli, L., Kumar, R. A., Castano, E. M., and Frangione, B. Beta-sheet breaker peptides inhibit fibrillogenesis in a rat brain model of amyloidosis: implications for Alzheimer's therapy. *Nat. Med.* **4**, 822–826 (1998).
31. Sigurdsson, E. M., Permanne, B., Soto, C., Wisniewski, T., and Frangione, B. *In vivo* reversal of amyloid-beta lesions in rat brain. *J. Neuropathol. Exp. Neurol.* **59**, 11–17 (2000).
32. Permanne, B., Adessi, C., Saborio, G. P., Fraga, S., Frossard, M.-J., Dorpe, J. V., Dewachter, I., Banks, W. A., Leuven, F. V., and Soto, C. Reduction of amyloid load and cerebral damage in a transgenic mouse model of Alzheimer's disease by treatment with a b-sheet breaker peptide. *FASEB J.* **16**, 860–862 (2002).
33. Szegedi, V., Fulop, L., Farkas, T., Rozsa, E., Robotka, H., Kis, Z., Penke, Z., Horvath, S., Molnar, Z., Datki, Z., Soos, K., Toldi, J., Budai, D., Zarandi, M., and Penke, B. Pentapeptides derived from Abeta 1-42 protect neurons from the modulatory effect of Abeta fibrils—an in vitro and in vivo electrophysiological study. *Neurobiol. Dis.* **18**, 499–508 (2005).
34. Datki, Z., Papp, R., Zádori, D., Soós, K., Fülöp, L., Juhász, A., Laskay, G., Hetényi, C., Mihalik, E., Zarándi, M., and Penke, B. In vitro model of neurotoxicity of A$\beta 1-42$ and

neuroprotection by a pentapeptide: Irreversible events during the first hour. *Neurobiol. Dis.* **17**, 507–515 (2004).

35. Juhasz, G., Marki, A., Vass, G., Fulop, L., Budai, D., Penke, B., Falkay, G., and Szegedi, V. An intraperitoneally administered pentapeptide protects against Abeta (1-42) induced neuronal excitation in vivo. *J. Alzheimers Dis.* **16**, 189–196 (2009).
36. Harrison, C. F., Lawson, V. A., Coleman, B. M., Kim, Y.-S., Masters, C. L., Cappai, R., Barnham, K. J., and Hill, A. F. Conservation of a glycine-rich region in the prion protein is required for uptake of prion infectivity. *J. Biol. Chem.* **285**, 20213–20223 (2010).
37. Zhang, J. and Zhang, Y. Molecular dynamics studies on 3D structures of the hydrophobic region PrP(109-136). *Acta Biochim. Biophys. Sin.* **45**, 509–519 (2013).
38. Cunningham, F., Poulsen, B. E., Ip, W., and Deber, C. M. Beta-branched residues adjacent to GG4 motifs promote the efficient association of glycophorin A transmembrane helices. *Biopolymers.* **96**, 340–347 (2011).
39. Horschitz, S., Lau, T., and Schloss, P. Glycine residues G338 and G342 are important determinants for serotonin transporter dimerisation and cell surface expression. *Neurochem. Int.* **52**, 770–775 (2008).
40. Johnson, R. M., Rath, A., and Deber, C. M. The position of the Gly-xxx-Gly motif in transmembrane segments modulates dimer affinity. *Biochem. Cell Biol.* **84**, 1006–1012 (2006).
41. Sato, T., Tang, T. C., Reubins, G., Fei, J. Z., Fujimoto, T., Kienlen-Campard, P., Constantinescu, S. N., Octave, J. N., Aimoto, S., and Smith, S. O. A helix-to-coil transition at the epsilon-cut site in the transmembrane dimer of the amyloid precursor protein is required for proteolysis. *Proc. Natl. Acad. Sci. U. S. A.* **106**, 1421–1426 (2009).
42. Liu, W., Crocker, E., Zhang, W., Elliott, J. I., Luy, B., Li, H., Aimoto, S., and Smith, S. O. Structural role of glycine in amyloid fibrils formed from transmembrane α-helices. *Biochemistry.* **44**, 3591–3597 (2005).
43. Sato, T., Kienlen-Campard, P., Ahmed, M., Liu, W., Li, H., Elliott, J. I., Aimoto, S., Constantinescu, S. N., Octave, J. N., and Smith, S. O. Inhibitors of amyloid toxicity based on beta-sheet packing of Abeta40 and Abeta42. *Biochemistry.* **45**, 5503–5516 (2006).
44. Wiesehan, K., Buder, K., Linke, R. P., Patt, S., Stoldt, M., Unger, E., Schmitt, B., Bucci, E., and Willbold, D. Selection of D-Amino-Acid peptides that bind to Alzheimer's disease amyloid peptide A beta(1-42) by mirror image phage display. *Chembiochem.* **4**, 748–753 (2003).
45. Wiesehan, K. and Willbold, D. Mirror-image phage display: Aiming at the mirror. *Chembiochem.* **4**, 811–815 (2003).
46. Wiesehan, K., Stoehr, J., Nagel-Steger, L., van Groen, T., Riesner, D., and Willbold, D. Inhibition of cytotoxicity and amyloid fibril formation by a D-amino acid peptide that specifically binds to Alzheimer's diseaseamyloid peptide. *Protein Eng. Des. Sel.* **21**, 241–246 (2008).
47. van Groen, T., Wiesehann, K., Funke, S. A., Kadish, I., Nagel-Steger, L., and Willbold, D. Reduction of Alzheimer's disease amyloid plaque load in transgenic mice by D3, a D-enantiomeric peptide identified by mirror image phage display. *ChemMedChem.* **3**, 1848–1852 (2008).

48. van Groen, T., Kadish, I., Funke, A., Bartnik, D., and Willbold, D. Treatment with Abeta42 binding D-amino acid peptides reduce amyloid deposition and inflammation in APP/PS1 double transgenic mice. *Adv. Protein Chem. Struct. Biol.* **88**, 133–152 (2012).
49. Funke, S. A., van Groen, T., Kadish, I., Bartnik, D., Nagel-Steger, L., Brener, O., Sehl, T., Batra-Safferling, R., Moriscot, C., Schoehn, G., Horn, A. H. C., Muller-Schiffmann, A., Korth, C., Sticht, H., and Willbold, D. Oral treatment with the D-enantiomeric peptide D3 improves the pathology and behavior of Alzheimer's disease transgenic mice. *Acs Chem. Neurosci.* **1**, 639–648 (2010).
50. Frydman-Marom, A., Rechter, M., Shefler, I., Bram, Y., Shalev, D. E., and Gazit, E. Cognitive-performance recovery of Alzheimer's disease model mice by modulation of early soluble amyloidal assemblies. *Angew. Chem. Int. Ed.* **48**, 1981–1986 (2009).
51. Karle, I. L. and Balaram, P. Structural characteristics of .alpha.-helical peptide molecules containing Aib residues. *Biochemistry.* **29**, 6747–6756 (1990).
52. Gilead, S. and Gazit, E. Inhibition of amyloid fibril formation by peptide analogues modified with alpha-aminoisobutyric acid. *Angew. Chem. Int. Ed.* **43**, 4041–4044 (2004).
53. Sievers, S. A., Karanicolas, J., Chang, H. W., Zhao, A., Jiang, L., Zirafi, O., Stevens, J. T., Munch, J., Baker, D., and Eisenberg, D. Structure-based design of non-natural amino-acid inhibitors of amyloid fibril formation. *Nature.* **475**, 96–100 (2011).
54. Pei, J. J., Grundke-Iqbal, I., Iqbal, K., Bogdanovic, N., Winblad, B., and Cowburn, R. F. Accumulation of cyclin-dependent kinase 5 (cdk5) in neurons with early stages of Alzheimer's disease neurofibrillary degeneration. *Brain Res.* **797**, 267–277 (1998).
55. Ahlijanian, M. K., Barrezueta, N. X., Williams, R. D., Jakowski, A., Kowsz, K. P., McCarthy, S., Coskran, T., Carlo, A., Seymour, P. A., Burkhardt, J. E., Nelson, R. B., and McNeish, J. D. Hyperphosphorylated tau and neurofilament and cytoskeletal disruptions in mice overexpressing human p25, an activator of cdk5. *Proc. Natl. Acad. Sci. U. S. A.* **97**, 2910–2915 (2000).
56. Noble, W., Olm, V., Takata, K., Casey, E., Mary, O., Meyerson, J., Gaynor, K., LaFrancois, J., Wang, L., Kondo, T., Davies, P., Burns, M., Veeranna, Nixon, R., Dickson, D., Matsuoka, Y., Ahlijanian, M., Lau, L.-F., and Duff, K. Cdk5 is a key factor in tau aggregation and tangle formation in vivo. *Neuron.* **38**, 555–565 (2003).
57. Shukla, V., Zheng, Y.-L., Mishra, S. K., Amin, N. D., Steiner, J., Grant, P., Kesavapany, S., and Pant, H. C. A truncated peptide from p35, a Cdk5 activator, prevents Alzheimer's disease phenotypes in model mice. *FASEB J.* **27**, 174–186 (2013).
58. Rangasamy, S. B., Corbett, G. T., Roy, A., Modi, K. K., Bennett, D. A., Mufson, E. J., Ghosh, S., and Pahan, K. Intranasal delivery of NEMO-binding domain peptide prevents memory loss in a mouse model of Alzheimer's disease. *J. Alzheimers Dis.* **47**, 385–402 (2015).
59. Gordon, D. J., Tappe, R., and Meredith, S. C. Design and characterization of a membrane permeable N-methyl amino acid-containing peptide that inhibits Abeta1-40 fibrillogenesis. *J. Pept. Res.* **60**, 37–55 (2002).
60. Findeis, M. A., Musso, G. M., Arico-Muendel, C. C., Benjamin, H. W., Hundal, A. M., Lee, J. J., Chin, J., Kelley, M., Wakefield, J., Hayward, N. J., and Molineaux, S. M. Modified-peptide inhibitors of amyloid β-peptide polymerization. *Biochemistry.* **38**, 6791–6800 (1999).
61. Taylor, M., Moore, S., Mayes, J., Parkin, E., Beeg, M., Canovi, M., Gobbi, M., Mann, D. M., and Allsop, D. Development of a proteolytically stable retro-inverso peptide

inhibitor of beta-amyloid oligomerization as a potential novel treatment for Alzheimer's disease. *Biochemistry.* **49**, 3261–3272 (2010).
62. Lin, C. Y., Cheng, Y. S., Liao, T. Y., Lin, C., Chen, Z. T., Twu, W. I., Chang, C. W., Tan, D. T., Liu, R. S., Tu, P. H., and Chen, R. P. Intranasal administration of a polyethylenimine-conjugated scavenger peptide reduces Aβ accumulation in a mouse model of Alzheimer's disease. *J. Alzheimers Dis.* **3**(53), 1053–1067 (2016).
63. Hughes, E., Burke, R. M., and Doig, A. J. Inhibition of toxicity in the beta-amyloid peptide fragment beta -(25-35) using N-methylated derivatives: a general strategy to prevent amyloid formation. *J. Biol. Chem.* **275**, 25109–25115 (2000).
64. Adessi, C., Frossard, M. J., Boissard, C., Fraga, S., Bieler, S., Ruckle, T., Vilbois, F., Robinson, S. M., Mutter, M., Banks, W. A., and Soto, C. Pharmacological profiles of peptide drug candidates for the treatment of Alzheimer's disease. *J. Biol. Chem.* **278**, 13905–13911 (2003).
65. Gordon, D. J., Sciarretta, K. L., and Meredith, S. C. Inhibition of beta-amyloid(40) fibrillogenesis and disassembly of beta-amyloid(40) fibrils by short beta-amyloid congeners containing N-methyl amino acids at alternate residues. *Biochemistry.* **40**, 8237–8245 (2001).
66. Lowe, T. L., Strzelec, A., Kiessling, L. L., and Murphy, R. M. Structure-function relationships for inhibitors of beta-amyloid toxicity containing the recognition sequence KLVFF. *Biochemistry.* **40**, 7882–7889 (2001).
67. Poduslo, J. F., Curran, G. L., Kumar, A., Frangione, B., and Soto, C. (Beta-sheet breaker peptide inhibitor of Alzheimer's amyloidogenesis with increased blood-brain barrier permeability and resistance to proteolytic degradation in plasma. *J. Neurobiol.* **39**, 371–382 (1999).
68. Findeis, M. A., Lee, J. J., Kelley, M., Wakefield, J. D., Zhang, M. H., Chin, J., Kubasek, W., and Molineaux, S. M. Characterization of cholyl-leu-val-phe-phe-ala-OH as an inhibitor of amyloid beta-peptide polymerization. *Amyloid.* **8**, 231–241 (2001).
69. Kokkoni, N., Stott, K., Amijee, H., Mason, J. M., and Doig, A. J. N-Methylated peptide inhibitors of beta-amyloid aggregation and toxicity. Optimization of the inhibitor structure. *Biochemistry.* **45**, 9906–9918 (2006).
70. Amijee, H., Madine, J., Middleton, David A., and Doig, Andrew J. Inhibitors of protein aggregation and toxicity. *Biochem. Soc. Trans.* **37**, 692 (2009).
71. Amijee, H., Bate, C., Williams, A., Virdee, J., Jeggo, R., Spanswick, D., Scopes, D. I. C., Treherne, J. M., Mazzitelli, S., Chawner, R., Eyers, C. E., and Doig, A. J. The N-methylated peptide SEN304 powerfully inhibits Aβ(1–42) toxicity by perturbing oligomer formation. *Biochemistry.* **51**, 8338–8352 (2012).
72. Haag, R. and Kratz, F. Polymer therapeutics: concepts and applications. *Angew. Chem. Int. Ed.* **45**, 1198–1215 (2006).
73. Ambrosio, E., Barattin, M., Bersani, S., Shubber, S., Uddin, S., van der Walle, C. F., Caliceti, P., and Salmaso, S. A novel combined strategy for the physical PEGylation of polypeptides. *J. Control. Release* **226**, 35–46 (2016).
74. Qi, Y. and Chilkoti, A. Protein-polymer conjugation-moving beyond PEGylation. *Curr. Opin. Chem. Biol.* **28**, 181–193 (2015).
75. Rocha, S., Cardoso, I., Borner, H., Pereira, M. C., Saraiva, M. J., and Coelho, M. Design and biological activity of beta-sheet breaker peptide conjugates. *Biochem. Biophys. Res. Commun.* **380**, 397–401 (2009).

76. Chafekar, S. M., Malda, H., Merkx, M., Meijer, E. W., Viertl, D., Lashuel, H. A., Baas, F., and Scheper, W. Branched KLVFF tetramers strongly potentiate inhibition of beta-amyloid aggregation. *Chem.Bio.Chem.* **8**, 1857–1864 (2007).
77. Zagorski, M. G., Yang, J., Shao, H., Ma, K., Zeng, H., and Hong, A. Methodological and chemical factors affecting amyloid beta peptide amyloidogenicity. *Methods Enzymol.* **309**, 189–204 (1999).
78. Ban, T., Hamada, D., Hasegawa, K., Naiki, H., and Goto, Y. Direct observation of amyloid fibril growth monitored by Thioflavin T fluorescence. *J. Biol. Chem.* **278**, 16462–16465 (2003).
79. Morimoto, A., Irie, K., Murakami, K., Masuda, Y., Ohigashi, H., Nagao, M., Fukuda, H., Shimizu, T., and Shirasawa, T. Analysis of the secondary structure of β-amyloid (Aβ42) fibrils by systematic proline replacement. *J. Biol. Chem.* **279**, 52781–52788 (2004).
80. Krebs, M. R. H., Bromley, E. H. C., and Donald, A. M. The binding of thioflavin-T to amyloid fibrils: localisation and implications. *J. Struct. Biol.* **149**, 30–37 (2005).
81. Wolfe, L. S., Calabrese, M. F., Nath, A., Blaho, D. V., Miranker, A. D., and Xiong, Y. Protein-induced photophysical changes to the amyloid indicator dye Thioflavin T. *Proc. Natl. Acad. Sci. U. S. A.* **107**, 16863–16868 (2010).
82. Stsiapura, V. I., Maskevich, A. A., Tikhomirov, S. A., and Buganov, O. V. Charge transfer process determines ultrafast excited state deactivation of Thioflavin T in low-viscosity solvents. *J. Phys. Chem. A.* **114**, 8345–8350 (2010).
83. Khurana, R., Coleman, C., Ionescu-Zanetti, C., Carter, S. A., Krishna, V., Grover, R. K., Roy, R., and Singh, S. Mechanism of Thioflavin T binding to amyloid fibrils. *J. Struct. Biol.* **151**, 229–238 (2005).
84. Naiki, H., Higuchi, K., Hosokawa, M., and Takeda, T. Fluorometric determination of amyloid fibrils in vitro using the fluorescent dye, Thioflavin T1. *Anal. Biochem.* **177**, 244–249 (1989).
85. Groenning, M., Olsen, L., van de Weert, M., Flink, J. M., Frokjaer, S., and Jorgensen, F. S. Study on the binding of Thioflavin T to beta-sheet-rich and non-beta-sheet cavities. *J. Struct. Biol.* **158**, 358–369 (2007).
86. Chen, A. K., Lin, R. Y., Hsieh, E. Z., Tu, P. H., Chen, R. P., Liao, T. Y., Chen, W., Wang, C. H., and Huang, J. J. Induction of amyloid fibrils by the C-terminal fragments of TDP-43 in amyotrophic lateral sclerosis. *J. Am. Chem. Soc.* **132**, 1186–1187 (2010).
87. Liao, T. Y., Lee, L. Y., and Chen, R. P. Leu-138 in the bovine prion peptide fibrils is involved in the seeding discrimination related to codon-129 M/V polymorphism in the prion peptide seeding experiment. *FEBS J.* **278**, 4351–4361 (2011).
88. Hsu, J. C., Chen, H. L., Snoeberger, R. C., Luh, F. Y., Lim, T. S., Hsu, C., and Chen, R. P. (2013) Thioflavin T and its photo-irradiative derivatives: exploring their spectroscopic properties in the absence and presence of amyloid fibrils. *J. Phys. Chem. B.* **117**, 3459–3468.
89. Shearman, M. S., Hawtin, S. R., and Tailor, V. J. The intracellular component of cellular 3-(4,5-dimethylthiazol-2-yl)-2, 5-diphenyltetrazolium bromide (MTT) reduction is specifically inhibited by beta-amyloid peptides. *J. Neurochem.* **65**, 218–227 (1995).
90. Shearman, M. S., Ragan, C. I., and Iversen, L. L. Inhibition of PC12 cell redox activity is a specific, early indicator of the mechanism of beta-amyloid-mediated cell death. *Proc. Natl. Acad. Sci. U. S. A.* **91**, 1470–1474 (1994).

91. Bruno, B. J., Miller, G. D., and Lim, C. S. Basics and recent advances in peptide and protein drug delivery. *Ther. Delivery* **4**, 1443–1467 (2013).
92. Kawarabayashi, T., Younkin, L. H., Saido, T. C., Shoji, M., Ashe, K. H., and Younkin, S. G. Age-dependent changes in brain, CSF, and plasma amyloid (beta) protein in the Tg2576 transgenic mouse model of Alzheimer's disease. *J. Neurosci.* **21**, 372–381 (2001).
93. Kamat, P. K., Tota, S., Rai, S., Swarnkar, S., Shukla, R., and Nath, C. A study on neuroinflammatory marker in brain areas of okadaic acid (ICV) induced memory impaired rats. *Life Sci.* **90**, 713–720 (2012).
94. Webster, S. J., Bachstetter, A. D., Nelson, P. T., Schmitt, F. A., and Van Eldik, L. J. Using mice to model Alzheimer's dementia: an overview of the clinical disease and the preclinical behavioral changes in 10 mouse models. *Front. Genet.* **5**, 88 (2014).
95. Weksler, M. E., Gouras, G., Relkin, N. R., and Szabo, P. The immune system, amyloid-beta peptide, and Alzheimer's disease. *Immunol. Rev.* **205**, 244–256 (2005).
96. Tanzi, R. and Bertram, L. Twenty years of the Alzheimer's disease amyloid hypothesis: a genetic perspective. *Cell* **120**, 545–555 (2005).
97. De Jonghe, C., Esselens, C., Kumar-Singh, S., Craessaerts, K., Serneels, S., Checler, F., Annaert, W., Van Broeckhoven, C., and De Strooper, B. Pathogenic APP mutations near the γ-secretase cleavage site differentially affect Aβ secretion and APP C-terminal fragment stability. *Hum. Mol. Genet.* **10**, 1665–1671 (2001).
98. Borchelt, D. R., Thinakaran, G., Eckman, C. B., Lee, M. K., Davenport, F., Ratovitsky, T., Prada, C. M., Kim, G., Seekins, S., Yager, D., Slunt, H. H., Wang, R., Seeger, M., Levey, A. I., Gandy, S. E., Copeland, N. G., Jenkins, N. A., Price, D. L., Younkin, S. G., and Sisodia, S. S. Familial Alzheimer's disease-linked presenilin 1 variants elevate Abeta1-42/1-40 ratio in vitro and in vivo. *Neuron.* **17**, 1005–1013 (1996).
99. Howlett, D. R. and Richardson, J. C. The pathology of APP transgenic mice: a model of Alzheimer's disease or simply overexpression of APP? *Histol. Histopathol.* **24**, 83–100 (2009).
100. Duyckaerts, C., Potier, M.-C., and Delatour, B. Alzheimer's disease models and human neuropathology: similarities and differences. *Acta Neuropathol.* **115**, 5–38 (2008).
101. Newman, M., Musgrave, F. I., and Lardelli, M. Alzheimer disease: amyloidogenesis, the presenilins and animal models. *Biochim. Biophys. Acta, Mol. Basis Dis.* **1772**, 285–297 (2007).
102. Guénette, S. Y. and Tanzi, R. E. Progress toward valid transgenic mouse models for Alzheimer's disease. *Neurobiol. Aging.* **20**, 201–211 (1999).
103. Chin, J. (2011) Selecting a mouse model of Alzheimer's disease. In ed. Roberson, D. E., *Alzheimer's Disease and Frontotemporal Dementia: Methods and Protocols*. Totowa, NJ: Humana Press; 2011. pp 169–189.
104. Hsiao, K., Chapman, P., Nilsen, S., Eckman, C., Harigaya, Y., Younkin, S., Yang, F., and Cole, G. Correlative memory deficits, Aβ elevation, and amyloid plaques in transgenic mice. *Science.* **274**, 99–103 (1996).
105. Bornemann, K. D. and Staufenbiel, M. Transgenic mouse models of Alzheimer's disease. *Ann. N. Y. Acad. Sci.* **908**, 260–266 (2000).
106. Qing, H., He, G., Ly, P. T. T., Fox, C. J., Staufenbiel, M., Cai, F., Zhang, Z., Wei, S., Sun, X., Chen, C.-H., Zhou, W., Wang, K., and Song, W. Valproic acid inhibits Aβ

production, neuritic plaque formation, and behavioral deficits in Alzheimer's disease mouse models. *J. Exp. Med.* **205**, 2781–2789 (2008).

107. Dewachter, I., Van Dorpe, J., Smeijers, L., Gilis, M., Kuiperi, C., Laenen, I., Caluwaerts, N., Moechars, D., Checler, F., Vanderstichele, H., and Van Leuven, F. Aging increased amyloid peptide and caused amyloid plaques in brain of old APP/V717I transgenic mice by a different mechanism than mutant presenilinl. *J. Neurosci.* **20**, 6452–6458 (2000).
108. Willis, M., Hutter-Paier, B., Wietzorrek, G., Windisch, M., Humpel, C., Knaus, H. G., and Marksteiner, J. Localization and expression of substance P in transgenic mice overexpressing human APP751 with the London (V717I) and Swedish (K670M/N671L) mutations. *Brain Res.* **1143**, 199–207 (2007).
109. Jankowsky, J. L., Slunt, H. H., Ratovitski, T., Jenkins, N. A., Copeland, N. G., and Borchelt, D. R. Co-expression of multiple transgenes in mouse CNS: a comparison of strategies. *Biomol. Eng.* **17**, 157–165 (2001).
110. Podlisny, M. B., Citron, M., Amarante, P., Sherrington, R., Xia, W., Zhang, J., Diehl, T., Levesque, G., Fraser, P., Haass, C., Koo, E. H. M., Seubert, P., St. George-Hyslop, P., Teplow, D. B., and Selkoe, D. J. Presenilin proteins undergo heterogeneous endoproteolysis between Thr291and Ala299 and occur as stable N- and C-terminal fragments in normal and Alzheimer brain tissue. *Neurobiol. Dis.* **3**, 325–337 (1997).
111. Thinakaran, G., Borchelt, D. R., Lee, M. K., Slunt, H. H., Spitzer, L., Kim, G., Ratovitsky, T., Davenport, F., Nordstedt, C., Seeger, M., Hardy, J., Levey, A. I., Gandy, S. E., Jenkins, N. A., Copeland, N. G., Price, D. L., and Sisodia, S. S. Endoproteolysis of presenilin 1 and accumulation of processed derivatives in vivo. *Neuron.* **17**, 181–190 (1996).
112. Ratovitski, T., Slunt, H. H., Thinakaran, G., Price, D. L., Sisodia, S. S., and Borchelt, D. R. Endoproteolytic processing and stabilization of wild-type and mutant presenilin. *J. Biol. Chem.* **272**, 24536–24541 (1997).
113. Garcia-Alloza, M., Robbins, E. M., Zhang-Nunes, S. X., Purcell, S. M., Betensky, R. A., Raju, S., Prada, C., Greenberg, S. M., Bacskai, B. J., and Frosch, M. P. Characterization of amyloid deposition in the APPswe/PS1dE9 mouse model of Alzheimer disease. *Neurobiol. Dis.* **24**, 516–524 (2006).
114. Reiserer, R. S., Harrison, F. E., Syverud, D. C., and McDonald, M. P. Impaired spatial learning in the APPSwe +PSEN1ΔE9 bigenic mouse model of Alzheimer's disease. *Genes Brain Behav.* **6**, 54–65 (2007).
115. van Groen, T., Kadish, I., Funke, S. A., Bartnik, D., and Willbold, D. (Treatment with D3 Removes amyloid deposits, reduces inflammation, and improves cognition in aged A beta PP/PS1 double transgenic mice. *J. Alzheimers Dis.* **34**, 609–620 (2013).
116. Oakley, H., Cole, S. L., Logan, S., Maus, E., Shao, P., Craft, J., Guillozet-Bongaarts, A., Ohno, M., Disterhoft, J., Van Eldik, L., Berry, R., and Vassar, R. Intraneuronal beta-amyloid aggregates, neurodegeneration, and neuron loss in transgenic mice with five familial Alzheimer's disease mutations: potential factors in amyloid plaque formation. *J. Neurosci.* **26**, 10129–10140 (2006).
117. Oddo, S., Caccamo, A., Kitazawa, M., Tseng, B. P., and LaFerla, F. M. Amyloid deposition precedes tangle formation in a triple transgenic model of Alzheimer's disease. *Neurobiol. Aging* **24**, 1063–1070 (2003).

118. Andorfer, C., Kress, Y., Espinoza, M., de Silva, R., Tucker, K. L., Barde, Y. A., Duff, K., and Davies, P. Hyperphosphorylation and aggregation of tau in mice expressing normal human tau isoforms. *J. Neurochem.* **86**, 582–590 (2003).
119. Polydoro, M., Acker, C. M., Duff, K., Castillo, P. E., and Davies, P. Age-dependent impairment of cognitive and synaptic function in the htau mouse model of tau pathology. *J. Neurosci.* **29**, 10741–10749 (2009).
120. Duff, K., Knight, H., Refolo, L. M., Sanders, S., Yu, X., Picciano, M., Malester, B., Hutton, M., Adamson, J., Goedert, M., Burki, K., and Davies, P. Characterization of pathology in transgenic mice over-expressing human genomic and cDNA tau transgenes. *Neurobiol. Dis.* **7**, 87–98 (2000).
121. Kowall, N. W., Beal, M. F., Busciglio, J., Duffy, L. K., and Yankner, B. A. An in vivo model for the neurodegenerative effects of beta amyloid and protection by substance P. *Proc. Natl. Acad. Sci. U. S. A.* **88**, 7247–7251 (1991).
122. Tajima, H., Kawasumi, M., Chiba, T., Yamada, M., Yamashita, K., Nawa, M., Kita, Y., Kouyama, K., Aiso, S., Matsuoka, M., Niikura, T., and Nishimoto, I. A humanin derivative, S14G-HN, prevents amyloid-beta-induced memory impairment in mice. *J. Neurosci. Res.* **79**, 714–723 (2005).
123. Mogi, M., Li, J.-M., Tsukuda, K., Iwanami, J., Min, L.-J., Sakata, A., Fujita, T., Iwai, M., and Horiuchi, M. Telmisartan prevented cognitive decline partly due to PPAR-γ activation. *Biochem. Biophys. Res. Commun.* **375**, 446–449 (2008).
124. Tsukuda, K., Mogi, M., Iwanami, J., Min, L.-J., Sakata, A., Jing, F., Iwai, M., and Horiuchi, M. Cognitive deficit in amyloid-β–injected mice was improved by pretreatment with a low dose of telmisartan partly because of peroxisome proliferator-activated receptor-γ activation. *Hypertension* **54**, 782–787 (2009).
125. Nitta, A., Itoh, A., Hasegawa, T., and Nabeshima, T. β-Amyloid protein-induced Alzheimer's disease animal model. *Neurosci. Lett.* **170**, 63–66 (1994).
126. O'Hare, E., Weldon, D. T., Mantyh, P. W., Ghilardi, J. R., Finke, M. P., Kuskowski, M. A., Maggio, J. E., Shephard, R. A., and Cleary, J. Delayed behavioral effects following intrahippocampal injection of aggregated Aβ (1–42). *Brain Res.* **815**, 1–10 (1999).
127. De Ferrari, G. V., Chacon, M. A., Barria, M. I., Garrido, J. L., Godoy, J. A., Olivares, G., Reyes, A. E., Alvarez, A., Bronfman, M., and Inestrosa, N. C. Activation of Wnt signaling rescues neurodegeneration and behavioral impairments induced by [beta]-amyloid fibrils. *Mol. Psychiatry* **8**, 195–208 (2003).
128. Chacon, M. A., Barria, M. I., Soto, C., and Inestrosa, N. C. Beta-sheet breaker peptide prevents Abeta-induced spatial memory impairments with partial reduction of amyloid deposits. *Mol. Psychiatry.* **9**, 953–961 (2004).
129. Illum, L. Nasal drug delivery — possibilities, problems and solutions. *J. Control. Release.* **87**, 187–198 (2003).
130. Illum, L. Is nose-to-brain transport of drugs in man a reality? *J. Pharm. Pharmacol.* **56**, 3–17 (2004).
131. Illum, L. Nasal drug delivery: new developments and strategies. *Drug Discov. Today.* **7**, 1184–1189 (2002).
132. Futami, J., Kitazoe, M., Murata, H., and Yamada, H. Exploiting protein cationization techniques in future drug development. *Expert. Opin. Drug. Discov.* **2**, 261–269 (2007).
133. Kitazoe, M., Murata, H., Futami, J., Maeda, T., Sakaguchi, M., Miyazaki, M., Kosaka, M., Tada, H., Seno, M., Huh, N. H., Namba, M., Nishikawa, M., Maeda, Y., and

Yamada, H. Protein transduction assisted by polyethylenimine-cationized carrier proteins. *J. Biochem.* **137**, 693–701 (2005).

134. Herve, F., Ghinea, N., and Scherrmann, J. M. CNS delivery via adsorptive transcytosis. *AAPS J.* **10**, 455–472 (2008).

135. Loftus, L. T., Li, H. F., Gray, A. J., Hirata-Fukae, C., Stoica, B. A., Futami, J., Yamada, H., Aisen, P. S., and Matsuoka, Y. In vivo protein transduction to the CNS. *Neuroscience* **139**, 1061–1067 (2006).

Statistical Mechanics and Kinetics of Amyloid Fibrillation

4

Liu Hong, Chiu Fan Lee† and Ya Jing Huang**

**Zhou Pei-Yuan Center for Applied Mathematics, Tsinghua University, Beijing 100084, P. R. China*
†Department of Bioengineering, Imperial College London, South Kensington Campus, London SW7 2AZ, UK

Amyloid fibrillation is a protein self-assembly phenomenon that is intimately related to well-known human neurodegenerative diseases. During the past few decades, striking advances have been achieved in our understanding of the physical origin of this phenomenon and they constitute the contents of this review. Starting from a minimal model of amyloid fibrils, we explore systematically the equilibrium and kinetic aspects of amyloid fibrillation in both dilute and semidilute limits. We then incorporate further molecular mechanisms into the analyses. We also discuss the mathematical foundation of kinetic modeling based on chemical mass-action equations, the quantitative linkage with experimental measurements, as well as the procedure to perform global fitting.

1. Introduction

The importance of understanding amyloid fibrillation comes not only from its intimate relation to amyloid diseases, such as the well-known Alzheimer's, Huntington's, and Parkinson's diseases,[1,2] but also from its physical simplicity and universality as a typical self-assembling phenomenon of linear biomolecules.[3] Various thermodynamic and kinetic approaches borrowed from classical polymer statistical mechanics, the kinetics of chemical reactions, as well as nonequilibrium

L. H. and Y. J. H. would like to dedicate this paper to the memory of Prof. Chia-Chiao Lin (1916–2013), a great applied mathematician, a beloved advisor and a dear friend, on his 100-year anniversary.

processes, have been developed and applied to experimentally and biologically relevant amyloid systems with great success.[4–6] Related fruitful results, developments, and applications in the past decades constitute the focus of our current paper: a self-contained review on the thermodynamic and kinetics of amyloid fibrillation.

Thermodynamics and kinetics are two sides of the same coin. The latter deals with time-dependent fibrillation processes in general; while the former is more focused on the final time-independent properties of the amyloid system — the equilibrium state. In the current review, we present the thermodynamics and kinetics of amyloid fibrillation separately in order to keep each part clear and self-contained. But, readers should bear in mind of the intrinsic correlations between those two descriptions, like requirements on reaction rate constants for various fibrillation processes in order to guarantee the existence of a genuine thermodynamic equilibrium state.[7]

The whole review is organized into three major sections. The first one is focused on the thermodynamics of amyloid fibrillation by using the language of statistical mechanics; the next two are devoted to kinetic descriptions based on chemical mass-action equations. To be specific, the former provides a systematic exploration of various amyloid fibrillation processes, including both model formulation and analysis; while the latter is about the mathematical foundation of kinetic modeling, as well as its linkage with experimental observations.

2. Statistical Mechanics of Amyloid Fibrillation

Although polymer physics is a well-established field of science, novel physics governing the behavior of the systems is still being uncovered. In the case of self-assembling biopolymers, the novelty comes from the fact that the binding energy driving the polymerization process is relatively low compared to covalently bonded polymers. Hence, polymer breakage and rejoining can potentially contribute to the polymerization kinetics at an experimentally and physiologically relevant scale. The polymeric system is thus called "living" since every polymer can shrink through breakage and grow through elongation *via* monomer additions and through end-to-end joining with another polymer. Taking these processes into account is important for the complete description of the kinetics of self-assembling polymers. In this section, we focus purely on how a system of living semiflexible polymers behaves at thermal equilibrium.

2.1. *How to construct a minimal model*

We start by considering a minimal model of polymerizing monomers as depicted in Fig. 1a. Namely, the monomers are purely spherical particles (beads) with two sticky patches at two opposite ends. We assume that the beads are in an over-damped environment and so their movement is Brownian. The "stickiness" is short ranged and is quantitatively described by two quadratic energy functions as follows: one controls the distance between the connected beads and the other enforces the rigidity of the resulting polymers (Fig. 1c and d). We denote the distance between two connected beads as $l_0 + \Delta l$. Given any consecutive segment of three monomers in a polymer, if the three monomers are not colinear, we denote the angular deviation as $\Delta\theta$ (Fig. 1c). A polymer thus consists of a

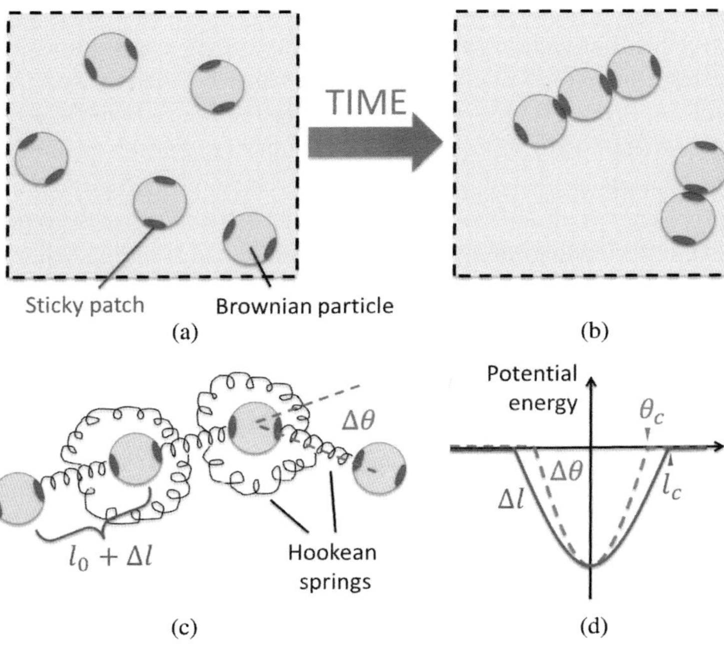

Figure 1: (a) A system of spherical particles (beads) with sticky patches on polar ends will self-assemble into polymers if the patches are stickiness enough (b). (c) In the minimal model considered here, the interactions within polymer act as Hookean springs between the beads to enforce extensile restriction and rigidity. The size of the beads has been shrunken to show the springs. (d) The specific potential energy functions governing the deviation in extension Δl (solid line) and angle $\Delta\theta$ (broken line) are assumed to be quadratic.

series of beads such that for all consecutive pairs of beads, the absolute values of the deviations, $|\Delta l|$, are smaller than the distance cutoff l_c; and that all deviation angles, are also less than the cutoff θ_c. The energy function for such an s-polymer is then

$$U(\{x\}) = \sum_{k=1}^{s-1} A\Delta l_k^2 + \sum_{h=1}^{s-2} B\Delta\theta_h^2 - E(s-1), \quad (1)$$

where k enumerates the number of distant bonds, h enumerates the number of angular bonds, and E is the binding energy between the patches that promotes aggregation.

Let us now imagine that at $t = 0$, we put these monomers in an inert solvent of a certain temperature T (Fig. 1a). The solvent is inert in the sense that their role in the system is purely to provide the thermalizing effects of a heat bath. The whole system is further connected to a much larger thermal bath of temperature T in such a way that heat, but not the beads, can flow back and forth between the system and the large heat bath. In other words, we are investigating the system from the perspective of a canonical ensemble.[8] The setup of this thought experiment corresponds to a typical experimental procedure in which polymerizing monomers are first dissolved in an appropriate solvent and then left unperturbed in the course of the experiment. In our case, if the binding energy is strong enough (i.e., E is large), then we expect that these monomers will self-assemble into polymers (Fig. 1b). Here, we assume that the threshold angle θ_c is small enough that we do not need to worry about the interactions of distant parts of the same polymer beyond what is already considered in our energy function. In particular, we can ignore the formation of loops in the system.

Although highly simplified, the model presented here is of relevance to some colloidal systems studied experimentally.[9] But as we show, the greatest virtue of this minimal model is its analytical tractability.

2.2. *How to deal with the dilute limit*

By the dilute limit, we mean that the concentration of solute and the resulting polymers in the system are dilute enough that we can ignore all solute interactions except those that lead to polymerization as described in the previous section. Given this assumption, the free energy density of the overall system can be calculated with the mean-field method. Specifically, we consider a system of N monomers in a volume V. The total partition function can be written in terms of the internal partition function of a single s-mer with its first bead's position fixed in space, z_s, in the following manner.[10]

$$Z_{\text{tot}} = \prod_{s}^{\prime} \frac{1}{N_s!} \left(\frac{V z_s}{\Lambda^3} \right)^{N_s} \tag{2}$$

where N_s denotes the number of s-mers in the system and the prime in the product denotes the number conservation of monomers: $\sum_{s=1}^{\infty} sN_s = N$. Since we are dealing with "classical" (i.e., not quantum mechanical) objects, the kinetic part of the partition function (resulting from momentum integrations) is irrelevant[11] and so Z_{tot} corresponds to the configurational partition function, with Λ being an arbitrary constant of dimension length to make Z_{tot} dimensionless. The denominator $N_s!$ is in Eq. 2 because the s-mers in the system are all indistinguishable and that the free energy is extensive. Note that these polymers are indistinguishable purely because we have chosen not to distinguish them in our analysis, which is typically the case in experiments.[12]

The total partition function in Eq. 2 follows from a mean-field approximation in the sense that the sequence $\{N_s\}$ is fixed by minimizing the free energy of the system.

$$F_{\text{tot}} = -k_B T \log Z_{\text{tot}}. \tag{3}$$

In other words, fluctuations away from the minimizing sequence $\{N_s\}$ are ignored. Such an approximation is expected to be qualitatively correct away from any critical points,[8] which, as we shall see, this system does not possess.

To proceed further analytically, we still need to calculate the s-mer partition function, z_s, which is of the form:

$$\begin{aligned} z_s &= \frac{4\pi l_0^2 \mathrm{e}^{(s-1)\beta E}}{\Lambda^{3(s-1)}} \left(\int_{-l_c}^{l_c} d\Delta l \mathrm{e}^{-\beta A \Delta l^2} \right)^{s-1} \left(l_0^2 \int_0^{\theta_c} d\Delta\theta \sin(\Delta\theta) \mathrm{e}^{-\beta B \Delta\theta^2} \right)^{s-2} \\ &= \frac{4\pi^{3/2} l_0^2 \mathrm{e}^{\beta E}}{\Lambda^3 \sqrt{\beta A}} \left(\frac{l_0^2 \sqrt{\pi} \mathrm{e}^{\beta E}}{\Lambda^3 \beta^{3/2} \sqrt{AB}} \right)^{s-2}, \end{aligned} \tag{4}$$

for $s > 1$, while $z_1 = 1$. Note that in Eq. 4, the factor $4\pi l_0^2$ comes from integrating over the orientation of the polymer given that the first bead is fixed in space, the integrals in the first brackets stem from the longitudinal degrees of freedom and those in the second brackets from the angular degrees of freedom along the polymer chain. To arrive at Eq. 4, we have taken the limits of integration to infinity, which is legitimate since $\beta A, \beta B$ are typically high where $\beta \equiv k_B T$.

The total free energy can now be expressed as follows:

$$
\begin{aligned}
F_{\text{tot}} &= -k_B T \sum_s{}' \left\{ N_s \log\left(\frac{V z_s}{\Lambda^3}\right) - \log N_s! \right\} \\
&= -k_B T \sum_s{}' \left\{ N_s \log\left(\frac{V z_s}{\Lambda^3}\right) - N_s \log N_s + N_s \right\} \\
&= \beta^{-1} \left\{ N_1 \left[\log N_1 - \log \frac{V}{\Lambda^3} - 1 \right] \right. \\
&\quad \left. + \sum_s{}' N_s \left[\log N_s - \log \frac{V}{\Lambda^3} - \chi s - \xi - 1 \right] \right\},
\end{aligned} \tag{5}
$$

and

$$\xi = \log \frac{4\pi^{1/2}\Lambda^3\beta^{5/2}A^{1/2}B^2}{l_0^2} - \beta E, \tag{6}$$

$$\chi = \log \frac{l_0^2\pi^{1/2}}{\Lambda^3\beta^{3/2}A^{1/2}B} + \beta E. \tag{7}$$

Given Eq. 5, we can finally minimize F_{tot} with respect to N_s using the Lagrange multiplier method to enforce the conservation $\sum_s sN_s = N$. To do so, we minimize the following summation with λ being the Lagrange multiplier.

$$F_{\text{tot}} + \lambda(\sum_s sN_s - N) \tag{8}$$

with respect to the set $\{N_s\}$, which leads to:

$$N_1 = \frac{V}{\Lambda^3} e^{\lambda}, \tag{9}$$

$$N_s = \frac{V}{\Lambda^3} e^{(x+\lambda)s+\xi} \text{ for } s > 1, \tag{10}$$

or for $s > 1$,

$$\frac{N_s}{V} = K\left(\frac{N_1}{V}\frac{1}{m_F^*}\right)^s, \tag{11}$$

where,

$$m_F^* = \frac{\beta^{3/2} A^{1/2} B}{l_0^2 \pi^{1/2} e^{\beta E}}, \tag{12}$$

$$K = \frac{4\pi^{1/2} \beta^{5/2} A^{1/2} B^2}{l_0^2 e^{\beta E}}. \tag{13}$$

Equation 11 expresses the s-mer concentration $n_s \equiv N_s/V$ in terms of monomer concentration $m = n_1 \equiv N_1/V$. Since s can be as big as we want, by the conservation of mass, we know that m can never exceed m_F^* for otherwise the terms in the brackets in Eq. 11 will blow up with s. Indeed, we shall see that m asymptotically approaches m_F^* as m_{tot} increases. For this reason, we ($\sum_{s \geq 1} s n_s = N/V \equiv m_{\text{tot}}$) shall call m_F^* the critical fibrillar concentration (CFC).[10,13] Note that although the system transition from being monomer dominated to fibril dominated as m_{tot} increases, it never goes through a phase transition in the thermodynamic sense[11] since the derivatives of the free energy are always continuous. This is also reflected, for example, by the lack of discontinuities in m_{tot} (Fig. 2). In the regime where $m_{tot} \gg m_F^*$, $m \simeq m_F^*$, and Eq. 11 shows that the size distribution of polymers is exponential, with the average size given by $\sqrt{m_{\text{tot}}/K}$.[10,14–16]

Let us now try to substitute in experimentally motivated parameters to see how our model corroborates with observation. Since we are primarily interested in protein aggregation, we take the average size l_0 to be 1 nm, and the binding energy E to be 25 $k_B T$. To estimate the spring constant A and B, we make the assumption

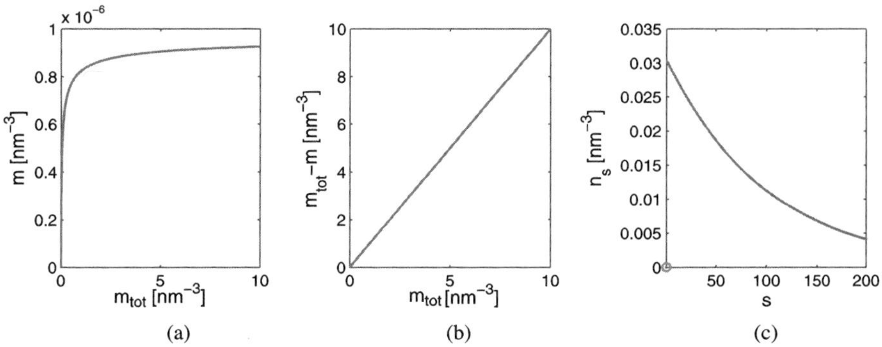

Figure 2: Using the parameters in the text, (a) depicts how the monomer concentration changes as monomer concentration increases; and correspondingly for the concentration of the fibrillar species (b). (c) The size (length) distribution of the system at $m_{tot} \simeq 10\,\text{nm}^{-3}$. The distribution is exponential, but note the discontinuity at $s = 1$ (highlighted by the red circle). The unit of the concentration is nm^{-3}.

that each monomer within a polymeric chain has a wriggle room of around 10% of its size, that is, $l_c \sim l_0/10$ and $\theta_c \sim 0.1$ rad. From this we can estimate A as $100E/l_0^2$ and B as $100\,E$. Using these parameters, we find that $m_F^* \simeq 9.8 \times 10^{-7}\,\text{nm}^{-3}$ or around $9.8\,\mu$M. The corresponding fibrillation behavior of this system is shown in Fig. 2.

With regard to experimental observation, the predicted exponential length distribution seems to deviate from some experimental studies.[17,18] Van Raaij *et al.* has interpreted the observed peaks as a result of the finite resolution of the atomic force microscopy imaging and length measurement procedure.[19] Besides this explanation, it is also known that it can take on the order of months for mature fibrils to form.[20] Therefore, the appearance of the peaks observed may also reflect the fact that the self-assembled systems have not yet reached thermal equilibrium.

2.3. *How to generalize to the semidilute limit*

In the dilute limit, mutual interactions between the solutes beyond the polymerizing interactions are ignored. What if we now increase the concentration of the solute so that such an approximation is no longer valid? This takes us to the semidilute limit. We first discuss the simplest and ubiquitous type of interactions: volume exclusion interactions.

2.3.1. *Pure volume exclusion interactions*

To take steric interactions into account, we start again with the noninteracting free energy density (see Eq. 5):

$$f_0 = \beta^{-1} \int ds n(s)[\log n(s) + \log \Lambda^3 - \chi s - \xi - 1], \tag{14}$$

where we have ignored the monomeric contribution, and pass to the continuum description in s since we are primarily interested in the fibril-dominant regime ($m_{\text{tot}} \gg$ CFC). We then add to f_0 the following interaction term:

$$f_{\text{int}} = \int ds ds' n(s) n(s') B(s, s'). \tag{15}$$

In the earlier equation, $B(s, s')$ is the second virial coefficients corresponding to the steric interactions of two semiflexible polymers of length s and s', and is of the form:[21]

$$B(s, s') = \frac{2\pi}{3} l_0^3 + \frac{\pi(s+s')l_0^3}{2} + 2ss' l_0^3 |\sin\phi|, \tag{16}$$

where ϕ is the angle between the two polymers. To deal with the additional variable ϕ, we ask ourselves what would happen to the system given the steric interactions. From the physics of liquid crystals,[22,23] we expect that as the polymer concentration increases, the system can become nematic, that is, the semiflexible polymers will be aligned.

In other words, we anticipate that similar to liquid crystals,[23] as the solute concentration increases, the system will first go through a phase separation where the regions of the system can be partitioned into two phases, one nematic and the other isotropic. If the concentration increases further, the system will become fully nematic. This is indeed what is observed experimentally in, for instance, a system of fibrilizing hens lysozyme[24] (Fig. 3). We now incorporate this expected picture into our free energy minimization. Specifically, we consider both free energy densities in the isotropic and nematic phases, f_I and f_N, respectively. For f_I, we can simply add to Eq. 5 the average over the angle that two randomly oriented semiflexible make (since the system is isotropic) in Eq. 16, hence,

$$f_I = f_0(\{n_I(s)\}) + \int ds ds' n(s) n(s') \left[\frac{2\pi}{3} l_0^3 + \frac{\pi(s+s')l_0^3}{2} + \pi ss' l_0^3 \right]. \tag{17}$$

Using again the Lagrange multiplier method to enforce protein number conservation, one finds that the distribution $n_I(s)$ that minimizes f_I is exponential as before, although the mean size is now $\sqrt{m_I e^{8\psi_I/3}/K}$, where m_I is the total protein concentration and ψ_I is the volume fraction of proteins in the isotropic phase.[25]

In the nematic phase, the picture is more complicated. Since the polymers can elongate, it was found that the flexibility of the polymer has to be taken into account in order to stop the unrealistic lengthening of the polymers in the nematic phase.[26] Here, we again quote the results in,[25] in which it was found that by minimizing

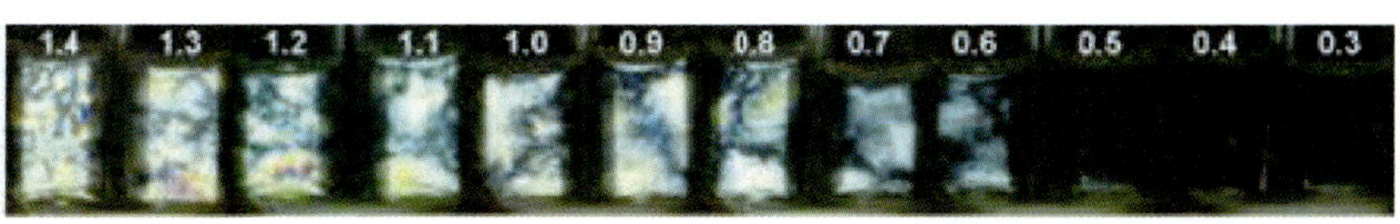

Figure 3: Glass vials (1 cm wide) with hen lysozyme fibril containing solutions imaged between crossed polars. Concentrations are indicated in mM. The lit up regions correspond to the nematic phase. This figure is reprinted with permission from.[24]

the free energy in the nematic phase, the length distribution is again exponential, and the mean size is now approximately $\sqrt{m_N e^{4\psi_I}/K}$. Here, m_N is the total protein concentration in the nematic phase. These results are in the regime where the mean polymer length is much greater than the persistence length of the polymer.

Minimizing f_I and f_N separately is not the whole story since there is also the possibility of phase separation. Namely, the volume of the system can be partitioned into isotropic and nematic regions. In the thermodynamic limit, we usually ignore the surface energy coming from the interface separating the isotropic and nematic regions. The minimization is thus performed on the following free energy density:

$$f_{\rm tot}(v_{\rm tot}, m_{\rm tot}) = \frac{v_I f_I(m_I) + v_N f_N(m_N)}{v_{\rm tot}}, \tag{18}$$

where $v_I(v_N)$ is the total volume of the isotropic (nematic) regions and $c_I(c_N)$ is solute concentration in the isotropic (nematic) regions. The conservation laws are, therefore, $v_I + v_N = v_{tot}$ and $v_I m_I + v_N m_N = v_{\rm tot} m_{\rm tot}$. Minimizing f_{tot} with respect to v and m finally enables us to confirm the expectation we had from the beginning. Namely, as solute concentration increases, regions of nematic phase appear in the system and coexists with the isotropic phase. The polymer length distributions in both phases remain exponential, although the average length in the nematic is higher than that in the isotropic region.[27] As the solute concentration increases further, regions of isotropic phase disappear and the whole system will be in the nematic phase. As mentioned, these theoretical findings are corroborated by experimental study on self-assembling hens lysozyme[24,28] (see Fig. 3).

2.3.2. *With lateral interactions*

We have seen how the isotropic–nematic phase transition in a system of semiflexible polymers is typically accompanied by phase separation. In this section, we show that if the polymers are mutually attractive (beyond the end-to-end binding that leads to the joining of the two polymers), then the tendency for phase separation is even stronger. Specifically, using again our minimal model system shown in Fig. 1, we imagine that besides the strong directional attractive interactions schematically depicted by the red (dark grey) patches, the beads are weakly attractive ($\sim k_B T$) toward each other, that is, the green (light grey) areas of the beads are also weakly sticky.

To appreciate conceptually how interacting polymers behave at thermal equilibrium, the Flory–Huggins theory is a good starting point since it is conceptually simple and can be easily analyzed numerically. Here, we ignore the rigidity of the polymers and focus solely on the effects of attractive interactions on the system's phase behavior. Using the lattice model where each lattice site can be either occupied by one solvent molecule and by one monomer, the following Flory–Huggins free energy density of mixing can be derived:[29]

$$f_m(\phi) = \frac{1}{\bar{n}}\phi \ln \phi + (1-\phi)\ln(1-\phi) + \chi\phi(1-\phi), \tag{19}$$

where ϕ is the volume fraction of the polymers in the system and $\bar{n}$ is the number of monomers in the polymer. We note that in the previous formula, all polymers are of the same length, that is, we have ignored the disperse length distribution here for simplicity. The parameter χ summarizes the interactions of the monomers and is of the form:

$$\chi \equiv \frac{z}{2k_B T}[2e_{ms} - e_{mm} - e_{ss}], \tag{20}$$

where e_{mm}, e_{ss}, e_{ms} are the interaction energy for the monomer–monomer, solvent–solvent, and monomer–solvent pairs on the lattice, respectively. Here, we assume that the monomers are weakly attractive ($e_{mm} \sim -k_B T$) and for simplicity, we set $e_{ss} = 0 = e_{ms}$. Moreover z is the coordination number of the lattice, which for a three-dimensional (3D) cubic lattice is six. As a result, the interaction parameter χ is $-3e_{mm}/k_B T$.

The first two terms in Eq. 19 promotes phase separation, while for positive χ,f_m becomes concave down when χ is large enough, this is a signature of phase separation, which means that we need to minimize the total free energy by considering the possibility of having the system partition into two parts of distinct phases as in the previous section (see Eq. 18). A typical phase diagram is shown in Fig. 4.

The key feature here is that as the polymers elongate, the tendency to phase separate gets stronger even if the polymers are only weakly attractive toward each other. Experimentally, it is observed that many amyloid fibrils aggregate in solution and thus *in vitro* and *in vivo* phase separation of amyloid fibrils may be expected to occur. Indeed, the observed clustering of Sup35 fibrils in the cytoplasm of yeast cells may be a signature of such phase separation (Fig. 5).

2.4. *How to incorporate oligomers*

We have so far focused on the behavior of fibrilizing system in which there are only monomers or fibrils. In the case of amyloid fibrils, the situation is more complex. Indeed, mounting evidence has indicated that proteins in the monomeric form and oligomeric form (potentially amorphous aggregates of tens of proteins), instead of proteins in the fibrillar form, are predominantly responsible for cell death.[32] In this section, we incorporate the oligomeric species into our analysis. To model the

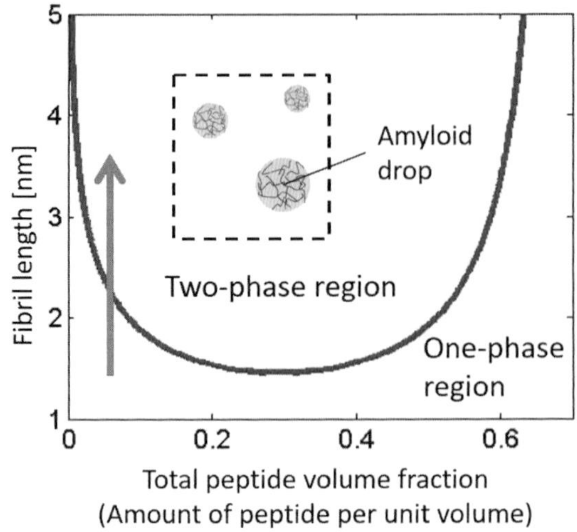

Figure 4: A typical phase diagram of a phase separating polymeric system according to the classic Flory–Huggins theory.[30] Phase separated polymer drops (inset) form in the two-phase region is bounded by the blue (dark grey) curve. At fixed peptide volume fraction, the tendency for the system to phase separate increases as the self-assembled fibrils elongate (indicated by the red [light grey] arrow).

presence of oligomers, we borrow the treatment of spherical micelles formation in a solution of surfactants.[13] Specifically, we assume that the monomers can aggregate together to form an amorphous cluster. However, there is an optimal number of monomers, W, in the cluster in the sense that the corresponding cluster partition function is greatest. In this system, a monomer can be classified into three categories: (1) monomeric, (2) part of an oligomer (which we call a micelle), and (3) part of a fibril. The total partition function in this system can be written as follows:[10]

$$Z_{\text{tot}} = \prod_{s,p}{}' \frac{V^{N_1}}{\Lambda^{3N_1} N_1!} \frac{(V z_s^{(f)})^{N_s}}{\Lambda^{3N_s} N_s!} \frac{(V z_p^{(m)})^{M_p}}{\Lambda^{3M_p} M_p!}, \tag{21}$$

where $z_s^{(f)}$ and $z_p^{(m)}$ are the internal partition functions for the fibrillar and micellar species. We have also singled out the monomeric contribution to the partition to highlight the fact that there are three different species in the system.

We now model the micellar partition function as follows:

$$z_p^{(m)} = \delta_{pW} \left(\frac{\alpha}{\Lambda^3}\right)^{W-1} e^{WE_m} \tag{22}$$

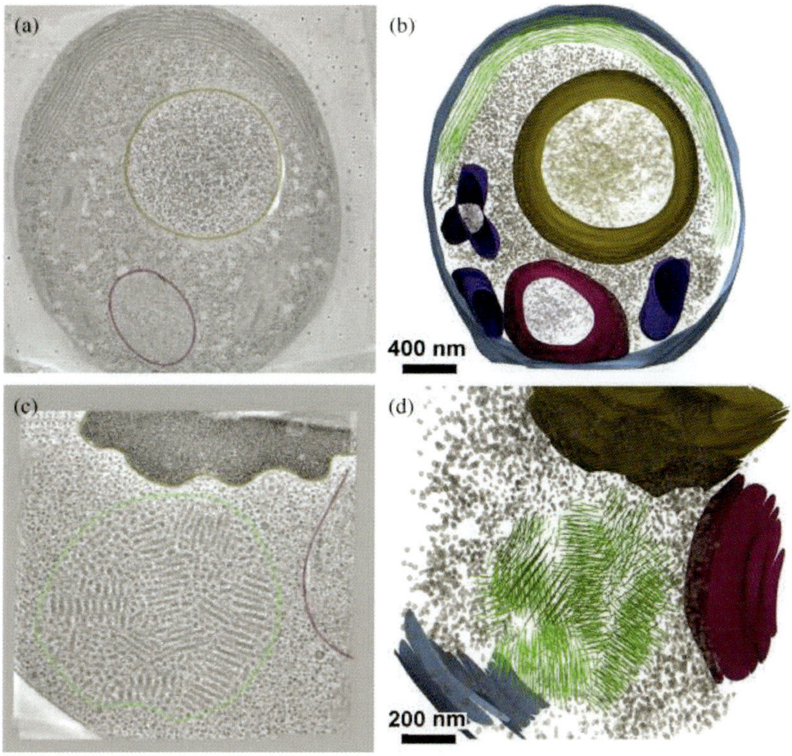

Figure 5: (a) Section through a tomogram of a yeast cell in which amyloid fibrils formed from Sup35 are close to the cell membrane. The nucleus is outlined in magenta and the vacuole in brown. (b) Rendered 3D model from serial tomograms of this cell. Amyloid fibrils are in green; membrane, blue; nucleus, magenta; vacuole, brown; mitochondria, purple; and large complexes (presumably ribosomes) as gray dots. (c) Section through of a cell in which the aggregate of amyloid fibrils is of a form of a drop (green). (d) Rendered 3D model of the dot from serial tomograms (coloring as in B). This figure is reprinted with permission from reference Saibil *et al.*[31]

where α is of the order of the dimension of the monomer. Specifically, all micelles are assumed to be of size W for simplicity and the clustering is driven by the binding energy E_m per monomer. If we ignore the fibrillar species for the time being, then using again the Lagrange multiplier method as before, we find that we can again relate the micellar concentration to the monomer concentration M:

$$m_W = \frac{1}{\alpha}\left(\frac{m}{m_M^*}\right)^W \tag{23}$$

where $m_M^* = (\alpha e^{\beta E_m})^{-1}$. For $W \sim \mathcal{O}(10)$, we can see that m is bounded above at around m_M^*, that is, most monomers are in the micellar form at $m_{\text{tot}} > m_M^*$. On the other hand, for $m \ll m_M^*$, m_K is negligible. For this reason, we call m_M^* the critical micellar concentration (CMC).

If we now incorporate the fibrillar species into the picture, the system is effectively partitioned into multiple regimes: if m_{tot} is smaller than both the CMC and the CFC, then the system is dominated by m_{tot}. If the CMC is lower than the CFC and $m_{\text{tot}} >$ then the micellar species will dominate the system; while if the CFC is lower than the CMC, fibrils will be dominant if $m_{\text{tot}} >$ (Fig. 6a).[10] Note that even though if the CFC is lower than the CMC, the micelles would still be transiently present in the system, and their existence may have important implications in the fibrillation kinetics as proposed in Lomakin *et al.*[33] (Fig. 6b). A similar model has also been recently studied using molecular dynamics simulation in Šarić *et al.*[34]

2.5. *How to incorporate internal monomeric structures*

Another natural generalization of our minimal bead-and-stick model is the incorporation of the internal monomeric structure into our analysis. A peptide can in general be in multiple states, for example, in the form of a beta sheet, random coil, alpha helix. In the specific case of Aβ peptides, the monomeric state is in the random-coil

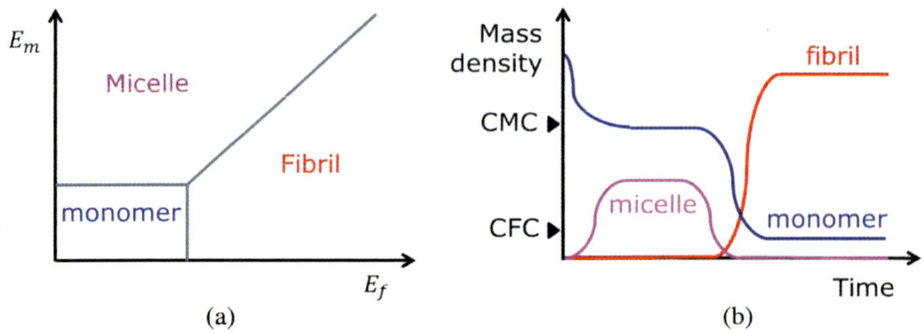

Figure 6: (a) At fixed m_{tot}, the system can be dominated by distinct species depending on the strength of the fibrillar binding energy E_F and the micellar binding energy E_M. (b) A schematic of a potential nucleation pathway of fibrillation proposed in Lomakin *et al.*[33] Starting with a pool of monomers, micelles form quickly due to the fast (potentially diffusion-limited) formation kinetics. Within the micelles, the protein concentration is high and thus facilitates the slow nucleation of nuclei of fibrils. As fibrils elongate, the monomers in the system will be eventually depleted below the CMC and thus the micelles will disappear in the system.

configuration, while the peptides are predominately in the β-sheet state in the fibrillar form.[35] In terms of our minimal model, we can account for this modification by assigning a nonzero value to the monomeric partition function $z_1^{(R)} = \gamma > 0$ corresponding to the random coil state in Eq. 2; while for monomers in the β-sheet form, the monomeric partition function $z_1^{(\beta)}$ is again set to one. Here, γ is positive means that the monomer in the random coil state is preferred over the beta sheet state in the fibrillar form, which originates from the fact that the random coil is entropically more favorable. At thermal equilibrium, we expected have $m^{(\beta)} = \mathrm{e}^{-\gamma} m^{(R)}$. The total monomer concentration m is therefore $m^{(\beta)} + m^{(R)} = m^{(R)}(1 + \mathrm{e}^{-\gamma})$. Incorporating the fibrillar species into the picture and writing the concentration of s-mers in terms of that of the monomer as done in Eq. 11, we have,

$$n_s = K\left(\frac{m^{(R)}}{\mathrm{e}^{\gamma} m_F^*}\right)^s = K\left(\frac{m}{\hat{m}_F^*}\right)^s \tag{24}$$

where $\hat{m}_F^*$ is $(1 + \mathrm{e}^{\gamma})m_F^*$ (see Eq. 11). In other words, most of qualitative analyses as before remains the same, except now the CFC is increased by the factor $(1 + \mathrm{e}^{\gamma})$, which signifies that the monomeric concentration m is generally increased when the random coil-beta sheet transition[36,37] is taken into account at the monomeric level. We note that a more detailed analysis of the effects of the degree of freedom from the monomeric conformation can be found in Schreck *et al.*[3]

2.6. *How to incorporate multiple fibril morphologies*

Besides the possibility of micelle formation, real biopolymers, and amyloid fibrils in particular, are likely to consist of two or more filaments. In addition, multiple morphologies may coexist in the system.[38] The simplest way to model such a system is to imagine that each bead in Fig. 1a also possesses a sticky patch on the equator. If the new "patch on the side" is small in area, then a two-filament fibril will form naturally (Fig. 7). In fact, many amyloid fibrils seemed to consist of filaments twisted together, which could be incorporated into our minimal model by twisting the location of the side patch along the axial direction (Fig. 7). The formalism employed so far can again be generalized to consider this system. Interestingly, as far as minimizing the total free energy is concerned, the only effects of the lateral association of filaments are to double the capping energy and E, where the factor 2 comes from the additional axial bond due to the bundling of the two filaments.[27] Because of this increase of capping energy *via* bundling, the species that dominate the system will always be the two-filament polymers. So how do these theoretical predictions square with experimental observations such that multiple morphology

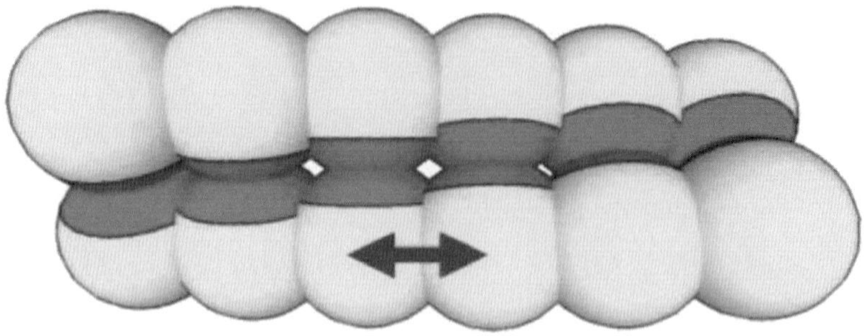

Figure 7: A schematic of a twisting fiber consisting of two filaments. In this model, besides the axial bonds (blue arrows) along the longitudinal direction of the fiber, there are lateral bonds (red patched) that bind the two filaments together.

exist? The simplest resolution again points to the conclusion that under typical experimental conditions, the self-assembly of amyloid fibril has not yet reached the thermal equilibrium state.

2.7. *Summary*

Here we summarize the key conclusions one can draw from statistical mechanics with regards to biopolymer self-assembly.

1. The existence of a CFC below which fibrillar mass is negligible. The emergence of such a critical concentration comes purely from the fact that the aggregates concerned consist of a large number of monomers.
2. Above the CFC, the length distribution of the fibrils is exponential. This remains so even in the semidilute limit if the fibril–fibril interactions are purely steric.
3. When micellar or other oligomeric species are taken into account, each distinct type of aggregates will have their specific critical concentrations, and the one with the lowest critical concentration will have the dominating mass density at high protein concentration (Fig. 6).
4. If fibrils of distinct widths exist, the mass density of the widest fibrils will dominate the system at thermal equilibrium.
5. If the fibrils exhibit attractive lateral interactions, there will be a strong tendency for phase separation of fibrils to occur.

We note that the existence of CFC is a well-established experimental observation. However, experimentally measured fibrillar length distributions seem generally to deviate from the predicted exponential distribution. In addition, multiple morphologies of fibrils seem to coexist in typical experimental conditions. These observations indicate that protein amyloid formation does not reach thermal equilibrium at the typical experimental timescale. Therefore, to account quantitatively for the experimental observation, we generally need a kinetic description of amyloid fibrillization tailored for the specific experimental condition.

3. Kinetics of amyloid fibrillation

In the first part, we have presented a general review on the thermodynamics of amyloid fibrillation. However, due to the lack of a mature nonequilibrium thermodynamics theory, currently we can only deal with the equilibrium states and the phase transition between them. In order to understand many unsolved puzzles concerning with the time evolution of an amyloid system, such as how fibrils grow, how they replicate, how they interact with cells, and be cytotoxic, we need to turn to a kinetic theory. The mathematical modeling based on chemical mass-action equations provides us a unified framework, as well as a quantitative linkage, between experimental observations and underlying molecular mechanisms. Fruitful results and interesting physical insights have been obtained based on this formulation in the past several years. In what follows, we focus on two aspects: one is a systematic exploration of the kinetic formulation of amyloid fibrillation, including both molecular mechanism basis and model analysis (Fig. 8); the other is the mathematical foundation of kinetic modeling as well as its linkage with experimental facts.

3.1. *How fibrils grow*

When examining a given amyloid system, a first question coming to mind usually would be how amyloid fibrils are able to grow? Regardless of various possible explanations for other self-assembling phenomena in nature,[39–42] we are really astonished at the fact that there is one truly simple answer valid for almost all amyloid fibrils. That is elongation, the process of which incorporates free protein molecules (or monomers as stated in literature) into existing fibrillar aggregates by monomer association and dissociation at fibril ends in a linear sequential way. Elongation not only is a geometrical consequence of the intrinsic one-dimensional structure of amyloid fiber, but also has a deep root in both thermodynamics and kinetics, that is, monomer association is more favorable than that of oligomers in general.[43,44] We come back to this point later.

As an elementary step in amyloid fibrillation, elongation has been paid great attention to in the past studies. Pioneer works dated back to Oosawa and his

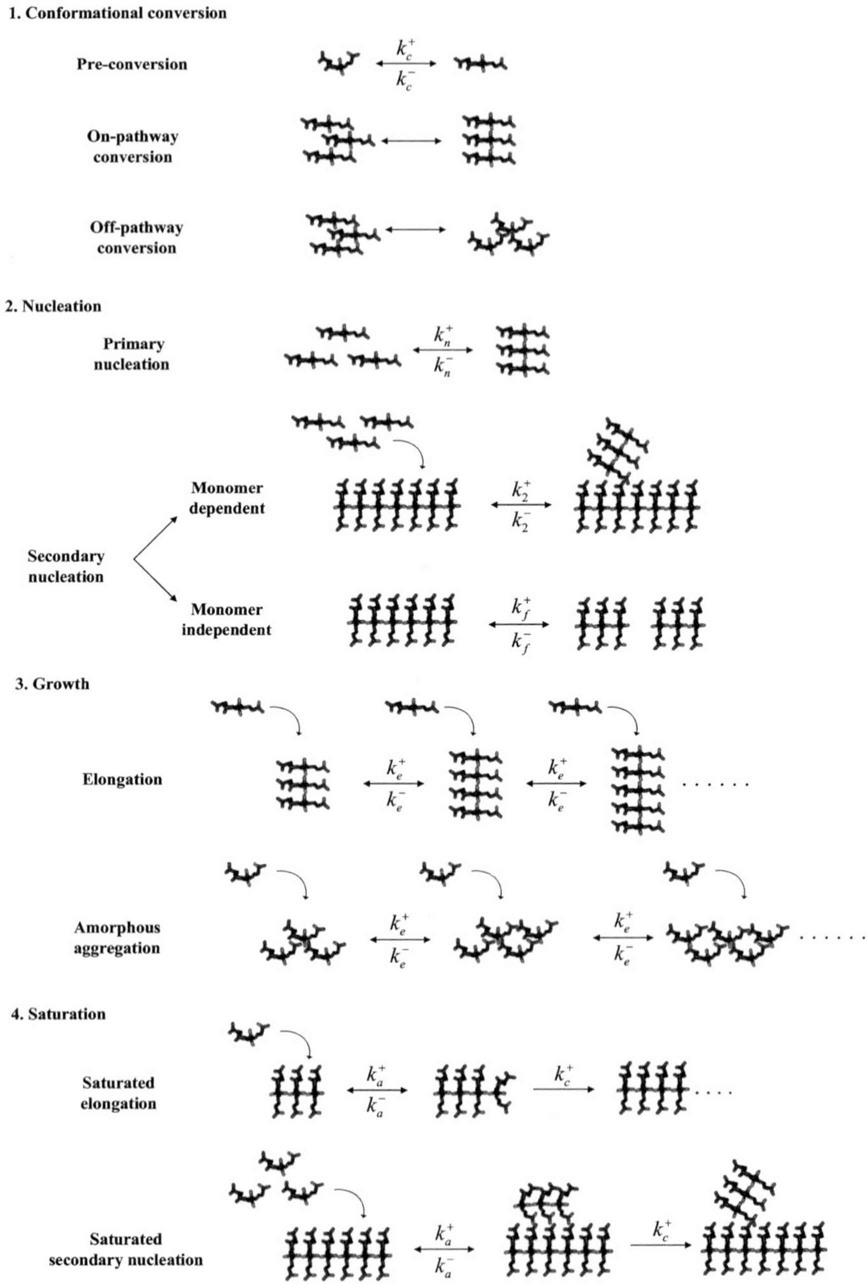

Figure 8: A cartoon depicting various microscopic mechanisms of amyloid fibrillation discussed in the current paper.

colleagues for their preliminary examination on actin formation in the late 1950s.[45] They are among the first ones who wrote down the explicit reaction schemes and rate laws for protein self-assembling, that is, once formed by nucleation processes (which will be discussed in next section), actins will extend or shrink by association or dissociation of monomeric units at one end mostly. They further verified that the initial rate of actin growth varies linearly with the monomer concentration, which implies that monomeric, rather than oligomeric, subunits have been added to actin filaments and make them to grow,[43] since the latter will give rise to a nonlinear dependence on the monomer concentration.

The linear dependence of elongation rate on the monomer concentration has been further investigated by Collins *et al.* through a combination of kinetic modeling and single-molecule fluorescence measurements on the N-terminal and highly charged middle (NM) domain of yeast prion protein Sup35.[44] Again, the initial rate of fiber growth was found to be directly proportional to the concentration of soluble NM over a range of $0 \sim 1\ \mu$M. However, at a higher NM concentration ($>0\ \mu$M), the rate of fiber growth shows a weaker-than-linear law. The reason for this is believed to be "a conformational rearrangement of NM after binding to fiber ends becomes rate limiting at high NM concentration",[44] a specific case of the saturation phenomenon. A quantitative treatment will be presented in the section of saturation.

Recently, Knowles and his colleagues developed a novel technique to measure the rate of fiber elongation directly,[46,47] which to some extent avoids the indistinguishability of several different fibrillation processes presented simultaneously in traditional methods. According to their setup, prepared fragments of amyloid fibrils have been attached to the surface of a quartz transducer. So that, once new monomers add to the fibrils, the resulting nanogram mass changes will be monitored through a shift in the resonance frequency of the quartz oscillator. Again, two different regimes in the fiber elongation rate are highlighted: an initial linear dependence on the monomer concentration, and a subsequent saturation of the growth rate at high monomer concentrations.

In literature, the possibility of fiber elongation through oligomer addition has been discussed from time to time.[48–51] For example, Serio *et al.* proposed a oligomer-based mechanism called "Nucleated Conformational Conversion",[48] in which "structurally fluid oligomeric complexes appear to be crucial intermediates in de novo amyloid nucleus formation" and "rapid assembly ensues when these complexes conformationally convert upon association with nuclei", to explain the assembling of prion protein Sup35. However, since in general "monomer addition is more rapid and efficient" than oligomer addition,[44] nowadays a common conclusion has been reached that fiber elongation has a first-order concentration dependence on both monomeric and fibril species, revealing a bimolecular mechanism of growth through monomer addition to both ends of existing fibrils.

Now we are going to formulate previous discussions into a quantitative model. If only the elongation process is included, it is straightforward to show that the mass concentration and number concentration of aggregates (see their mathematical definitions and physical meanings in section of how to quantify fibrillation kinetics) evolve according to following equations:

$$\frac{\mathrm{d}}{\mathrm{dt}}P = 0, \tag{25}$$

$$\frac{\mathrm{d}}{\mathrm{dt}}M = k_e^+ mP - k_e^- P, \tag{26}$$

where k_e^+ and k_e^- denote the reaction rate constants for monomer association and dissociation, respectively. Clearly, the first term on the right-hand side of Eq. 26 represents the desired process of fiber elongation by monomer addition at fibril ends (here the geometrical factor 2 is not directly written out); while the second one is the corresponding inverse process, whose necessity lies on "the maintenance of monomer pool" in the equilibrium state.[52]

It is noted that, according to Eq. 25, the number concentration is conserved, that is, $P(t) = P(0)$. This is a significant feature of elongation, in contrast to all kinds of nucleation processes which will be introduced in the following section. Toward the mass concentration, obviously its growth rate is maximal at the beginning of the reaction and then decreases monotonically. Actually, we have:

$$M(t) = (M_0 - \epsilon)e^{-\kappa t} + \epsilon, \tag{27}$$

where $\kappa = k_e^+ P_0, \epsilon = m_{\mathrm{tot}} - k_e^-/k_e^+, M(0) = M_0$ and $P(0) = P_0$ are the initial mass and number concentrations of aggregates. Accordingly, the half-time and apparent fiber growth rate (see section on how to quantify fibrillation kinetics for their definitions) are given by

$$t_{1/2} = \frac{\ln 2}{\kappa} \propto P_0^{-1}, \tag{28}$$

$$k_{\mathrm{app}} = \frac{\epsilon + M_0}{2m_{\mathrm{tot}}}\kappa \propto P_0. \tag{29}$$

here, both $t_{1/2}$ and k_{app} are only concerned with the number concentration of seeds P_0 rather than the mass concentration M_0. This is another significant feature of the elongation process. Contrarily, as we show, the speed of secondary nucleation generally depends on the mass concentration of seeds M_0; while that of primary nucleation does not rely on seeds at all. Finally, from the static solution $M(\infty) = \epsilon$, we can exactly see that the inverse reaction — monomer dissociation is essential to maintain a genuine thermodynamic equilibrium as well as an observable "monomer pool" at the end of reaction.[52]

In Eq. 26, the rate constants for monomer association and dissociation are assumed to be independent of fibril length. An indirect support comes from the argument that they are "characterized by the local interaction of a monomer with the end of a fibril", so that "the total length of the fibril is likely to introduce only a minor effect through the reduction of the diffusive encounter rate when the size of the fibril increases".[47] Under the assumption of fibril length independence, there are plenty of studies attempting to determine exact values of the rate constants for monomer association and dissociation. For example, by atomic-force microscopy (AFM) analysis Collins *et al.* reported the rate constant for fiber elongation of yeast prion protein Sup35 is approximately $k_e^+ \approx 2\times10^5 M^{-1}s^{-1}$ at a soluble NM concentration of 2.5 μM and can be two times bigger at higher concentrations.[44] Knowles *et al.* found an elongation rate of $(9.2 \pm 0.3) \times 10^3 M^{-1}s^{-1}$ for insulin at a concentration of 0.17 mM by quartz crystal microbalance, equivalent to one molecule attaching every $3.1 \pm 1.2s$ on average.[46,53] At 2.5 mg/ml $A\beta_{25-35}$ peptide, Kellermayer *et al.* determined $k_e^+ \approx 10^6 M^{-1}s^{-1}$ and $k_e^- \approx 10\, s^{-1}$ through scanning-force kymograph, given the equilibrium monomer concentration to be 10 μM.[54] In a similar way, referring to a critical concentration of α-synuclein at 2 μM, Pinotsi *et al.*[55] gave an average growth rate of $k_e^+ \approx (1 \pm 0.375) \times 10^3 M^{-1}s^{-1}$ as well as a lower bound for the average dissociation rate $k_e^- \approx (2.0 \pm 0.8) \times 10^{-3} s^{-1}$ by two-color single-molecule localization microscopy.

From the aforementioned far-from-complete list, we can clearly see that the rate constant for fiber elongation varies for at least three orders of magnitude from protein to protein and could also be influenced by the pH condition, salt concentration, temperature, and so on.[55] This fact to some extent reveals the intrinsic high complexity and heterogeneity of fiber elongation processes. Actually, for the growth of individual filament, an intermittent, stop-and-go behavior has been widely confirmed,[56,57] which shows fiber elongation operates "in a way analogous to the landscape models of protein folding defined by stochastic dynamics on a characteristic energy surface".[27,47] Therefore, the current single-reaction-rate-based picture turns out to be a rough average of various underlying stochastic processes concerning with the fiber growth. Readers should be aware of this point.

3.2. *How fibrils replicate*

Generally speaking, the fibrillation of amyloid proteins is constituted by two aspects: nucleation and growth. In the previous section, we have shown how elongation provides a simple way to let fibrils grow. But, we still lack a knowledge on where those templates (or seeds) for elongation come from and how they evolve during the fibrillation procedure. As far as we know, there are at least three basic processes contributing to the generation of new seeds, that is, primary nucleation, surface-catalyzed secondary nucleation and fragmentation. There is a long story for the study of primary nucleation in crystal and liquid formation. Surface-catalyzed secondary nucleation and fragmentation present two alternative ways to bypass the slow seeding procedure of primary nucleation by making use of existing fibrils in the system, so they are also named as "secondary nucleation" in literature. A major difference between them is that the former dependents on the monomer concentration, while the latter does not.

Secondary nucleation plays a key role in the fibrillation of amyloid proteins.[6,58] In the presence of secondary nucleation, the fibrillation speed will be dramatically accelerated and time courses for fiber mass concentration will be changed into a sigmoidal shape with prominent lag phase. Furthermore, in contrast to primary nucleation and surface-catalyzed secondary nucleation, fragmentation will not only cause a global change of the fiber length distribution, but also largely enhance the cytotoxicity by generating more harmful oligomers. Detailed discussions could be found in following sections correspondingly.

3.2.1. *Primary nucleation: rate-limiting step*

For different amyloid proteins under different conditions, time courses for fibrillation may vary considerably from each other, but in general a sigmoidal-like behavior with a prominent lag phase is observed if initial proteins are all in monomeric form.[59] Furthermore, by introducing certain amount of preseeded filaments into the system at the very beginning of the reaction, the lag phase can be completely removed.[5] This evidence reveals that amyloid fibrillation involves a process called primary nucleation, in which nuclei of aggregates are formed from monomeric proteins directly. In contrast, the formation of amorphous aggregates, which often acts as competitive pathways against regular amyloid fibrils, is generally believed to follow a nonnucleated polymerization (or random polymerization).[60–62]

The idea of primary nucleation has been extensively explored in various natural phenomena, such as the formation of snow flakes, clouds and bubbles, the crystallization of mineral, metal, protein, DNA, and so on.[63–69] It is the first step in the formation of either a new thermodynamic phase or a new structure via self-assembly, and typically determines how long we have to wait before the new phase or self-organized structure appears. According to whether the process is catalyzed by particles of foreign substance (like surface and substrate) or not, primary nucleation could be further divided into two categories: homogeneous nucleation and heterogeneous nucleation. The latter usually occurs much more often and faster than the former. This behavior could be understood by classical nucleation theory (CNT), which predicts the nucleation rate:[70–72]

$$R = N_S Zj \exp\left(\frac{-\Delta G^*}{k_B T}\right), \tag{30}$$

where ΔG^* is the free energy barrier for forming a critical nucleus, which will be explicitly defined later. $k_B T$ represents the thermal energy with the Boltzmann constant k_B and the absolute temperature T. N_S is the number of nucleation sites. j is the rate for monomers attaching to the nucleus. Z is called the Zeldovich factor, which is the forward probability for a critical nucleus to grow diffusively into a larger nucleus rather than shrink back to nothing.

To further model the free energy barrier, CNT treats the microscopic nucleus as a macroscopic droplet, so that the free energy ΔG for forming a nucleus can be generally written as the sum of a bulk term proportional to the volume of the nucleus and a surface term proportional to its surface area. Especially for homogeneous nucleation, the nucleus modeled by a sphere of radius r gives[70–72]:

$$\Delta G = \frac{4}{3}\pi r^3 \Delta g + 4\pi r^2 \sigma. \tag{31}$$

The first term stands for the volume contribution, in which Δg is the free energy difference per unit volume between the nucleated and nonnucleated phases and usually negative. The second term comes from the interface between the nucleus and its surroundings. σ is the surface tension and always positive. As a consequence, for small r, the surface term dominates and $\Delta G(r) > 0$; while for large r, the volume term dominates and $\Delta G(r) < 0$. Especially at some intermediate value of r, the free energy goes through a maximum, corresponding to a least probability for nucleus occurring. This is called the critical nucleus and occurs at $\mathrm{d}G/\mathrm{d}r = 0$, which gives a critical nucleus radius:

$$r^* = -\frac{2\sigma}{\Delta g}. \tag{32}$$

Adding new monomers to nuclei larger than this critical radius decreases the free energy, so the overall nucleation rate is then limited by the probability of forming the critical nucleus, which is $\Delta G^* = 16\pi\sigma^3/[3(\Delta g)^2]$. This is exactly the free energy barrier needed in the CNT expression for the nucleation rate R.

The reason for heterogeneous nucleation occurring much easier than homogeneous nucleation is that the nucleation barrier ΔG^* is much lower at a surface. For homogeneous nucleation, the nucleus is approximated by a sphere. However, for heterogeneous nucleation, when a nucleus is formed at the surface, its form is not completely spherical and depends on the contact angle.[73,74] This geometrical factor reduces the interfacial area and so the interfacial free energy, which in turn reduces the nucleation barrier. In principle, we expect nucleation to be fastest when the nucleus forms a small contact angle on its surface. However, detailed calculation is not straightforward and will not be listed here.

From the kinetic aspect, primary nucleation, including both homogeneous and heterogeneous nucleation, is often modeled with the following formula[33,75]:

$$\frac{dP}{dt} = k_n(m - m_F^*)^{n_c}, \tag{33}$$

where k_n is the macroscopic reaction rate constant. With respect to the microscopic nucleation rate R considered in previous thermodynamic picture, we expect $\lim_{V\to\infty} R/N_s = k_n$ in the limit of sufficiently large volume. m_F^* is the (CFC) and $(m - m_F^*)$ is also known as supersaturation. Supersaturation is the driving force for both the initial nucleation step and the following growth, both of which could not occur in saturated or undersaturated conditions. n_c stands for the critical nucleus size, which can be as large as 10, but generally ranges between 2 and 4.

In the 1960s, Oosawa *et al.* borrowed this idea to study of the polymerization of actin[43,76] and wrote down a reaction scheme consisting of three basic process — primary nucleation, monomer association and dissociation. However, they did not include the supersaturation effect explicitly, since mostly the protein concentration used for fibrillation is much higher than that required for saturation. According to Oosawa's model, the number concentration and mass concentration of aggregates evolve according to the equation:

$$\frac{d}{dt}P = k_n m^{n_c}, \tag{34}$$

$$\frac{d}{dt}M = k_e^+ mP \underline{- k_e^- P + n_c k_n m^{n_c}}. \tag{35}$$

In previous model, the inverse process for primary nucleation has been omitted. A major reason is that it is accounted by a boundary term $k_n^-[A_{n_c}]$ (where k_n^- is the rate constant and $[A_{n_c}]$ is the concentration of critical nucleus), not compatible with other terms constituted by moments M and P (the moment-closure method we introduced in the section of how to coarse-grain model can systematically solve this problem). The last two terms underlying in Eq. 35 are generally negligible too, due to the fact that under proper fibrillation conditions, fiber elongation is more energetically favorable and proceeds faster than primary nucleation and monomer dissociation.[43] This leads to an important conclusion — the major role of primary

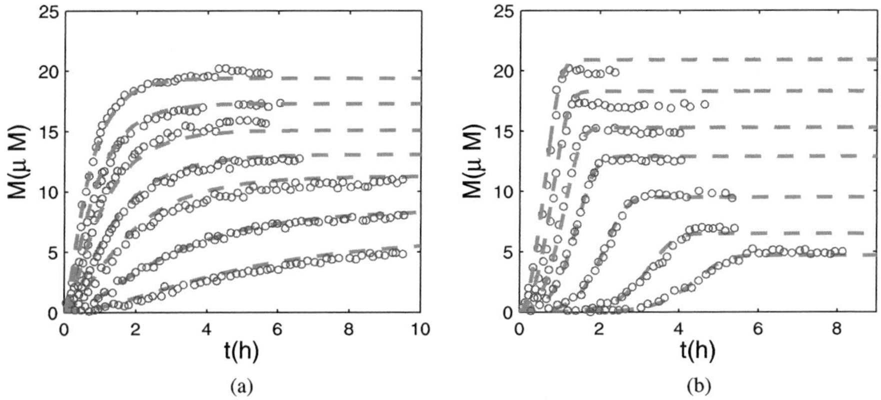

Figure 9: Actin growth in the presence of KCl vs $MgCl_2$, which highlights two different dominated mechanisms.[81] (a) Action samples were prepared with 40 mM KCl at monomer concentrations $m_{\text{tot}} = 7.4, 9.6, 12.4, 14.2, 16.2, 18.4$, and $20.5\,\mu$M separately (blue circles). Predictions based on the NE model (red dashed lines) in Eqs. 34 and 35 were performed with $n_c = 4$, $k_n = 3 \times 10^9\,\text{M}^{-3}\text{s}^{-1}$, $k_e^+ = 9 \times 10^2\,\text{M}^{-1}\text{s}^{-1}$, $k_e^- = 10^{-3}\text{s}^{-1}$. (b) Action samples were prepared with 0.6 mM $MgCl_2$ and 0.5 mM EGTA at monomer concentrations $m_{\text{tot}} = 6.7, 8.5, 11.5, 14.9, 17.3, 20.3$, and $22.9\,\mu$M. Predictions based on the nucleation-elongation-fragmentation (NEF) model in Eqs. 47 and 48 were performed with $n_c = 6$, $k_n = 2 \times 10^{16}\,\text{M}^{-5}\text{s}^{-1}$, $k_e^+ = 9 \times 10^4\,\text{M}^{-1}\text{s}^{-1}$, $k_e^- = 0.18\,\text{s}^{-1}$, $k_f^+ = 6 \times 10^{-7}\text{s}^{-1}$, $k_f^- = 0$.

nucleation in fibrillation is to generate new seeds rather than directly consuming free monomers like elongation. Similar argument also applies to secondary nucleation introduced in the next section. Typical amyloid systems, which follow the aforesaid mechanism without referring to secondary nucleation, include actins in KCl solvents[77] (Fig. 9a), γC−crystallin,[78] Apo C-II,[79] and so on.

According to,[80] Oosawa's model admits analytical solutions as follows:

$$M(t) = m_{\text{tot}} - (m_{\text{tot}} - M_0)[\mu \sec h(\nu + \lambda\beta\mu t)]^{2/n_c}, \tag{36}$$

$$P(t) = P_0 + k_n(m_{\text{tot}} - M_0)^{n_c}\mu(\beta\lambda)^{-1}[\tanh(\nu + \beta\lambda\mu t) - \tanh(\nu)], \tag{37}$$

where $\lambda = \sqrt{k_n k_e^+(m_{\text{tot}} - M_0)^{n_c}}$, $\beta = \sqrt{n_c/2}$, $\gamma = \beta k_e^+ P_0/\lambda$, $\mu = \sqrt{1+\gamma^2}$ and $\nu = \text{arcsinh}(\gamma)$. $m_{\text{tot}}, M(0) = M_0$ and $P(0) = P_0$ are the total protein concentration, initial mass, and number concentrations of aggregates, respectively. Especially, in the absence of initial seeds $M_0 = P_0 = 0$, we have $M(t) = m_{\text{tot}}[1 - \text{sech}^{2/n_c}(\lambda\beta t)]$, which recovers the classical Oosawa's result.[76]

Insight into the early time behavior of fiber mass concentration can be obtained by expanding Eq. 36 around $t = 0$

$$M(t) = M_0 + k_e^+ P_0(m_{\text{tot}} - M_0)t + \frac{1}{2}(m_{\text{tot}} - M_0)$$
$$\times[\lambda^2 - (k_e^+ P_0)^2]t^2 + O(t^3), \quad t \to 0. \tag{38}$$

This expression recovers the characteristic t^2 dependence relating to the primary nucleation in Oosawa's theory.[76] An additional term linear in time is raised by the growth of preadded seeds. This difference actually provides a simple way to distinguish the seeding contributions from primary nucleation and preadded seeds.

After some calculation, the half-time and the apparent fiber growth rate are given, respectively, as follows:

$$t_{1/2} = \frac{1}{\lambda\beta\mu}[\text{arccosh}(2^{n_c/2}\mu) - \nu], \tag{39}$$

$$k_{\text{app}} = \frac{\lambda\beta\mu}{n_c}\frac{m_{\text{tot}} + M_0}{m_{\text{tot}}}\sqrt{2^{n_c}\mu^2 - 1}. \tag{40}$$

According to the aforementioned integrated rate laws, we can easily see that, in the absence of initial seeds $M_0 = P_0 = 0$, the half time of fibrillation is proportional to λ^{-1}, which means $t_{1/2} \propto m_{\text{tot}}^{-n_c/2}$, a prominent feature of seeding through primary nucleation; on the contrary, if high concentrations of preformed seeds are added to the system, $\gamma \gg 1$ and we recover results for pure elongation $t_{1/2} \propto (k_e^+ P_0)^{-1}$.

Again, we can take advantage of this difference in scaling relations to separate the elongation process from primary nucleation. At first, without any initial seeds, the critical nucleus size for primary nucleation n_c could be directly read out by examining the scaling dependence of half-time under different initial monomer concentrations. Furthermore, parameter λ provides a combined knowledge of both primary nucleation and elongation. Finally, by adding high concentrations of performed seeds into the system, seeding through primary nucleation will be completely suppressed and the elongation rate could be extracted from the fibrillation kinetics solely. Similar argument is also applicable to secondary nucleation.

3.2.2. *Monomer-dependent secondary nucleation: self-catalyzed process*

As indicated by the early time expansion of solutions for primary nucleation, a linear growth in the presence of initial seeds and a growth depending on t^2 for purely monomeric proteins are expected,[82] which cannot account for the apparent high cooperativity of fibrillation observed in many amyloid systems. For example, in 1980 Ferrone *et al.*[83] observed an exponential growth at the early stage of aggregation for the aberrant gelation of sickle hemoglobin. Later, they studied the kinetics of polymerization of hemoglobin S and found a transition time much shorter than the lag phase predicted by primary nucleation only.[84,85] Miranker *et al.* observed a similar process in Islet amyloid polypeptide (IAPP).[86] Intriguingly, they found that both the reaction order and the activation enthalpy of two nucleation processes are identical, which made them to conclude that both primary nucleation and monomer-dependent secondary nucleation are "alternative manifestations of the same, surface-catalyzed nucleation event".[87] The aforesaid observations confirmed the existence of a secondary pathway for nucleation, which generates new nuclei by making use of the surface of existing aggregates in the system. Therefore, this kind of mechanism is called surface-catalyzed secondary nucleation. Since it depends on both monomer concentration and fiber concentration, it is also known as monomer-dependent secondary nucleation, in contrast to a different kind of secondary nucleation mechanism — fragmentation that is monomer independent. Just as Miranker claimed, monomer-dependent secondary nucleation is a special kind of heterogeneous nucleation and takes the surface of amyloid fibrils as catalyst. Therefore, it proceeds much faster than primary nucleation and exhibits an exponential growth due to the self-catalyzed nature. Besides sickle-hemoglobin,[83] hemoglobin S[84,85] and IAPP,[87] Aβ40[88] and Aβ42[89,90] (Fig. 10) have also been revealed to adopt the surface-catalyzed secondary nucleation mechanism.

A model, incorporating both primary nucleation and monomer-dependent secondary nucleation, could be expressed as follows:

$$\frac{d}{dt}P = k_n m^{n_c} + k_2 m^{n_2} M, \tag{41}$$

$$\frac{d}{dt}M = k_e^+ mP - k_e^- P \underline{+ n_c k_n m^{n_c} + n_2 k_2 m^{n_2} M}, \tag{42}$$

where the new term $k_2 m^{n_2} M$ accounts for the contribution of secondary nucleation on seeding, which is proportional to the surface area of existing aggregates (scale as M). The parameter n_2 stands for the critical nucleus size for secondary nucleation, analogous to n_c for primary nucleation. It is noted that when $n_2 = 0$, the previous equation offers a good approximation to the model of fragmentation, an alternative secondary nucleation mechanism going to be shown in the next section. Therefore, we can use the same model to describe both monomer dependent and independent secondary nucleation by just tuning the parameter n_2. As we have stated in the section of primary nucleation that the major role of various nucleation processes is to create

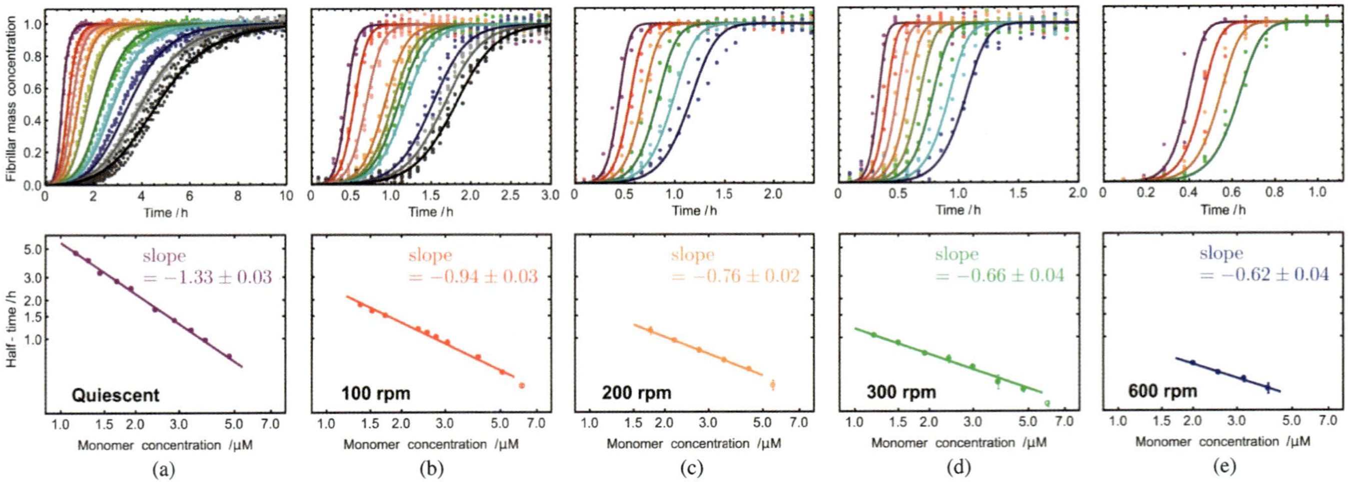

Figure 10: Kinetics of A42 aggregation is shifted from surface catalyzed secondary nucleation dominated mechanism to fragmentation dominated by continuously increasing agitating speeds. The upper plots show the time profile of A42 aggregation under different shear rates generated by agitating the sample under different speeds; and the lower ones show the corresponding power-law relationships between the half-time and the initial monomer concentration of A42. In (a), the rate parameters $\sqrt{k_e^+ k_n} = 42.4\,\mathrm{M}^{-1}\mathrm{s}^{-1}$ and $\sqrt{k_e^+ k_2} = 2.8 \times 10^5\ \mathrm{M}^{-3/2}s^{-1}$, $n_c = n_2 = 2$ are fixed and $k_f^+ = k_f^- = 0$. In (b)–(f), $\sqrt{k_e^+ k_f^+} = 0.6, 0.9, 1.4, 1.9\ \mathrm{M}^{-1/2}\mathrm{s}^{-1}$ are taken, respectively, under each shear rate in replace of $\sqrt{k_e^+ k_2}$. Figure is adapted with permission from Ref. 90

new sites for elongation rather than increasing the mass of aggregates directly, the terms underlying in Eq. 42 could be neglected with respect to elongation.

Under this condition, first-order self-iterative solutions have been obtained through fixed-point analysis[91] by Cohen *et al.*[92] as follows:

$$M(t) = M(\infty) - [M(\infty) - M_0]e^{-k_\infty t} \times \left(\frac{B_- + C_+ e^{\kappa t}}{B_+ + C_+ e^{\kappa t}} \cdot \frac{B_+ + C_+}{B_- + C_+}\right)^{k_\infty^2/(\kappa \bar{k}_\infty)} \tag{43}$$

$$P(t) = \frac{P_l(t)}{1 + P_l(t)/P(\infty)}, \tag{44}$$

in which $P_l(t) = (C_+\kappa e^{\kappa t} + C_-\kappa e^{-\kappa t})/k_e^+$ is the linearized solution for early time. $\kappa = \sqrt{(k_e^+ m_0 - k_e^+)k_2 m_0^{n_2}}$, $\lambda = \sqrt{k_e^+ k_n m_0^{n_c}}$, $c_\pm = k_e^+ P_0/(2\kappa) \pm k_e^+ M_0/[2(m_0 k_e^+ - k_e^-)] \pm \lambda^2/(2\kappa^2)$ $k_\infty = k_e^+ P(\infty)$, $\bar{k}_\infty = \sqrt{k_\infty^2 - 4C_+C_-\kappa^2}$, $B_\pm = (k_\infty \pm \bar{k}_\infty)/(2\kappa)$. $M(0) = M_0$, $P(0) = P_0$ and $M(\infty) = m_{\text{tot}} - k_e^-/k_e^+$, $P(\infty) = (k_e^+)^{-1} \sqrt{2\kappa^2/[n_2(n_2+1)] + 2\lambda^2/n_c + 2M_0\kappa^2/(n_c m_0) + (k_e^+ P_0)^2}$ are the aggregates number concentration and mass concentration at the beginning and in the equilibrium, respectively. $m(0) = m_0 = m_{tot} - M_0$ and m_{tot} are the initial and total monomer concentrations. Alternative solutions for early time based on perturbation methods could be found in Refs. Ferrone *et al.*[85] and Ferrone[93] and we not go into them here.

According to self-iterative solutions, the half-time and apparent fiber growth rate can be extracted,

$$t_{1/2} \approx \kappa^{-1} \ln(1/c_+), \tag{45}$$

$$k_{\text{app}} \approx \frac{\kappa}{2} \frac{k_\infty^2/(\kappa \bar{k}_\infty)}{B_+ + 1}. \tag{46}$$

It is clear, in the absence of initial seeds $M_0 = P_0 = 0$, the half time of fibrillation is around $2\kappa^{-1} \ln(\kappa/\lambda)$, which means $t_{1/2} \propto m_{\text{tot}}^{-(n_2+1)/2}$ a feature of seeding through secondary nucleation. As primary nucleation only enters into the half-time through a logarithmic correction, its influence will be limited to early time, which is expectable since secondary nucleation is more effective in generating seeds. Furthermore, even if medium amount of preformed seeds are introduced to the system $\lambda \ll k_e^+ P_0 \ll \kappa$, the scaling dependence of half-time on monomer concentration will not be changed except for a logarithmic correction, showing only primary nucleation is screened out in this step. Unless high concentrations of preformed seeds are added $k_e^+ P_0 \gg \kappa$, we cannot eliminate the effect of secondary nucleation and recover the result for pure elongation $t_{1/2} \propto (k_e^+ P_0)^{-1}$.

Since the major role of primary nucleation and secondary nucleation is to generate new seeds, it is generally expected, by introducing preformed seeded into the system, various nucleation mechanisms could be partially or even completely screened out. The previous examination confirms this point from an analytical point. In particular, we see that firstly primary nucleation and then secondary nucleation are suppressed with an increase of seeds concentration. As long as two nucleation processes are well separated in time, given by $\sqrt{k_e^+ k_n m_{\text{tot}}^{n_c}} \ll \sqrt{k_e^+ k_2 m_{\text{tot}}^{n_2+1}}$, rates of primary nucleation, secondary nucleation and elongation could be extracted, respectively, by simply varying the seeds concentration. In this way, the significant role of seeds in fibrillation kinetics could be fully appreciated. All these discussions are applicable to the fragmentation case in the next section by just setting $n_2 = 0$.

3.2.3. *Monomer-independent secondary nucleation: fragmentation and annealing*

Another efficient way to generate seeds without involving primary nucleation is fragmentation, which means one filament breaks into two smaller fragments. In principle, breaking into three or more pieces simultaneously is also possible but extremely rare. The universality of fragmentation in amyloid formation has long been established in the famous book of Oosawa.[43] It is well known that single filament becomes mechanically unstable and tends to break when exceeding certain threshold in length.[44,58] Even for bundled fibrils, breakage is unavoidable in the presence of mechanical stress,[94] thermal motion,[95–97] or chaperons like Hsp104.[98]

Compared to surface-catalyzed secondary nucleation, fragmentation is monomer-independent and thus becomes predominant under the condition of low monomer concentration. It can dramatically accelerate the formation of breakable filaments and change time courses for mass concentration of aggregates into a sigmoidal like behavior. Fragmentation could further alter the fiber length distribution globally just like elongation and enhance the toxicity of fibril samples by generating more harmful oligomers in low molecular weight. Due to its significant roles in both amyloid fibrillation and cytotoxicity, fragmentation has been paid plenty of attention to in past studies. And many typical amyloid systems have been shown to follow this mechanism, including actins in $MgC1_2$ solvents[77] (Fig. 9b), Sup35 NW region,[44] Ure2p,[99] $CsgB_{\text{trunc}}$,[100] Stefin B,[101] $\beta 2-$microglobulin,[102] WW domain,[103] $\alpha -$ synucleins,[104] and insulin.[105]

Generally speaking, fragmentation rates are length-dependent. In literature, several assumptions have been proposed, including the random scission, the central scission, the Gaussian scission[106,107] and modified partially random scission,[108] and so on. The latter two got supports from fiber-like PI264-b-PFS48 micelles under sonication,[109] thermodynamically induced shear degradation of polystyrene in semiconcentrated solutions,[107] single-stranded random-coiled poly-(uridylic acid), double-stranded helical DNA, and triple-stranded helical poly-(adenylic acid) · 2poly (inosinic acid).[108] Hill suggested a formula in log form for polymer fragmentation and annealing based on statistical mechanics.[110] And it has been applied by Hong *et al.* to the study of several amyloid proteins.[111] Schreck and Yuan examined the length-dependent rates too. By numerical techniques, they calculated fiber length distribution at an arbitrary time.[112]

In the presence of fiber-length-dependent fragmentation, the derivation of self-closed models for fiber formation becomes a big trouble. Based on Maximum Entropy Principle,[113–115] a systematical and reliable way has been proposed to derive close-formed mass-action equations from microscopic length-dependent fragmentation models[111] (see section on how to coarse-grain model for details). In particular, under the assumption of totally random scission (or length-independent fragmentation), the kinetics of fiber formation can be expressed by the following model:

$$\frac{d}{dt}P = k_n m^{n_c} + k_f^+[M - (2n_c - 1)P] - k_f^- P^2, \tag{47}$$

$$\frac{d}{dt}M = k_e^+ mP - k_e^- P \underline{+ n_c k_n m^{n_c}}, \tag{48}$$

where k_f^+ is the rate constant for length-independent fragmentation, and k_f^- is that for the inverse process — filaments annealing. Terms for fiber fragmentation and annealing in Eq. 47 are explicit and can be obtained directly from the microscopic model. Since here fibrils are not allowed to break into monomers directly and so is annealing, no term for fragmentation or annealing will enter into the equation for $M(t)$. This is a dramatic feature of fiber fragmentation and annealing, which only affect the number concentration of aggregates and keep the mass concentration conserved.

As an inverse process of fragmentation, filaments annealing is crucial for the fiber length distribution, number concentration of aggregates as well as their average length. Based on Eq. 47, it is clearly seen that in the absence of annealing (by setting $k_f^- = 0$), the average length of fibrils will be around $2n_c - 1$, in contrast to the estimated lower bounds of at least hundreds of monomers.

Again, by fixed-point analysis, self-iterative solutions could be derived in exactly the same form as those in Eqs. 43 and 44, except for some internal parameters have to be changed correspondingly,[116] that is, the linearized solution for early time $P_l(t) = C_1 e^{k_1 t} + C_2 e^{k_2 t} - n_c k_f^+ \epsilon$, number concentration of aggregates in equilibrium $P(\infty) = 2M(\infty)(2n_c - 1 + \sqrt{(2n_c - 1)^2 + 4k_f^- M_0/k_f^+})^{-1}$, $\kappa = \sqrt{(k_e^+ m_0 - k_e^-)k_f^+}$, as well as $C_\pm = k_e^+ C_{1,2}/k_{1,2}$, where $k_{1,2} = -k_f^- P_0 \pm \sqrt{(k_f^- P_0)^2 + \kappa^2}$, $C_{1,2} = (1 - k_{2,1}/k_{1,2})^{-1}[P_0 - k_{2,1}(M_0 + k_n m_0^{n_c}/\kappa^2 + n_c k_f^+ k_f^- P_0 \epsilon/\kappa^2)]$ The same expressions for the half-time and apparent fiber growth rate could be obtained as in Eqs. 45 and 46 and will not be addressed here. Just note in current case, except for a logarithmic correction, $t_{1/2} \propto m_{\text{tot}}^{-1/2}$ and $k_{\text{app}} \propto m_{\text{tot}}^{1/2}$, a special case of monomer-dependent secondary nucleation with $n_2 = 0$ for fragmentation as we claimed.

Models constituted by previous four basic mechanisms: elongation, primary nucleation, surface-catalyzed secondary nucleation, and fragmentation (as well as monomer dissociation and filaments annealing as two typical inverse processes) could already explain most fibrillation kinetics observed in experiments. In the past several years, fruitful results have been obtained in this direction,[5,6] which greatly enhance our understandings on the underlying microscopic processes, the kinetics and thermodynamics of amyloid fibrillation, effects of pH value, temperature, and so on. Besides this main framework, several key issues need to be addressed to provide a complete picture. Firstly, we have mentioned that high monomer concentration could dramatically change the kinetics of amyloid formation, which is known as saturation. Luckily, the effect of saturation could be easily accounted through a Michaelis–Menton-like formula, though model analysis becomes far more difficult. Secondly, it is well known that the formation of amyloid fibrils is a direct consequence of protein misfolding, and the conformational conversion of monomers and oligomers plays an unreplaceable role in it. However, the step of conformational conversion is easily neglected in kinetics due to the interference of primary nucleation.

3.3. *How high concentration affect?*

In the section of fiber growth, we have mentioned that fiber elongation shows a sublinear dependence in the regime of high monomer concentration. In fact, this phenomenon is rather universal and has a deep physical basis on saturation. It is imaginable, in the presence of too many monomers competing for the same fibril end at the same time, the fiber end will appear to be "saturated" since the incorporation of each monomer requires certain amount of time and cannot be finished at once. As a consequence, the elongation process appears to be blind to the instantaneous monomer concentration in the system and shows a weaker-than-linear dependence.

Physically, the process of saturated elongation could be modeled through a "dock-lock" mechanism,[117,118] which includes two substeps — unspecific attachment and detachment of a monomer with the fibril end, and subsequent conformational change of the attached monomer to make the attachment specifically. Usually, the second process is the rate-limiting step. For example, Scheibel *et al.* studied how nuclei mediate the conversion of soluble NM domain of Sup35 to the amyloid form in the elongation phase of fiber formation.[119] By creating single-cysteine substitution mutants at different positions of NM domain to provide unique attachment sites for various probes, they established that elongation is a two-step process involving the capture of an intermediate, followed by its conformational conversion.

If we take P_{bound} and $P_{\text{free}} = P - P_{\text{bound}}$ as the number concentration of fibril ends which has and has not monomers unspecifically attached, the "dock-lock" mechanism is expressed through following equations:

$$\frac{d}{dt}M = k_c P_{\text{bound}}, \tag{49}$$

$$\frac{d}{dt}P_{\text{bound}} = k_a^+ m P_{\text{free}} - k_a^- P_{\text{bound}} - k_c P_{\text{bound}}, \tag{50}$$

where k_a^+, k_a^-, and k_c represent reaction rate constants for monomer unspecific attachment, detachment and conformational change at the fibril end. It is clear that the three terms on the right-hand side of Eq. 50 account for the contribution of each step mentioned in the "dock-lock" mechanism to fibril ends, respectively.

To eliminate the additional variable P_{bound}, we refer to the classical Quasi Steady-State Approximation,[120,121] which assumes the generation and consumption of P_{bound} are always in a dynamical equilibrium. Such that we can take the sum of terms on the right-hand side of Eq. 50 to be zero, that is,

$$0 = k_a^+ m P_{\text{free}} - k_a^- P_{\text{bound}} - k_c P_{\text{bound}},$$

which gives solutions in a form of the famous Michaelis–Menten equation for enzyme kinetics,[122]

$$P_{\text{free}} = P\left(1 + \frac{m}{K_m}\right)^{-1},$$

$$P_{\text{bound}} = P\frac{m}{K_m}\left(1 + \frac{m}{K_m}\right)^{-1}.$$

Here the Michaelis constant, $K_m = (k_a^- + k_c)/k_a^+$.

Inserting the aforementioned formula into Eq. 49, a simplified model incorporating the process of saturated elongation is reached,

$$\frac{d}{dt}M = k_e^+ mP\left(1 + \frac{m}{K_m}\right)^{-1}, \tag{51}$$

$$\frac{d}{dt}P = k_n m^{n_c} + k_2 m^{n_2} M, \tag{52}$$

in which the new rate constant for fiber elongation is defined as $k_e^+ = k_c/K_m = k_c k_a^+/(k_a^- + k_c)$. Compared to the formula in the section of fiber growth, a correction factor $(1 + m/K_m)^{-1}$ has been added. Therefore, by introducing an "effective" monomer concentration as $m/(1 + m/K_m)$, earlier model will recover the classical one without saturation. Furthermore, if monomer concentration is much higher than the critical saturation concentration (given by the Michaelis constant, K_m) $m \gg K_m$, the effective monomer concentration becomes a constant K_m; otherwise if the monomer concentration is much lower than the critical saturation concentration $m \gg K_m$, the effective monomer concentration approaches to the real monomer concentration. In this way, both the linear dependence of fiber elongation rate in the regime of low monomer concentration and sublinear dependence in the high regime could be explained by a unified picture based on saturated elongation. The Michaelis constant, K_m serves as a key to characterize the transition from linear to sublinear (Fig. 11a and b).

The model for saturated elongation is far more difficult to solve than the classical one without saturation. In the absence of initial seeds, a suggested practical solution is (unpublished result)

$$M(t) = m_{\text{tot}}\{1 - [1 + \frac{1}{\theta}y(1 + y)^{\alpha/(1+\alpha)}]^{-\theta}\}, \tag{53}$$

where $y = \epsilon e^{\kappa t/\sqrt{1+\alpha}}$, $\alpha = m_{\text{tot}}/K_m$, $\theta = \sqrt{2/[n_2(n_2+1)]}$, and $\kappa = \sqrt{k_e^+ k_2 m_{\text{tot}}^{n_2+1}}$. A critical concentration of fibrils $M(0)/m_{\text{tot}} = \epsilon = k_n^+ m_{\text{tot}}^{n_c - n_2 - 1}/(2k_2) \ll 1$ has been introduced to seed the system, so that the resulting expression for $M(t)$ matches the leading order term for early time. In the special case of $n_2 = 0$, which corresponds to fragmentation, the solution reduces to:

$$M(t) = m_{\text{tot}}\{1 - \text{e}[-y(1 + y)^{\alpha/(1+\alpha)}]\}, \tag{54}$$

by exploiting the identity $\lim_{b\to\infty}(1 + a/b)^b = e^a$. Meanwhile, we have:

$$P(t) = \frac{\sqrt{\theta^2 + \alpha\vartheta^2\kappa}}{2k_+}\left[1 - \left(\frac{m}{m_{\text{tot}}}\right)^{1/\theta}\right], \tag{55}$$

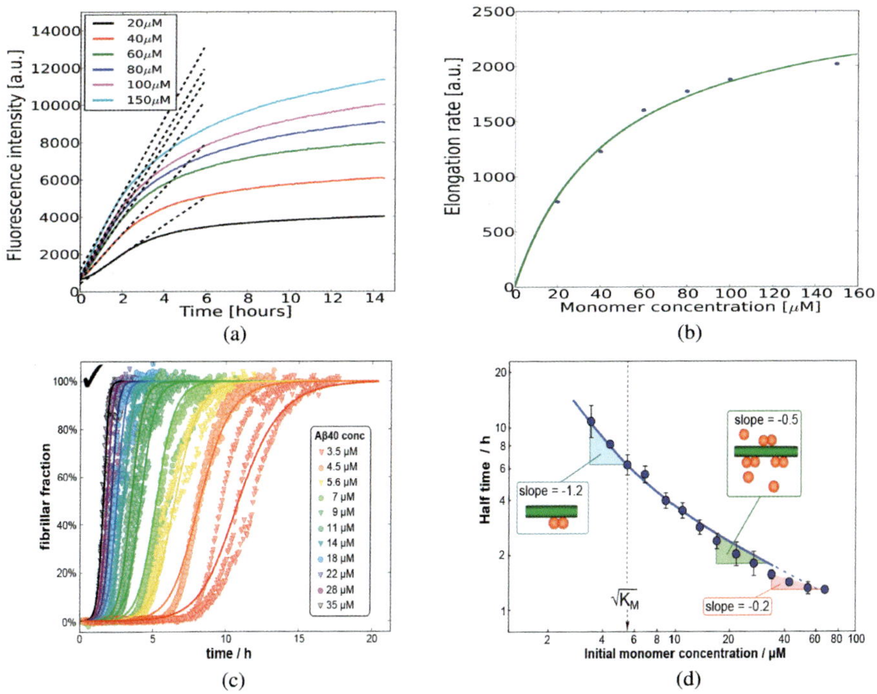

Figure 11: (a, b): Saturation of elongation rate for α-synuclein under different concentrations of soluble proteins and a constant 3.5 μM seeds. Images are taken from.[123] (c, d): Saturation of surface catalyzed secondary nucleation for Aβ40 under various monomer concentrations without initial seeds. Data fitting was performed according to Eqs. 59 and 60 with $k_e^+ = 6 \times 10^5\,\mathrm{M}^{-1}\mathrm{s}^{-1}$, $k_n = 2 \times 10^{-6}\,\mathrm{M}^{-1}\mathrm{s}^{-1}$, $k_2 = 3 \times 10^3\,\mathrm{M}^{-2}\mathrm{s}^{-1}$, $K_s = 6 \times 10^{-6}\mathrm{M}$, $n_c = n_2 = 2$. Figure is adapted with permission from Ref. 88

where $\vartheta = \sqrt{2/[(n_2+1)(n_2+2)]}$. Again, we need to pay attention to the case $n_2 = 0$, which gives:

$$P(t) = \frac{1}{2} k_n m_{\mathrm{tot}}^{n_c} t + \frac{k_2 m_{\mathrm{tot}}}{\kappa} \ln\left(\frac{1+y}{1+\epsilon}\right). \tag{56}$$

Based on the aforementioned solutions, we can roughly determine the half time and the apparent fiber growth rate as follows:

$$t_{1/2} \sim \ln(1/\epsilon)\sqrt{1+\alpha}/\kappa, \tag{57}$$

$$k_{\mathrm{app}} \sim \kappa/\sqrt{1+\alpha}. \tag{58}$$

It is easily seen that $t_{1/2} \propto m_{\rm tot}^{-(n_2+1)/2}$ in the regime of low monomer concentrations $m_{\rm tot} \ll K_m$ and $t_{1/2} \propto m_{\rm tot}^{-n_2/2}$ in the regime of high monomer concentrations $m_{\rm tot} \gg K_m$. Especially, when $n_2 = 0$ and $m_{\rm tot} \gg K_m$, we have $t_{1/2} \propto ln(1/m_{\rm tot})$, which well explains the observed sublinear dependence of fiber elongation under the condition of high monomer concentrations.

In principle, all monomer-dependent processes may get saturated once the monomer concentration exceeds certain threshold. And large amyloid proteins are more prone to get saturated than smaller ones under the same condition, since the former generally requires a longer time to fit itself to the fibrillar structure. In view of saturation, the model for surface catalyzed secondary nucleation Eq. 41 requires to be modified too (Fig. 11c and d). Following a similar "dock-lock" mechanism as well as derivations for saturated elongation, saturated secondary nucleation could be formulated as follows:

$$\frac{d}{dt}P = k_n m^{n_c} + k_2 \frac{m^{n_2}}{1 + (m/K_s)^{n_2}} M, \tag{59}$$

$$\frac{d}{dt}M = k_e^+ m P, \tag{60}$$

where K_s is the critical saturation concentration for secondary nucleation. Again, if monomer concentration is much lower than the critical saturation concentration $m \ll K_s$, we recover the classical model without saturation; contrarily, if monomer concentration is much higher $m \gg K_s$, only a constant concentration of monomers K_s could contribute to secondary nucleation due to saturation.

Self-iterative solutions in a similar form of Eqs. 43 and 44 could be derived for the aforementioned model through fixed-point analysis. Interested readers may refer to[88] for details. Basically, everything is the same except k_2 is replaced by $k_2/[1 + (m_{\rm tot}/K_s)^{n_2}]$. Of particular interest is the scaling behavior of half-time, which are solved within logarithmic corrections as $t_{1/2} \approx \kappa^{-1} \ln(1/C_+)$, where $\kappa = \sqrt{k_e^+ k_2 m_0^{n_2+1}/[1 + (m_0/K_s)^{n_2}]}$, $c_+ = k_e^+ P_0/(2\kappa) + k_e^+ M_0/(2m_0 k_e^+) + \lambda^2/(2\kappa^2)$, $\lambda = \sqrt{k_e^+ k_n m_0^{n_c}}$. And $m(0) = m_0$, $P(0) = P_0$, $M(0) = M_0$ are the monomer concentration, number and mass concentration of seeds at the start of reaction, respectively. Thus, in the presence of high monomer concentration $m_0 \gg K_s$, we have $t_{1/2} \propto m_{\rm tot}^{-1/2}$, similar as that for the model of fragmentation; while in the regime of low monomer concentration $m_0 \ll K_s$, the half-time $t_{1/2} \propto m_{\rm tot}^{-(n_2+1)/2}$, which recovers the unsaturated case as expected. The model combined both saturated elongation and saturated secondary nucleation is similar and will not be shown here.

3.4. *How to incorporate different conformations*

It is well known that the formation of amyloid fibrils is a direct consequence of the misfolding of amyloid proteins.[1,124] As an example, in one of the most well studied amyloid systems — prion, Prusiner identified that the conformational conversion of prion protein from PrP to PrP^{sc} gives rise to the famous mad cow disease and scrapie in sheep.[125] In human, prion causes Creutzfeldt–Jakob disease (CJD) and kuru. According to the molecular size, conformational conversion could be divided into either monomer conversion or oligomer conversion. While based on the position where the conformational conversion happens inside the whole fibrillation procedure, it can be classified into pre-, on-pathway and off-pathway conversion separately.

Among them, preconversion is a prerequisite for primary nucleation, elongation and surface catalyzed secondary nucleation. It is directly related to monomer conversion and means the conformation of monomers has to be adjusted in order to be incorporated into fibrillar structures.[126,127] The kinetic modeling of conformational preconversion is most straightforward. All terms concerning instantaneous monomer concentration $m(t)$ in previous models should be replaced by a new concentration of monomers in the unfolded (or partially unfolded) state $m_u(t)$ as required by the fibrillar structure.[128]

$$m_u(t) = \int_0^t k_c^+ m(\tau) e^{-(k_c^+ + k_c^-)(t-\tau)} d\tau + m_u(0) e^{-(k_c^+ + k_c^-)t}, \tag{61}$$

is the solution of

$$\frac{dm_u}{dt} = k_c^+(m - m_u) - k_c^- m_u, \tag{62}$$

where k_c^+ and k_c^- are unfolded states. In practice, the effect of conformational preconversion is easily neglected due to the existence of other rate-limiting steps, like primary nucleation. This fact is fully appreciated based on following argument. If conformational conversion of monomers is faster than fiber growth rate, which is generally limited by primary nucleation, we could assume monomers in the folded and unfolded states are in dynamic equilibrium and apply QSSA approximation to obtain $k_c^+(m - m_u) - k_c^- m_u = 0$. By redefining $\tilde{k}_n = [k_c^+/(k_c^+ + k_c^-)]^{n_c} k_n$, $\tilde{k}_e^+ = [k_c^+/(k_c^+ + k_c^-)] k_e^+$, $\tilde{k}_2 = [k_c^+/(k_c^+ + k_c^-)]^{n_2} k_2$, the model in Eqs. 41 and 42 is recovered without including conformational preconversion explicitly (see Fig. 12).

In contrast to preconversion, objects of on-pathway and off-pathway conversion are both oligomers. Their main difference lies on whether conformational conversion is necessary for fibril generation or not. In the on-pathway conversion, oligomers need to take some rearrangements in structure, for example, repacking from globular oligomeric conformation to linear fibril-like conformation, before growing into mature fibrils; while in the off-pathway conversion, conformational conversion of oligomers will leads to amorphous aggregates in competition with fibrillar aggregates. The kinetic modeling of on-pathway and off-pathway conversions is far more difficult than preconversion, due to the conformational variety of oligomers, complicated interactions between oligomers, fibrils, amorphous aggregates and so on. Preliminary results for on-pathway conversion leading to fibrillary aggregates in transthyretin (TTR),[129] tau proteins,[130] insulin stabilized by Zn^{2+},[131] and so on, and off-pathway conversion leading to the formation of amorphous aggregates[60–62] could be found in literature. But, details will be omitted with a pity.

Besides those well-formulated mechanisms discussed in current paper, there are still many processes, which play an important role in changing the morphology of fibrils, affecting fibril thermodynamic stability, modifying fibrillation kinetic profiles, and so on, worthy of exploring. For instance, Anderson *et al.* observed glucagon fibrils are able to generate new fibril ends by continuously branching, which prefers an angle of $35° - 40°$ along the forward direction of parent fibril and never occurs at the tip.[132] Murphy *et al.* further included lateral aggregation of filaments into fibrils to build up a complete description of Aβ fibrillation.[133] Most importantly, *in vivo* conditions are completely different from that *in vitro*,[134,135] Not only the cellular crowding environment,[136,137] but also synthesis,[138] degradation,[139] and transportation[140,141] of amyloid proteins may exert a great impact on the fibrillation kinetics. However, currently we still lack a quantitative characterization for most of them. These interesting topics have to be left to the future with regrets.

3.5. *How cytotoxicity arise?*

After looking into so many fibrillation mechanisms, we still have no idea about the most important issue, that is, how the aggregation of misfolded amyloid proteins affects normal cell function and gives rise to amyloidosis? To answer this question, we first need to make sure which species of aggregates is responsible for cytotoxicity. For quite a long time, mature fibrils have been taken for granted as the major cause of cell damage.[2] However, this view has been challenged again and again in recent years, with accumulated evidences coming from the morphology, atomic structure and functions of oligomers and fibrils both *in vitro* and *in vivo*.[142,143] Now heterogeneous oligomers are generally believed far more toxic than mature fibrils.[144–146] Meanwhile, solvable monomers are considered as less harmful to cells even in a misfolded state.[147,148]

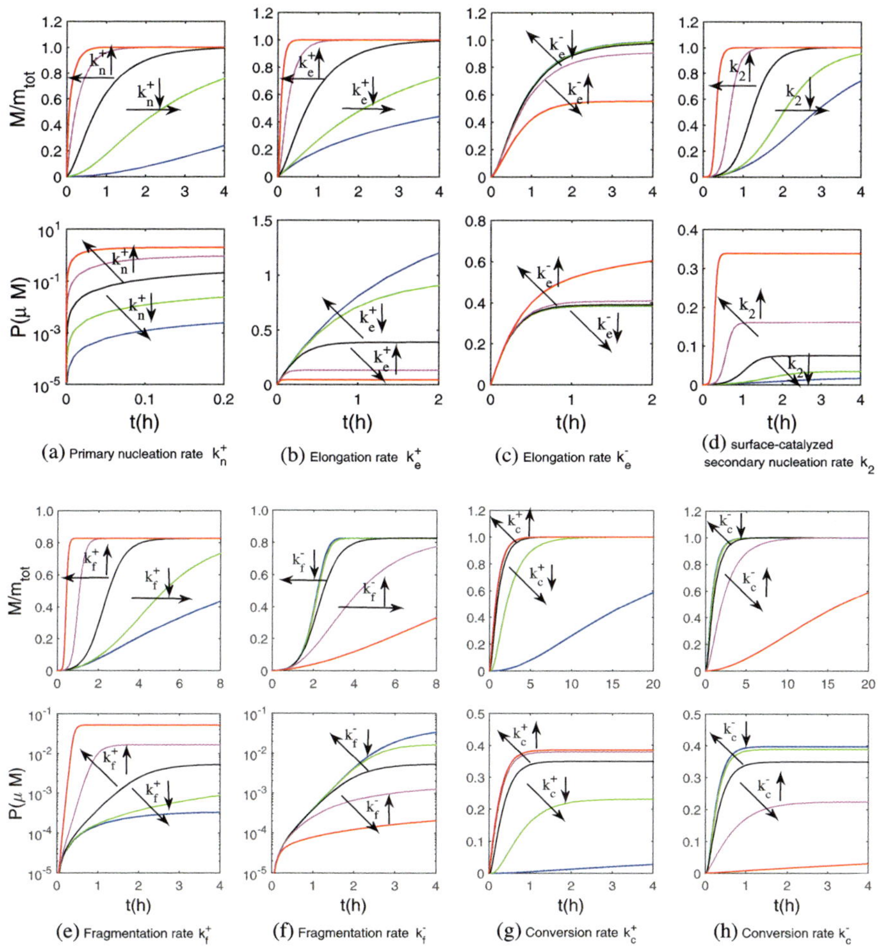

Figure 12: Influence of rate constants on the fibrillation kinetics. In (a–c), rates for primary nucleation, elongation and monomer dissociation are changed with respect to those in Fig. 9a ($m_{\rm tot} = 18.4\,\mu$M) by one or two orders of magnitude higher or lower. The base line is drawn in black. Lines in purple and red mean increasing the parameter, while lines in green and blue mean decreasing. In (d–h), similar procedures are performed on rates for monomer-dependent secondary nucleation, fragmentation, annealing and monomer conversion, respectively. To be exact, parameters in (d) are taken according to those for the first subplot of Fig. 10 with $m_{\rm tot} = 3.5\,\mu$M; parameters for (e) and (f) are the same as Fig. 9b with $m_{\rm tot} = 11.5\,\mu$M; (g) and (h) show a modified NE model by introducing an additional step of monomer conversion in Eq. 62. Besides $k_c^+ = 0.01\text{s}^{-1}$, $k_c^- = 0.001\text{s}^{-1}$, $m_{\rm tot} = 18.4\mu$ M other parameters are taken from Fig. 9a.

In literature, several mechanisms have been proposed for the molecular basis of cytotoxicity caused by oligomeric species. Most arguments are based on the interaction between oligomers and lipid membrane. It is generally believed that the binding of oligomers could dramatically affect the shape and permeability of lipid membrane,[149,150] induce undesired pores or ion-channel-like structures,[151,152] and give rise to fatal abnormal ion leakage therefore.[153,154] For instance, Demuro *et al.* observed in Xenopus oocytes that $A\beta(1-42)$ oligomers can lead to abnormal Ca^{2+} flux independent of ion-channel from 5 to 40 mer.[155] Schauerte *et al.* further pointed out that hexamer is the smallest stable oligomer that can penetrate cell membrane, whereas 12- to 14-mers give rise to the largest ion current.[156] By MD simulation, Jang *et al.* proposed that 16- to 24-mers of $A\beta$ arrange into pore-like structures,[157] which are compatible with the pores observed by atomic force microscopy.[158,159] Meanwhile, Shafrir *et al.* found that $A\beta$ pores is an assembly made up of six hexamers.[160]

Alternative hypothesis based on cellular regulatory network suggests that *in vivo* the exposed flexible hydrophobic surfaces of oligomers can promote aberrant protein interactions, deregulate cytosolic stress response,[161,162] trigger inflammatory responses, oxidative damage,[163] alter kinase and phosphatase activities, increase neurofibrillary tangles,[148,164,165] change synaptic plasticity[166–169] and so on.

Recently, Hong and his colleagues extended kinetic models for amyloid fibrillation and included the cell damage caused by oligomer formation too.[170] Four basic assumptions have been put forward as a general criterion for modeling, that is, (1) the basic procedure of amyloid formation is well formulated by kinetic models; (2) cell damage is mainly caused by oligomers, rather than mature fibrils or monomers, through their binding to lipid membrane; (3) cytotoxicity is quantified through the amount of membrane-bounded oligomers (or leaked ion concentration in original manuscript); (4) oligomer binding does not affect the kinetics of amyloid formation. Or, in other words, the consumption of oligomers during membrane binding could be neglected. The last assumption is generally unnecessary; however, mathematically it offers a simple way to get rid of the feed-back influence of oligomer consumption during membrane binding and keep previous well-formulated equations of fibrillation kinetics unaffected.

Besides two equations for number concentration and mass concentration of aggregates $P(t)$ and $M(t)$, which characterize the fibrillation process under difference mechanisms in previous sections, the concentration of cells in different states, classified based on the number of membrane-bounded oligomers, evolves according to

$$\frac{d}{dt}[C_i] = P_{\text{oli}}(t)\{k_b^+(i-1)[C_{i-1}] - k_b^+(i)[C_i]\}$$
$$- \{k_b^-(i)[C_i] - k_b^-(i+1)[C_{i+1}]\}, \quad i = 0, \ldots, N \qquad (63)$$

where $[C_i]$ is the concentration of cells in state i, in which the lipid membrane has been bounded with i oligomers. The current definition of cellular state is quite natural, since the condition of cell damage is positively correlated with the amount of bounded oligomers according to assumption (3). The total cell concentration $c_{tot} = \sum_{i=0}^{N}[C_i]$ is a constant, in which N stands for the maximal number of oligomers allowed to be bounded to the same membrane. $k_b^+(i)$ and $k_b^-(i)$ are rate constants for oligomer binding and unbinding, which potentially depend on the cellular state. According to Gregory *et al.* (2008) and Gregory *et al* (2009).[171,172] $k_b^+ \sim (2-5) \times 10^5 \text{M}^{-1}\text{s}^{-1}$ and $k_b^- \sim 5 - 300\,\text{s}^{-1}$ on average. $P_{\text{oli}}(t)$ is the number concentration of oligomers at time t and can be expressed as a function of $M(t)$ and $P(t)$ according to the moment-closure method shown in section of how to coarse-grain model. However, it is still uncertain whether M_{oli} (the mass concentration of oligomers) or even more complicated functions are required to include those toxic species and their relative damage to cells explicitly.

Especially, a simple two-state model including only normal and damaged cells is obtained when $N = 1$, based on which the fraction of damaged cells is given as follows:

$$[C_1]/c_{\text{tot}} = \int_0^t k_b^+ P_{\text{oli}}(\tau) \exp\left\{-\int_\tau^t [k_b^+ P_{\text{oli}}(\sigma) - k_b^-]d\sigma\right\} d\tau. \qquad (64)$$

Once the time course for $P_{\text{oli}}(t)$ is determined through the fibrillation kinetics, a full knowledge about the progress of cell damage caused by oligomers binding could be obtained. A similar approach by examining the leaked ion concentration has been adopt by Hong *et al.*[170] and applied to β_2m and IAPP fibrils induced membrane leakage with great success (Fig. 13).

3.6. *How to manipulate fibrillation*

A central aim for examining various fibrillation mechanisms is to manipulate the kinetics of amyloidosis. Based on different targets, there are basically two large groups of approaches to achieve this goal. The indirect approach is to control the environment, in which amyloid fibrillation takes place. Here the word "indirect" means we not manipulate amyloid proteins or fibrils directly. Alternatively, by

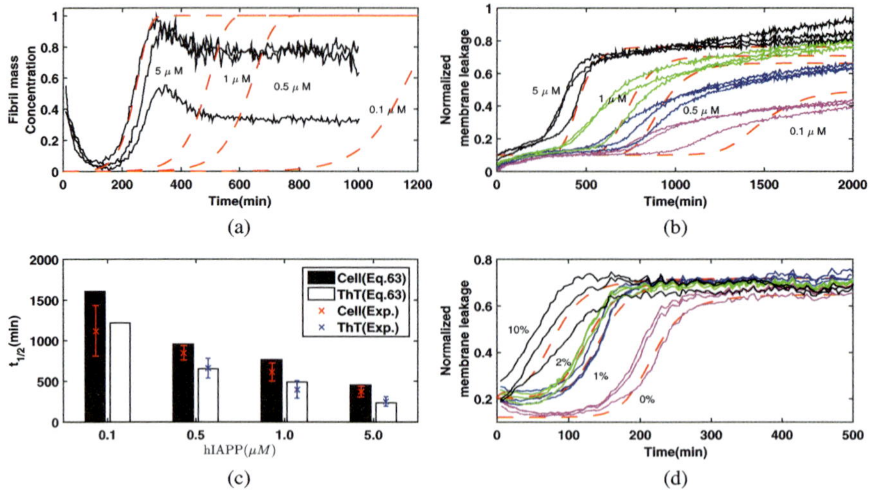

Figure 13: hIAPP fibril growth and hIAPP-induced membrane leakage: (a) fibril mass concentration measured by ThT fluorescence; (b) the amount of membrane leakage under different initial protein concentration; (c) the half-time for fibril formation and membrane leakage; and (d) the amount of membrane leakage under different concentrations of seeds. Figure is adapted with permission from Ref. 170 but all fittings are reperformed according to Eq. 63 with $k_n^+ = 4 \times 10^{-5}\ \mathrm{M}^{-1}\mathrm{s}^{-1}$, $k_e^+ = 1.2 \times 10^5 \mathrm{M}^{-1}\mathrm{s}^{-1}$, $k_e^- = 0\ \mathrm{s}^{-1}$, $k_f^+ = 7 \times 10^{-5}\ \mathrm{s}^{-1}$, $k_f^- = 1 \times 10^4\ \mathrm{M}^{-1}\mathrm{s}^{-1}$, $k_b^+ = 1.5 \times 10^5 \mathrm{M}^{-1}\mathrm{s}^{-1}$, $k_b^- = 2.5 \times 10^{-3}\mathrm{s}^{-1}$, $n = 1$, $n_c = 2$, $n_o = 40$. Particularly, in (d) $m_{\mathrm{tot}} = 2.5 \times 10^{-5}\mathrm{M}$, $k_b^+ = 6 \times 10^5 \mathrm{M}^{-1}\mathrm{s}^{-1}$, $k_b^- = 1 \times 10^{-3}\mathrm{s}^{-1}$ and $M(0)/P(0) = 400$.

varying the temperature, pH value, salt concentration, and so on, reaction rate constants for different processes will be changed accordingly. As a result, we can either accelerate or decelerate any given amyloid fibrillation, as well as increase or decrease the population of any oligomeric and fibrillar species. To be concrete, by decreasing the elongation rate, monomers will be overpopulated; otherwise, the population of mature fibrils will be enlarged. To increase the population of oligomers is a bit complex, which requires increasing the rate of primary nucleation and decreasing that of elongation at the same time.

According to the classical transition state theory (TST) for elementary chemical reactions, the rate constant for a given process is determined by the Arrhenius equation[173,174]

$$k = A \exp[-E_{\mathrm{act}}/(k_B T)], \tag{65}$$

where A is referred to as the frequency factor, and E is regarded as the activation energy, which in principle is a function of various experimental conditions, $E_{act} = E_{act}$ (temperature, pH value, salt concentration, etc.). Therefore, if we know how the activation energy for various elementary fibrillation processes relies on conditions quantitatively, we can control the fibrillation kinetics as well as the population of each species freely as we wish.

Although for many cases the indirect approach is quite useful, it suffers some intrinsic limitations, for example, it is difficult to make a big change of some fibrillation conditions in experiments; some conditions may play a complex role in determining the reaction rate and no quantitative or qualitative description is available; in many cases, changing one condition may affect almost all processes at the same time, which makes it almost impossible to perform analysis.

The direct approach means to manipulate the concentration of each species directly. Varying the initial monomer concentration and seeds concentration are two most popular and effective methods *in vitro*, whose effects on fibrillation kinetics and species populations have been fully appreciated in previous sections. *In vivo*, directly adding or removing monomers and seeds becomes less promising. Therefore, specific binding through antibodies and chaperons is introduced to control the fibrillation kinetics as well as species population.

Antibodies and chaperons are both well known for their specificity in binding to certain molecular structures. In principle, we can diminish any monomeric, oligomeric and fibrillar species by introducing proper antibody or chaperon into the system, which therefore allows us to manipulate each elementary fibrillation process independently. For example, nano particles have been widely used in literature to inhibit amyloid fibrillization, induce fibril dissociation and mitigate neurotoxicity.[175–177] Similarly, many chaperon molecules, like Hsp70 family members, are known for their ability to inhibit and reverse the formation of amyloid aggregates.[178–180] Now, searching for various promising antibodies and chaperons to inhibit amyloid fibrillation, or to prevent and cure amyloidosis as a potential goal, becomes very popular in this field (see Fig. 14 as an example). Though in many cases, the molecular basis for why it works has not yet been clarified.

To mathematically probing various potential binding effects of antibodies (or chaperons) on amyloid fibrillation is a systematic and laborious task, and has not been developed into a mature theoretical framework. Here we just look into one particular example to show how to do it in principle. The mechanism of antibody blocking fibril ends and stopping the elongation process could be formulated

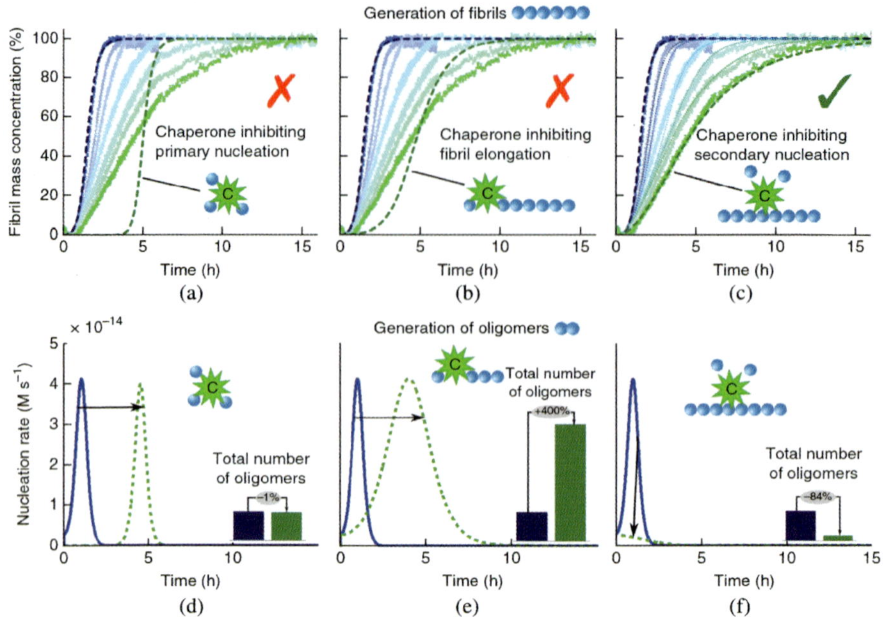

Figure 14: Brichos slows down Aβ42 aggregation by inhibiting the surface catalyzed secondary nucleation. (a–c): From left (blue) to right (green), 0%, 10%, 15%, 35%, 50%, 75%, and 100% Aβ42 monomer equivalents of Brichos have been added, respectively. The concentration of monomeric Aβ42 is 3 μM; (d–f): The blue line corresponds to the situation in the absence of Brichos. The green dashed lines in (d–f), respectively, show predictions for the cases in which primary nucleation, elongation and secondary nucleation are inhibited. Figure is adapted with permission from Ref. 181

into:

$$\frac{d}{dt}[B] = -k_b^+ P_{\text{free}}[B] + k_b^- P_{\text{bound}}, \tag{66}$$

$$\frac{d}{dt}P_{\text{free}} = k_n m^{n_c} + k_2 m^{n_2} M \underline{- k_b^+ P_{\text{free}}[B] + k_b^- P_{\text{bound}}}, \tag{67}$$

$$\frac{d}{dt}M = k_e^+ m P_{\text{free}} - k_e^- P_{\text{free}}, \tag{68}$$

where $P_{\text{free}} = P - P_{\text{bound}}$ and $P_{\text{bound}} = b_{\text{tot}} - [B]$ are concentrations of free fibril ends and antibody blocked fibril ends separately. $[B](t)$ is the free antibody concentration at time t, and b_{tot} is the total concentration of antibodies, which is conserved during reactions. k_b^+ and k_b^- are rate constants for antibody binding and unbinding, respectively. In a similar way, antibody binding oligomers and blocking primary nucleation or secondary nucleation, chaperons binding fibrils, reversing amyloid formation, and so on could be modeled and will not be addressed further.

In the presence of antibody or chaperon, model analysis becomes far more difficult. However, under one particular condition interested in experiments, that is, the rates for antibody (or chaperon) binding and unbinding are much faster than the fibrillation speed, we can safely apply the partial equilibrium approximation (PEA) to Eq. 66 and suppose terms on the right-hand side canceling each other at every moment, meaning $-k_b^+[B]P+k_b^-(b_{tot}-[B])=0$. Then class results for P and M without antibody binding are recovered. This leads to an important conclusion that, in order to manipulate amyloid fibrillation, the rate for antibody unbinding must be slower than that of fibrillation.

3.7. *Summary*

The fibrillation kinetics is a central issue in the study of amyloidosis and amyloid diseases. In the past several years, fruitful results and plenty of deep physical insights have been obtained in this direction. Among them, mathematical modeling based on chemical mass-action equations offers a unified framework to treat this problem. In the current section, we focus on how to apply kinetic modeling to explain various kinetic phenomena we have observed in real amyloid systems, which could be characterized and classified according to the half-time and fiber apparent growth rate as shown in Table 1. Actually, the scaling relations between half-time of fibrillation (or fiber apparent growth rate) and monomer concentration provide us a useful way in practice to pick out the proper model, to classify amyloid systems based on their own fibrillation mechanisms (Fig. 15), and so on.

4. Mathematical and Application Foundations

The former section is wholly based on macroscopic kinetic models for the mass concentration and number concentration of fibrils, but why we can adopt this simple picture and how its accuracy is compared to models at a molecular lever have never been addressed. For this purpose, here we are going to the mathematical foundation of kinetic modeling, in which the mathematical linkage between models at the microscopic scale (molecular lever) and macroscopic scale (what we adopted in the former section) will be clarified in quantity. Interestingly, this linkage provides us ways to reconstruct the full fiber length distribution from the knowledge of mass concentration and number concentration of fibrils in a high accuracy, which

Table 1: A summary of half-time and apparent fiber growth rate under different fibrillation mechanisms. Capital E stands for elongation, S for surface catalyzed secondary nucleation, F for fragmentation and $*$ for saturation of corresponding processes. "Seeded" means conditions with preadded seeds. In general, fragmentation can be regarded as a special case of secondary nucleation with $n_2 = 0$.

Mechanism	$t_{1/2}$	k_{app}	Parameters
NE	λ^{-1}	λ	$\lambda = \sqrt{k_n k_e^+ m_{\mathrm{tot}}^{n_c}}$
NES	$k_s^{-1} \ln (k_s/\lambda)$	k_s	$k_s = \sqrt{k_e^+ k_2 m_{\mathrm{tot}}^{n_2+1}}$
NEF	$k_f^{-1} \ln (k_f/\lambda)$	k_f	$k_f = \sqrt{k_e^+ k_f^+ m_{\mathrm{tot}}}$
NE*	$\sqrt{1+\alpha}\lambda^{-1}$	$\lambda/\sqrt{1+\alpha}$	$\alpha = m_{\mathrm{tot}}/K_m$
NE*S	$\sqrt{1+\alpha}k_s^{-1} \ln (k_s/\lambda)$	$k_s/\sqrt{1+\alpha}$	
NE*F	$\sqrt{1+\alpha}k_f^{-1} \ln (k_f/\lambda)$	$k_f/\sqrt{1+\alpha}$	
NES*	$\sqrt{1+\beta}k_s^{-1} \ln (k_s/\lambda)$	$k_s/\sqrt{1+\beta}$	$\beta = (m_{\mathrm{tot}}/K_s)^{n_2}$
NE*S*	$\sqrt{(1+\alpha)(1+\bar{\beta})}k_s^{-1} \ln (k_s/\lambda)$	$k_s/\sqrt{(1+\alpha)(1+\ )}$	
Seeded E	ν^{-1}	ν	$\nu = k_e^+ P_0$
Seeded E*	$\alpha\nu^{-1}$	ν/α	

is generally believed to be very difficult as an inverse problem. Finally, issues on how to convert model predictions into experimental observations, how to determine unknown model parameters and how to perform reliable global fittings are discussed too, in order to provide a relatively comprehensive review on the kinetic aspect.

4.1. *How to quantify fibrillation kinetics*

In order to quantify the kinetics of amyloid fibrillation, let's introduce the fiber length distribution $\{[A_i],\ i \geq 1\}$, each of whose components represents the concentration of aggregates containing exactly i protein molecules. Especially, $[A_1]$ stands for the concentration of monomers. According to different reaction schemes for amyloid aggregation which we have addressed in details in the main text, the time evolution of $\{[A_i](t)\}$ will be characterized through a group of coupled ordinary differential equations (without taking the spacial distribution into consideration) by using laws of mass-action. This leads to the so-called "microscopic chemical kinetic equations", since they contain a full knowledge of the fiber length distribution.

On the other hand, either due to resolution limitation of experiments which makes it impossible to obtain a full spectrum of fiber length distribution, or for a purpose to speed up simulation without taking too many details into consideration,

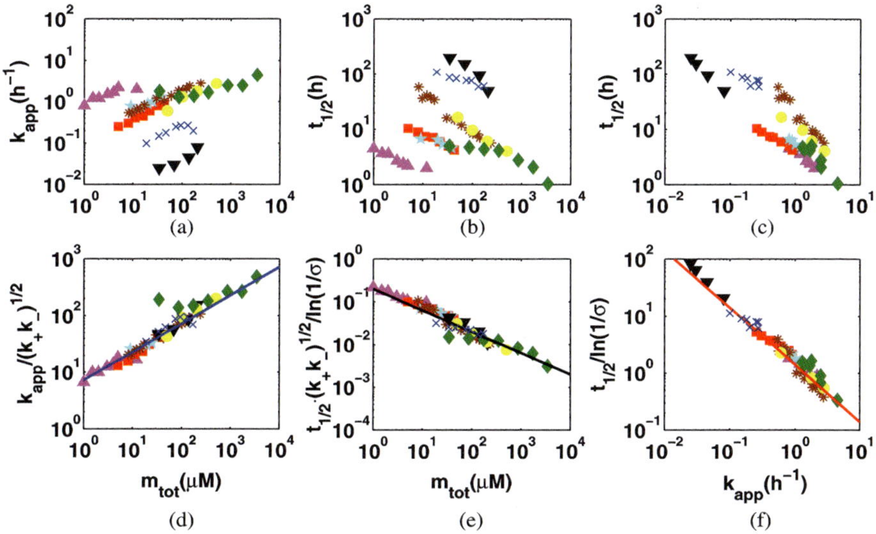

Figure 15: Scaling relationships between $k_{\rm app}$, $t_{1/2}$, and $m_{\rm tot}$ for eight fragmentation-dominated amyloid proteins, that is, the yeast prion Sup35 NW region (purple triangles up), Csg $B_{\rm trunc}$ (red squares), Ure2 protein (cyan pentacles), β2-microglobulin (brown stars), Stefin B (blue cross), α-synucleins (black triangles down), WW domain (yellow circles), and insulin (green diamonds). Figure is adapted with permission from Ref. 128

a simple coarse-grained description is preferred. Among various candidates, formulation involving two macroscopic quantities — the number concentration and mass concentration of aggregates (including both oligomers and fibrils)

$$P = \sum_{i=n_c}^{\infty} [A_i], \quad M = \sum_{i=n_c}^{\infty} i \cdot [A_i], \tag{69}$$

where n_c stands for the critical nucleus size, is the most popular and welcomed. A basic reason is that this formulation actually provides a simplest self-consistent way to examine the amount of amyloid fibrils formed inside a given system, a quantity we are most interested in.

From the aforesaid definition, we can see that P and M are actually the zeroth and first-order moments of fiber length distribution $\{[A_i]\}$ for $i \geq n_c$. The total mass concentration $m_{tot} = \sum_{i=1}^{\infty} i \cdot [A_i]$ is another often used first-order moment. In particular, if oligomer concentration $[A_i](2 \leq i \leq n_c - 1])$ could be neglected, we have $m_{\rm tot} = m(t) + M(t)$, an equality representing the mass conservation of total protein molecules during fibrillation. In principle, other high order moments could also be introduced into the formulation in order to achieve a high accuracy, but in many cases it is not very worthwhile due to the sacrifice of both simplicity and efficiency. In literature, the governing equations of moments are referred to as "macroscopic chemical kinetic equations" in contrast to microscopic formulation based on fiber length distribution.

The relation between macroscopic and microscopic chemical kinetic equations is an interesting question. In fact, a whole branch of statistical physics is dealing with such problems on how to derive macroscopic quantities and their governing kinetic equations from the knowledge of microscopic descriptions.[182] There are enormous investigations, fruitful results and endless debates in history which are far beyond the scope of this paper. But on current specific topic, almost all issues could be solved by some well-formulated moment-closure methods. We come back to this point with all necessary details in the next section.

At last, we want to introduce the half time of fibrillation $t_{1/2}$ and the apparent fiber growth rate k_{app},[128] that is,

$$M(t_{1/2}) = \frac{M(0) + M(\infty)}{2}, \quad k_{app} = \frac{1}{m_{tot}} \frac{dM}{dt}\bigg|_{t=t_{1/2}} \tag{70}$$

These two quantities are essential to characterize the kinetics of amyloid fibrillation in an empirical way (see Eq. 89), even without referring to any models or analytical solutions. Two additional quantities appear in literature from time to time too. One is the lag time, which is defined as $t_{lag} = t_{1/2} - 1/(2k_{app})$. The lag time has an intuitive physical meaning, which measures how long an amyloid system has to wait in order to accumulate enough seeds to pass the phase dominated by primary nucleation. In the presence of secondary nucleation, the lag time becomes prominent due to the sigmoid fibrillation profile, and could be greatly shortened by an introduction of initial seeds. There are also other ways to define the lag time in literature, for example, the time when the mass concentration of aggregates reaches 1% of its static value.[183] Obviously, our definition is more natural and meaningful in physics. The other quantity is related to the speed of fibril growth, its maximal value to be exact. However, in most cases the maximal fiber growth rate k_{max} relies on the whole fibrillation profile, which makes it difficult to determine and use.

4.2. *How to coarse-grain model*

In last section, we have introduced the fiber length distribution and its moments in different orders, like the number concentration and mass concentration of aggregates. We have also claimed there is a direct connection between the macroscopic and microscopic chemical kinetic equations. Now we are going to address this point based on the method of moment closure.

Without loss of generality, we consider the reaction scheme proposed for the length-dependent fragmentation.[111] According to laws of mass-action, the time evolution of fiber length distribution obeys following equations

$$\frac{d[A_1]}{dt} = -n_c k_n^+ [A_1]^{n_c} + n_c k_n^- [A_{n_c}] - k_e^+ [A_1] \sum_{j=n_c}^{\infty} [A_j] + k_e^- \sum_{j=n_c+1}^{\infty} [A_j], \tag{71}$$

$$\begin{aligned} \frac{d[A_i]}{dt} &= k_e^+ [A_1]([A_{i-1}] - [A_i]) - k_e^-([A_i] - [A_{i+1}]) \\ &\quad + 2 \sum_{j=n_c+i}^{\infty} k_f^+(i, j-i)[A_j] - \sum_{j=n_c}^{i-n_c} k_f^+(j,\ i-j)[A_i] \\ &\quad - 2 \sum_{j=n_c}^{\infty} k_f^-(i,j)[A_i][A_j] + \sum_{j=n_c}^{i-n_c} k_f^-(j,\ i-j)[A_j][A_{i-j}] \\ &\quad + (k_n^+ [A_1]^{n_c} - k_n^- [A_i] - k_e^+ [A_1][A_{i-1}] + k_e^- [A_i])\delta_{i,n_c},\ i \geq n_c. \end{aligned} \tag{72}$$

As a consequence, it is straightforward to show that the number concentration $P(t)$ and mass concentration $M(t)$ of aggregates evolve according to

$$\begin{aligned} \frac{d}{dt}P &= k_n^+ (m_{\mathrm{tot}} - M)^{n_c} - k_n^- [A_{n_c}] + \sum_{i=n_c}^{\infty} \sum_{j=i+n_c}^{\infty} k_f^+(i,\ j-i)[A_j] \\ &\quad - \sum_{i=n_c}^{\infty} \sum_{j=n_c}^{\infty} k_f^-(i,\ j)[A_i][A_j], \end{aligned} \tag{73}$$

$$\frac{d}{dt}M = n_c k_n^+ (m_{\mathrm{tot}} - M)^{n_c} - n_c k_n^- [A_{n_c}] + k_e^+ (m_{tot} - M)P - k_e^- P + k_e^- [A_{n_c}], \tag{74}$$

which clearly are not closed, as unknown variables $\{[A_i]\}$ have not been expressed through P and M.

To solve this problem, a systematic moment-closure method based on Maximum Entropy Principle[113,114] has been proposed and applied with great success[111,170,184] (as shown in Fig. 16). Namely, we seek for solutions of following

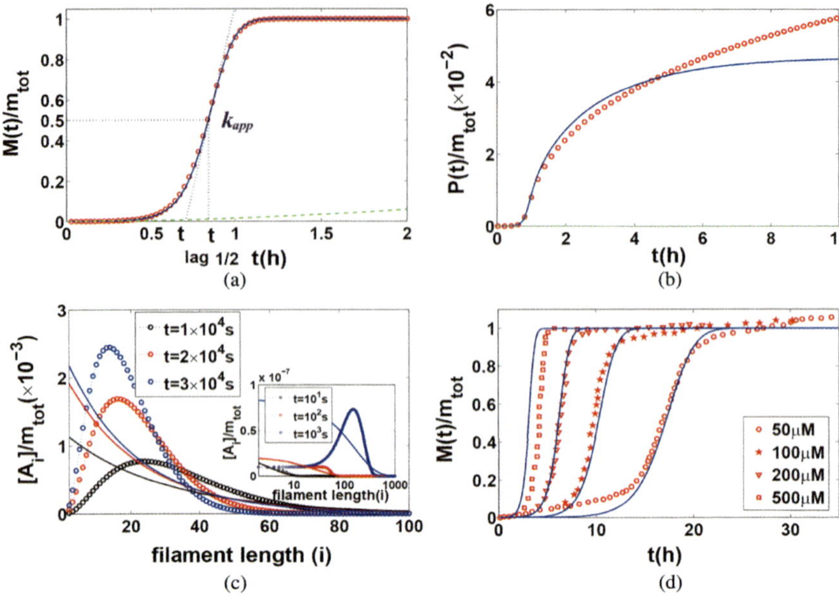

Figure 16: Accuracy of moment-closure method through comparisons on the fiber mass concentration, number concentration and fiber length distribution in (a–c). Values obtained from microscopic kinetic equations Eqs. 71 and 72 are drawn in circles and that from moment-closure methods in solid lines. (d) Application to the polymerization of WW domain. Figure is adapted with permission from Ref. 111

constrained optimization problem, that is,

$$\max\ S(\{[A_i]\}) = -k_B \sum_{i=n_c}^{\infty} ([A_i] \ln[A_i] - [A_i]) \tag{75}$$

$$\text{s.t.} \sum_{i=n_c}^{\infty} [A_i] = P, \ \sum_{i=n_c}^{\infty} i \cdot [A_i] = M, \ [A_1] + \sum_{i=n_c}^{\infty} i \cdot [A_i] = m_{\text{tot}}. \tag{76}$$

It could also be translated into an equivalent variational problem through the method of Lagrangian multiplier,

$$\frac{\delta}{\delta[A_i]} \left[S(A_i)/k_B + \lambda_1 \left(\sum_{i=n_c}^{\infty} [A_i] - P \right) + \lambda_2 \left(\sum_{i=n_c}^{\infty} i \cdot [A_i] - M \right) \right.$$
$$\left. + \lambda_3 \left([A_1] \sum_{i=n_c}^{\infty} i \cdot [A_i] - m_{\text{tot}} \right) \right] = 0, \tag{77}$$

where λ_1, λ_2 and λ_3 are Lagrangian multipliers. The solution of the aforesaid equation is given by

$$[A_1] = \exp(\lambda_3) \tag{78}$$

$$[A_i] = \exp[\lambda_1 + i(\lambda_2 + \lambda_3)], \tag{79}$$

based on which λ_1, λ_2 and λ_3 are related to $P(t)$, $M(t)$ and m_{tot} as follows:

$$\lambda_1 = \ln\left[\frac{P^2}{M - (n_c - 1)P}\right] - n_c \ln\left[\frac{M - n_c P}{M - (n_c - 1)P}\right], \tag{80}$$

$$\lambda_2 = \ln\left[\frac{M - n_c P}{M - (n_c - 1)P}\right] - \ln(m_{\text{tot}} - M) \tag{81}$$

$$\lambda_3 = \ln(m_{\text{tot}} - M) \tag{82}$$

Putting these formulas back into Eqs. 73 and 74, we obtain desired macroscopic chemical kinetic equations solely concerning with $P(t)$ and $M(t)$.

In Ref. Hong *et al.*,[111] it has been proven that above moment-closure method based on Maximum Entropy Principle is mathematically equivalent to PEA on fiber elongation, which assumes the elongation process is much faster than primary nucleation and fragmentation. Therefore, for given number concentration $P(t)$ and mass concentration $M(t)$ of aggregates, each component of fiber length distribution $\{[A_i]\}$ is considered in quasi-equilibrium with each other, which is exactly the way how we are able to express $[A_i](t)$ as a function of $P(t)$, $M(t)$ and m_{tot}.

4.3. *How to reconstruct fiber length distribution from moments*

In previous section, we have focused on how to simplify the model by coarse-graining, which turns out to be a very promising approach with tremendous successful applications. However, during coarse-graining, we are facing with an inevitable loss of information. Original knowledge of full fiber length distribution has been compressed into that about only two macroscopic moments — the number and mass concentration of aggregates to be exact. Is it possible to reconstruct original fiber length distribution based on a knowledge of these two moments? This is a question not only of mathematical interest, but also with great practical usage. In principle, inverse problems are generally very difficult to solve and do not admit a unique solution.[185–187] But, in current case without considering filaments fragmentation and annealing, as we are so lucky to have a complete understanding about underlying microscopic processes, the full fiber length distribution at any time could be explicitly extracted just from one moment — mass concentration of aggregates as well as some knowledge about the initial length distribution. A brief derivation is listed as follows.

Without loss of generality, let us start with following microscopic model, including primary nucleation, elongation, and secondary nucleation,[188]

$$\frac{d[A_i]}{dt} = k_e^+ m(t)([A_{i-1}] - [A_i]) + k_n m(t)^{n_c} \delta_{i,n_c} + k_2 m(t)^{n_2}[m_{\text{tot}} - m(t)]\delta_{i,n_2},\ i \geq n_c \tag{83}$$

where $n_2 \geq n_c > 0$. Introduce the generating function[189]:

$$C(z,\ t) = \sum_{j=n_c}^{\infty} z^j \cdot [A_j](t) \tag{84}$$

which is a natural mathematical generalization of the physical moments. Especially, we have $P(t) = C(z = 1,\ t)$ and $M(t) = \frac{\partial C(z,t)}{\partial z}|_{z=1}$. In fact, the full fiber length distribution could be recovered from the generating function, that is, $[A_j](t) = \frac{1}{j!}\frac{\partial^j C(z,t)}{\partial z^j}|_{z=0}$ for $i \geq n_c$ which means the generation function is a one-to-one mapping of the fiber length distribution.

It is straightforward to show that the generating function satisfies following equation:

$$\frac{\partial C(z,t)}{dt} = k_e^+ m(t)(z-1)C(z,\ t) + k_n m(t)^{n_c} z^{n_c} + k_2 m(t)^{n_2}[m_{\text{tot}} - m(t)]z^{n_2}. \tag{85}$$

Define a new time scale $\tau(t) = \int_0^t k_e^+ m(s)\text{d}s$, which acts as the characteristic time for fiber elongation, above-stated equation could be rewritten as follows:

$$\frac{\partial C(z,\tau)}{d\tau} = (z-1)C(z,\ \tau) + \{k_n m(\tau)^{n_c} z^{n_c} + k_2 m(\tau)^{n_2}[m_{\text{tot}} - m(\tau)]z^{n_2}\}\frac{\partial t}{\partial \tau}, \tag{86}$$

whose solution is given by

$$C(z,\ \tau(t)) = \int_0^t e^{-(z-1)[\tau(s)-\tau(t)]}\{k_n m(s)^{n_c} z^{n_c} + k_2 m(s)^{n_2}[m_{\text{tot}} - m(s)]z^{n_2}\}\text{d}s + C(z,\ 0)e^{(z-1)\tau(t)}. \tag{87}$$

Now the fiber-length distribution at any given time t could be calculated through following formula:

$$
\begin{aligned}
[A_j](t) &= \frac{1}{j!}\left.\frac{\partial^j C(z,t)}{\partial z^j}\right|_{z=0} \\
&= \int_0^t \Theta_{j-n_c}(t,\ s)k_n m(s)^{n_c}\mathrm{d}s + \int_0^t \Theta_{j-n_2}(t,\ s)k_2 m(s)^{n_2}[m_{\mathrm{tot}} - m(s)]\mathrm{d}s \\
&\quad + \sum_{k=n_c}^{j} \Theta_{j-k}(t,\ 0)[A_k](0),
\end{aligned}
\tag{88}
$$

where $\tau(t) = \int_0^t k_e^+ m(s)\mathrm{d}s$ and $\Theta_k(t,\ s) = e^{\tau(s)-\tau(t)}[\tau(t) - \tau(s)]^k/k!$. During the calculation, we use identities $0^0 = 1$ and $0! = 1$.

The aforementioned formula provides the mathematical foundation on how to extract the full fiber length distribution at any time just based on the time course of monomer concentration (or mass concentration of aggregates $M(t)$ as well as the initial fiber length distribution. This is a quite astonishing result, as we have mentioned that inverse problems are usually extremely difficult to solve and do not admit a unique solution in general. Our success in this case could be contributed to two reasons: one is we have a complete knowledge on the microscopic kinetics which governs the time evolution of fiber length distribution; the other is both primary nucleation and secondary nucleation can solely affect the concentration of single species $[A_{n_c}]$ or $[A_{n_2}]$. The only way to perform a global change of the fiber length distribution is elongation, which follows a Poisson process characterized by the integral kernel $\Theta_k(t,\ s)$ with intrinsic time scale $\tau(t) = \int_0^t k_e^+ m(s)ds$, since each monomer association is obviously random and independent of each other.

In the presence of filaments fragmentation and annealing, an additional global change of the fiber length distribution will be introduced besides elongation. How to include these two processes into earlier picture explicitly is an unsolved problem. In particular, Michaels *et al.* derived approximate solutions for length-independent fragmentation in open and close systems.[190] While for general models with length-dependent fragmentation and annealing, an empirical method has been proposed, whose theoretical foundation lies on the fact that during the procedure of moment closure, the microscopic fiber length distribution is expressed through macroscopic moments based on maximum entropy principle. The thus obtained fiber length distribution directly determines the accuracy of macroscopic equations, which as a consequence could be adopted as an empirical candidate to approximate the exact

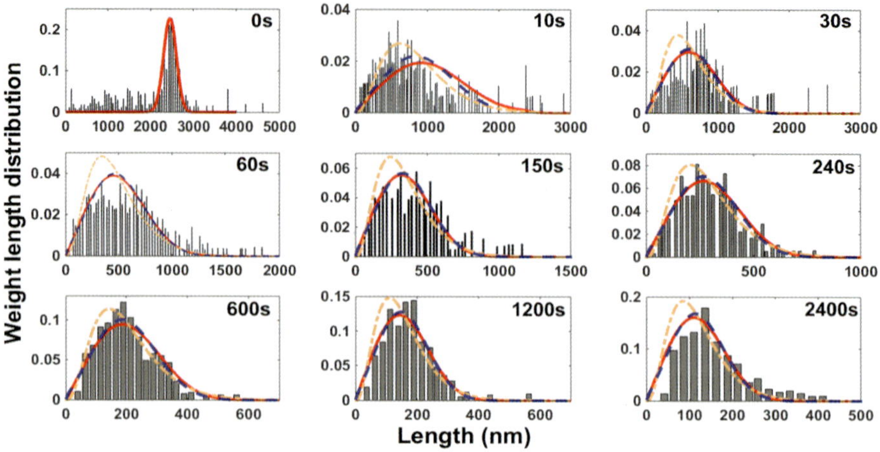

Figure 17: Accuracy of approximate fiber length distribution constructed from moment-closure method for a fragmentation only model. Comparisons were made among TEM measurements of fiber-like PI264-b-PFS48 micelles under sonication (black symbols), exact fiber length distribution in Eqs. 71 and 72 (red solid curves), and approximate fiber length distributions (brown and blue dashed curves). Figure is adapted with permission from Ref. 184

fiber length distribution without introducing new errors. This empirical approach has been applied to examine a fragmentation-only model and matches perfectly with numerical solutions and experimental data for fiber-like PI264-b-PFS48 micelles under sonication as shown in Fig. 17.[184] In addition, similar procedure has been shown effective for a more complicated model, including primary nucleation, elongation and length-dependent fragmentation.[191]

4.4. *How to make a connection to experiments*

According to Nilsson,[192] there are three criteria that define a protein aggregate as an amyloid fibril: green birefringence upon staining with Congo Red, fibrillar morphology, and β-sheet secondary structure. Based on these three criteria, plenty of novel instruments and techniques have been developed to probe amyloid fibrils and their kinetics. Instead of providing a comprehensive review on each technique, like its basic setup, procedure and protocols, advantages and limitations, we focus on the relation between those physical quantities we can measure and mathematical quantities we have defined for characterizing the fibrillation kinetics.

The most widely adopted technique to monitor the fibrillation kinetics is Thioflavin T(ThT) fluorescence. When it binds to β-sheet-rich structures, like those in amyloid aggregates, the dye displays enhanced fluorescence and a characteristic red shift of its emission spectrum.[193,194] It has been shown that: (1) the ThT fluorescence intensity increases nearly linearly with the total amount of amyloid fibrils for several orders of magnitude; (2) the fluorescence intensity is independent of number concentration of amyloid fibrils if mass concentration is constant; (3) if the number concentration is fixed as in the extension kinetic study, the increase of average length of amyloid fibrils corresponds to an increase in the fluorescence intensity too.[195,196] These features make ThT fluorescence an unusually sensitive and efficient reporter for the mass concentration of aggregates $M(t)$, despite of limitations that ThT is not perfectly specific for amyloid fibrils, as well as some amyloid fibrils do not affect the fluorescence. In experiments, following empirical formula is often used to interpret the fluorescence intensity observed during amyloid fibrillation.[104]

$$F(t) = F(0) + \frac{A}{1 + \exp[-k_{\text{app}}(t - t_{1/2})]}, \tag{89}$$

where $F(t)$ is the fluorescence intensity at time t, $F(0)$ is the background fluorescence intensity at the starting time, and A is a fitting constant to correlate the mass concentration of aggregates with the absolute fluorescence intensity. Based on previous formula, we can easily extract the apparent fiber growth rate k_{app} and half-time of fiber formation $t_{1/2}$ from data fitting.

The goal of any absorption spectroscopy, for example, Fourier transform infrared (FTIR) spectroscopy, ultraviolet–visible (UV) spectroscopy, circular dichroism (CD), and so on, is to detect the presence of certain molecular structure by measuring the light absorbed at each wavelength. It has been established that, in FTIR, a peak near 1645 cm^{-1} indicates random coil, 1655 cm^{-1} for α-helix, and 1620–1640 cm^{-1} for β-sheet;[197] while in far-ultraviolet (170–250 nm) circular dichroism, a pronounced double minimum at 208 and 222 nm indicates α-helical structure, and a single minimum at 204 or 217 nm reflects random-coil or β-sheet structure.[198,199] Thus we can make a quantitative correlation between the fraction of certain structure (say β-sheet as an indicator of amyloid fibrils) with absorbed light intensity at the corresponding wavelength. Here we take the CD spectroscopy as a simple example, in which the fraction of β-sheet structure could be expressed through the measured ellipticity as follows[200]:

$$\frac{C_{\text{beta}}}{C_{\text{beta}} + C_{\text{coil}}} = \frac{\Theta_{217}^{\text{obs}} - \Theta_{217}^{0}}{\Theta_{217}^{\text{max}} - \Theta_{217}^{0}}, \tag{90}$$

where Θ_{217}^{0} is the ellipticity at 217 nm at the beginning of time; while $\Theta_{217}^{\text{obs}}$ and $\Theta_{217}^{\text{max}}$ represent observed and maximal ellipticity, respectively, at 217 nm during the whole measurement. C_{beta} and C_{coil} represent the concentration of β-sheet and random coil structures separately.

Instead of light absorption, dynamic light scattering (DLS) is a technique that can be used to determine the size distribution profile of small particles in suspension or polymers in solution.[201] Although light scattering by particles in solution depends on a number of factors, the most relevant one is the ratio of particle size with respect to quantity $\lambda/[2\pi \sin(\theta/2)]$, where λ is the wavelength of incident light and θ is the angle of detection.[133,202] As the average length of amyloid fibrils is expected to get larger during amyloid fibrillation, the intensity of light scattered by fibrils will be directly proportional to the mass concentration of aggregates.[62,203]

$$I_{\text{scat}}(t) = \frac{Q\lambda M(t)}{4l \sin(\theta/2)}, \tag{91}$$

where l is the monomer size, Q is a constant depending on the setup.

Currently, transmission electron microscopy (TEM), AFM and scanning-force microscopy (SFM) are most advanced techniques with highest resolution. According to,[204] the smallest distance that can be resolved with a TEM is approximately 0.2−0.5 nm and, for STM, a typical resolution of several tenths of one nanometer can be achieved. In comparison, the diameter of typical filaments is around 3–4 nm and 8–10 nm for mature fibrils, while the length may vary from hundreds of nanometers to a few micrometers.[205–207] Therefore, we may use those instruments to directly observe the growth, inhibition, propagation and adaptation of single fibril and even its breakage and branching in real time.[132,208–210] With high-resolution images in hand, the fiber length distribution could be extracted to certain accuracy, based on which both the number concentration and mass concentration of aggregates could be obtained. As a conclusion, advanced high-resolution microscopies could provide us most detailed information about amyloid fibrils, both their morphology and kinetics, though at a great expense of money and time.

4.5. *How to determine model parameters*

How to determine unknown parameters is a key step in model application. Although it is as important as modeling itself, details behind parameter fitting have seldom been clarified. A major obstacle is that usually most model parameters are empirical and hard to be precisely determined by either experiments or fundamental principles in nature. Their accuracy and validity heavily depends on the experience of modelers.

Here, we face with the same problem. In current study, the parameters, we adopted for describing amyloid fibrillation, cell damage and antibody inhibition in the main text, can be roughly divided into four groups.

The initial protein concentration m_{tot}, mass concentration M_0 and number concentration P_0 of initial seeds, total cell concentration c_{tot} and antibody concentration b_{tot} are generally prespecified in the setup. They are not adjustable during the fitting and can be classified into Group I.

Group II contains those application-insensitive parameters, like the critical nucleus size n_c, the monomer disassociation rate k_e^- and filaments annealing rate k_f^- in some cases. Since their values have a minor influence on the model performance, according to Occam's Razor, they can set to a default value (usually $n_c = 2$ and $k_e^- = k_f^- = 0$). In mathematics, sensitivity analysis allows a systematical determination of model parameters in group II.

Group III contains those parameters which can be determined either by experiments or by some fundamental principles. For example, as discussed in section of how fibrils grow, the fiber elongation rate k_e^+ for different amyloid proteins has been measured by plenty of techniques in recent years. We can directly take their values from the literature giving the same amyloid protein under similar experimental conditions.

The last group includes those freely adjustable model parameters. In most cases, the primary nucleation rate k_n, critical nucleus size n_2 and rate k_2 for surface-catalyzed secondary nucleation, fiber fragmentation rate k_f^+ are essential for modeling the fibrillation kinetics, and oligomer binding and unbinding rates k_b^+ and k_b^- for modeling cell damage. The first four parameters can be gotten by performing global fitting of amyloid formation, while the last two are determined through data on cytotoxicity (a sensitivity analysis of model parameters could be learned from Fig. 18). In this sense, there will be no free tunable parameter left.

4.6. *How to perform global fitting*

As claimed by John von Neumann, "With four parameters I can fit an elephant, and with five I can make him wiggle his trunk". An over-fit of experimental data with too many undetermined model parameters is encountered from time to time. Although a great effort has been dedicated to eliminate unnecessary model parameters as shown in last section, we are still facing with the problem how to perform a fitting reasonably and robustly. Global fitting, as suggested by its name, provides a nice way to partially solve this difficulty and make the fitting more reliable and promising.

The central idea of global fitting is to try to fit all data at the same time with the same parameters. In such a way, the redundancy in model parameters could be eliminated as much as possible and those key parameters will be highlighted. Although, the requirement of global fitting is quite natural from a theoretical view,

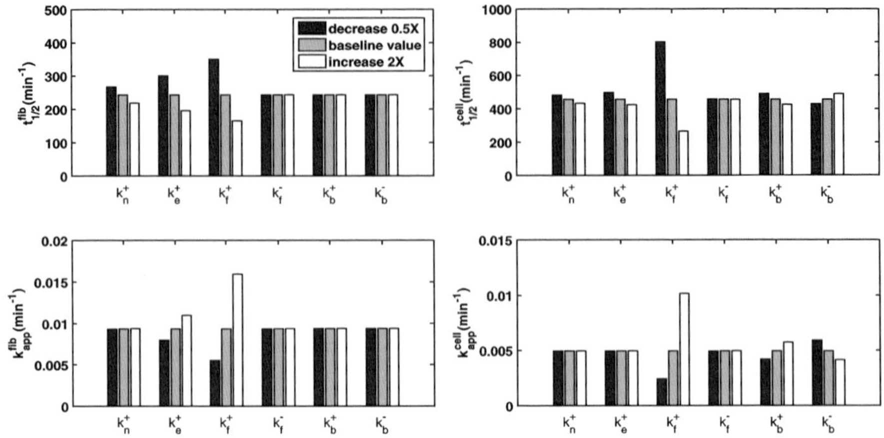

Figure 18: Model sensitivity on six adjustable parameters. The baseline values are taken in accordance with those in Fig. 13b. All six reaction rates are changed by either two times or one half with respect to their baseline values separately.

in practice it is not so easy to perform it. One obstacle is data noise. In the presence of large noise, a global fit becomes impossible, which means we have to either try best to eliminate noise source and perform measurements as precisely as possible, or include the influence of noise into modeling at the beginning (e.g., stochastic models). Another obstacle comes from the high dimensionality of the parameter space. Efficient global exploration methods, applicable to high dimensional space, have to be implemented during the procedure of fitting. In contrast, local exploration methods or those only valid for low dimensional space are not applicable. Finally, quantitative judgements or scores are needed in order to tell which group of model parameters gives the best fitting.

For this purpose, programs for global fitting have been developed for the kinetic models we have discussed in the main text.[211] Various global exploration methods, such as simulated annealing, genetic algorithm, particle swarm method, and so on, have been implemented into the program to avoid local trapping (private codes). Generally speaking, according to our own experience, simulated annealing is efficient and reliable in most cases; genetic algorithm consumes the longest CPU time among the three; while particle swarm is also quite slow, but in some cases it gives better results than simulated annealing. In these programs, nonlinear least-square regression is adopted to minimize the sum of squared errors between experimental data and those predicted by the model. Now we plan to take the influence of noises on global fitting into consideration. Related works are going on.

5. Conclusion

In the past decades, due to the increasing interest on amyloid related diseases, a variety of amyloid proteins and their fibrillation processes have been investigated in details. In this self-contained review, fruitful results on both thermodynamics and kinetics of amyloid fibrillation have been shown and discussed, with a purpose to provide a relatively comprehensive physical picture on what we know and what we do not know in this field. As a summary, facts we have learned from thermodynamic and kinetic modeling are:

1. The existence of a CFC below which fibrillar mass is negligible.
2. Above the CFC, the length distribution of the fibrils is exponential. This remains so even in the semidilute limit, if the fibril–fibril interactions are purely steric.
3. If the fibrils exhibit attractive lateral interactions, there will be a strong tendency for phase separation of fibrils.
4. Molecular mechanisms of amyloid fibrillation could be formulated into a group of microscopic chemical kinetic equations concerning with fiber length distribution.
5. Low-order moments, a function of fiber length distribution, evolves according to macroscopic chemical mass-action equations derived by moment-closure method.
6. Mass concentration and number concentration of aggregates are two most widely used moments and provide a well characterization of fibrils.
7. In many cases, fiber length distribution could be reconstructed from moments exactly or approximately once the underlying fibrillation kinetics is known.
8. Primary nucleation, elongation and secondary nucleation, including monomer-independent fragmentation and monomer-dependent surface catalyzed secondary nucleation, constitute a basic framework for amyloid fibrillation, which has been applied to many amyloid systems successfully.
9. Conformational conversion of monomeric, oligomeric and fibrillar structures is crucial for amyloid fibrillation but easily neglected in kinetic modeling due to other rate-limiting steps.
10. Effect of high monomer concentration on fiber elongation and surface catalyzed secondary nucleation could be explained by saturation.
11. Oligomers binding to lipid membrane play a key role in cytotoxicity and can easily include in kinetic models.
12. Quantitative knowledge on how to manipulate amyloid species and fibrillation kinetics could be learned from kinetic modeling, which provides us a power method to probe amyloidosis.

In our opinion, answering the following questions would constitute fruitful research direction:

1. How to incorporate various morphologies of amyloid fibrils into the thermodynamic picture?
2. How to construct a complete picture of phase separation and transition for amyloid fibrillation?
3. How to include oligomeric species explicitly in kinetic models?
4. How to model lateral association of protofibrils or filaments into mature fibrils, like coiling and twisting?
5. How to determine various reaction rate constants under a given fibrillation condition?
6. How to systematically quantify the effect of antibody and chaperon in order to manipulate amyloid fibrillation?
7. How to correlate amyloidosis with their fibrillation mechanisms at a molecular level?
8. How to probe amyloid diseases with kinetic modeling?

Acknowledgment

This work was supported by the National Natural Science Foundation of China (Grants 11204150), Tsinghua University Initiative Scientific Research Program (Grants 20151080424) and the program of China Scholarships Council (CSC). Ya Jing Huang also acknowledges the Postdoctoral Science Foundation of China (Grants 2015M581050).

List of Symbols

$[A_i]$	Concentration of filaments of size i
$[C_i]$	Concentration of cells in state i
$N_s(n_s)$	Number (concentration) of s-mer
$N_s^{(\beta)}(n_s^{(\beta)})$	Number (concentration) of monomers in the β-sheet configuration
$N_s^R(n_s^R)$	Number (concentration) of monomers in the random coil configuration
m	Monomer concentration
P	Number concentration of total aggregates

M	Mass concentration of total aggregates
$P_{\rm oli}$	Number concentration of oligomers
$M_{\rm oli}$	Mass concentration of oligomers
$m_{\rm tot}$	Total concentration of proteins
m_0	Initial monomer concentration
P_0	Initial number concentration of aggregates
M_0	Initial mass concentration of aggregates
n_c	Critical nucleus size for primary nucleation
n_2	Critical nucleus size for secondary nucleation
K_m	Critical saturation concentration for elongation
K_s	Critical saturation concentration for secondary nucleation
m_F^*	Critical fibrillar concentration
m_M^*	Critical micellar concentration
k_e^+	Rate constant for monomer association
k_e^-	Rate constant for monomer dissociation
k_n	Rate constant for primary nucleation
k_2	Rate constant for surface catalyzed secondary nucleation
k_f^+	Rate constant for filaments fragmentation
k_f^-	Rate constant for filaments annealing
k_c^+	Forward reaction rate constant for conformation conversion
k_c^-	Backward reaction rate constant for conformation conversion
k_b^+	Rate constant for membrane binding
k_b^-	Rate constant for membrane unbinding
$k_{\rm app}$	Apparent fiber growth rate
$k_{\rm max}$	Maximal fiber growth rate
$t_{1/2}$	Half-time for fibrillation
$t_{\rm lag}$	Lag-time for fibrillation

References

1. Dobson, C. M. Protein folding and misfolding. *Nature.* **426**(6968), 884–890 (2003).
2. Chiti, F. and Dobson, M. C. Protein misfolding, functional amyloid, and human disease. *Annu. Rev. Biochem.* **75**, 333–366 (2006).
3. Schreck, J. S. and Yuan, J. M. Statistical mechanical treatments of protein amyloid formation. *Int. J. Mol. Sci.* **14**(9), 17420–17452 (2013).
4. Morris, A. M., Watzky, M. A., and Finke, R. G. Protein aggregation kinetics, mechanism, and curve-fitting: a review of the literature. *Biochim. Biophys. Acta.* **1794**(3), 375–397 (2009).

5. Cohen, S. I., Vendruscolo, M., Dobson, C. M., and Knowles, T. P. From macroscopic measurements to microscopic mechanisms of protein aggregation. *J. Mol. Biol.* **421**(2), 160–171 (2012).
6. Gillam, J. E. and Mac Phee C. E. Modelling amyloid fibril formation kinetics: mechanisms of nucleation and growth. *J. Phys. Condens. Mat.* **25**(37), 373101 (2013).
7. Hong, L., Qi, X., and Zhang, Y. A lattice-gas model for amyloid fibril aggregation. *Europhys. Lett.* **94**(6), 68006 (2011).
8. Callen, H. B. *Thermodynamics and an Introduction to Thermostatistics*. 4.John Wiley & Sons, New York (1985).
9. Wang, Y., Wang, Y., Breed, D. R., Manoharan, V. N., Feng, L., Hollingsworth, A. D., Weck, M., and Pine D. J. Colloids with valence and specific directional bonding. *Nature*. **491**(7422), 51–55 (2012).
10. Lee, C. F. Self-assembly of protein amyloids: a competition between amorphous and ordered aggregation. *Phys. Rev. E*. **80**(3), 031922 (2009).
11. Huang, K. *Statistical Mechanics*, 2nd Ed. New York: John Wiley & Sons (1987).
12. Frenkel, D. Why colloidal systems can be described by statistical mechanics: some not very original comments on the gibbs paradox. *Mol. Phys.* **112**(17), 2325–2329 (2014).
13. Israelachvili, J. N. *Intermolecular and Surface Forces*, 3rd Ed. Academic Press, New York; 2011.
14. Cates, M. E. and Candau, S. J. Statics and dynamics of worm-like surfactant micelles. *J. Phys.* **2**(33), 6869 (1990).
15. van Gestel, J. and de Leeuw, S. W. A statistical–mechanical theory of fibril formation in dilute protein solutions. *Biophys. J.* **90**(9), 3134–3145 (2006).
16. Schmit, J. D., Ghosh, K., and Dill, K. What drives amyloid molecules to assemble into oligomers and fibrils? *Biophys. J.* **100**(2), 450–458 (2011).
17. Rogers, S. R., Venema, P., Sagis, L. M. C., van der, E., Linden., and Donald, A. M. Measuring the length distribution of a fibril system: a flow birefringence technique applied to amyloid fibrils. *Macromolecules.* **38**(7), 2948–2958 (2005).
18. Xue, W. F., Homans, S. W., and Radford, S. E. Amyloid fibril length distribution quantified by atomic force microscopy single-particle image analysis. *Protein Eng. Des. Sel.* **22**(8), 489–496 (2009).
19. Van Raaij, M. E., Gestel, J. V., Segers-Nolten, I. M. J., De Leeuw, S. W., and Subramaniam, V. Concentration dependence of α-synuclein fibril length assessed by quantitative atomic force microscopy and statistical-mechanical theory. *Biophys. J.* **95**(10), 4871–4878.
20. Morel, B., Varela, L. Azuaga, A. I., and Conejero-Lara, F. Environmental conditions affect the kinetics of nucleation of amyloid fibrils and determine their morphology. *Biophys. J.* **99**(11), 3801–3810 (2010).
21. Schoot, P. V. D. and Cates, M. E. Growth, static light scattering, and spontaneous ordering of rodlike micelles. *Langmuir.* **10**(3), 670–679 (1994).
22. de Gennes, P. D. and Prost, J. *The physics of Liquid Crystals* (International Series of Monographs on Physics 83). Oxford University Press, New York; 1995.
23. Doi, M. and Edwards, S. F. *The Theory of Polymer Dynamics*. Clarendon Press, Oxford (1986).

24. Corrigan, A. M., Müller, C., and Krebs, M. R. H. The formation of nematic liquid crystal phases by hen lysozyme amyloid fibrils. *J. Am. Chem. Soc.* **128**(46), 14740–14741 (2006).
25. Schoot, P. V. D. and Cates, M. E. The isotropic-to-nematic transition in semi-flexible micellar solutions. *EPL.* **25**(7), 515 (1994).
26. Odijk, T. Effect of micellar flexibility on the isotropic–nematic phase transition in solutions of linear aggregates. *J. Phys. France.* **48**(1), 125–129 (1987).
27. Lee, C. F., Loken, J., Jean, L., and Vaux D. J. Elongation dynamics of amyloid fibrils: a rugged energy landscape picture. *Phys. Rev. E.* **80**(4), 041906 (2009).
28. Lee, C. F. Isotropic–nematic phase transition in amyloid fibrilization. *Phys. Rev. E.* **80**(3),031902 (2009).
29. Barrat, J.-L. and Hansen, J. P. *Basic Concepts for Simple and Complex Liquids*. Cambridge University Press, New York; 2003.
30. Doi, M. *Introduction to Polymer Physics*. Oxford University Press, New York; 1996.
31. Saibil, H. R., Seybert, A., Habermann, A., Winkler, J., Eltsov, M., Perkovic, M., Castaño-Diez, D., Scheffer, M. P., Haselmann, U., and Chlanda, P. Heritable yeast prions have a highly organized three-dimensional architecture with interfiber structures. *Proc. Natl. Acad. Sci.* **109**(37), 14906–14911 (2012).
32. Lambert, M. P., Barlow, A. K., Chromy, B. A., Edwards, C., Freed, R., Liosatos, M., Morgan, T. E., Rozovsky, I., Trommer, B., and Viola K. L. Diffusible, nonfibrillar ligands derived from Aβ1–42 are potent central nervous system neurotoxins. *Proc. Natl. Acad. Sci.* **95**(11), 6448–6453 (1998).
33. Lomakin, A., Teplow, D. B., Kirschner, D. A., and Benedek, G. B. Kinetic theory of fibrillogenesis of amyloid β-protein. *Proc. Natl. Acad. Sci.* **94**(15), 7942–7947 (1997).
34. Šarić, A., Chebaro, Y. C., Knowles, T. P. J., and Frenkel, D. Crucial role of nonspecific interactions in amyloid nucleation. *Proc. Natl. Acad. Sci.* **111**(50), 17869–17874 (2014).
35. Petkova, A. T., Ishii, Y., Balbach, J. J., Antzutkin, O. N., Leapman, R. D., Delaglio, F., and Tycko, R. A structural model for Alzheimer's β-amyloid fibrils based on experimental constraints from solid state NMR. *Proc. Natl. Acad. Sci.* **99**(26), 16742–16747 (2002).
36. Hong, L. and Lei, J. Statistical mechanical model for helix-sheet-coil transitions in homopolypeptides. *Phys. Rev. E.* **78**(5), 051904 (2008).
37. Hong, L. A statistical mechanical model for antiparallelsheet/coil equilibrium. *J. Chem. Phys.* **129**(225101), 1–225101 (2008).
38. Usov, I., Adamcik, J., and Mezzenga, R. Polymorphism complexity and handedness inversion in serum albumin amyloid fibrils. *ACS Nano.* **7**(12), 10465–10474 (2013).
39. Whitesides, G. M. and Laibinis, P. E. Wet chemical approaches to the characterization of organic surfaces: self-assembled monolayers, wetting, and the physical-organic chemistry of the solid–liquid interface. *Langmuir.* **6**(1), 87–96 (1990).
40. Ulman, A. Formation and structure of self-assembled monolayers. *Chem. Rev.* **96**(4), 1533–1554 (1996).
41. Schreiber, F. Structure and growth of self-assembling monolayers. *Prog. Surf. Sci.* **65**(5), 151–257 (2000).

42. Ulman, A. *An Introduction to Ultrathin Organic Films: From Langmuir–Blodgett to Self–Assembly*. Academic Press, Boston; 2013.
43. Oosawa, F. and Asakura, S. *Thermodynamics of the Polymerization of Protein*. Academic Press, New York, 1975.
44. Collins, S. R., Douglass, A., Vale, R. D., Weissman, J. S. Mechanism of prion propagation: amyloid growth occurs by monomer addition. *PLoS Biol.* **2**, 1582–1590 (2004).
45. Oosawa, F., Asakura, S., Hotta, K., Imai, N., and Ooi, T. G-F transformation of actin as a fibrous condensation. *J. Polym. Sci.* **37**(132), 323–336 (1959).
46. Knowles, T. P. J., Shu, W., Devlin, G. L., Meehan, S., Auer, S., Dobson, C. M., and Welland, M. E. Kinetics and thermodynamics of amyloid formation from direct measurements of fluctuations in fibril mass. *Proc. Natl. Acad. Sci.* **104**(24), 10016–10021 (2007).
47. Buell, A. K., Blundell, J. R., Dobson, C. M., Welland, M. E., Terentjev, E. M., and Knowles, T. P. J. Frequency factors in a landscape model of filamentous protein aggregation. *Phys. Rev. Lett.* **104**(22), 228101 (2010).
48. Serio, T. R., Cashikar, A. G., Kowal, A. S., Sawicki, G. J., Moslehi, J. J., Serpell, L., Arnsdorf, M. F., and Lindquist, S. L. Nucleated conformational conversion and the replication of conformational information by a prion determinant. *Science.* **289**(5483), 1317–1321 (2000).
49. Souillac, P. O., Uversky, V. N., and Fink, A. L. Structural transformations of oligomeric intermediates in the fibrillation of the immunoglobulin light chain LEN. *Biochemistry.* **42**(26), 8094–8104 (2003).
50. Caughey, B. and Lansbury, P. T. Jr. Protofibrils, pores, fibrils, and neurodegeneration: separating the responsible protein aggregates from the innocent bystanders. *Annu. Rev. Neurosci.* **26**(1), 267–298 (2003).
51. Bitan, G., Kirkitadze, M. D., Lomakin, A., Vollers, S. S., Benedek, B., and Teplow, D. B. Amyloid beta-protein (abeta) assembly: abeta40 and abeta42 oligomerize through distinct pathways. *Proc. Natl. Acad. Sci.* **100**(1), 330–335 (2003).
52. O'Nuallain, B., Shivaprasad, S., Kheterpal, I., and Wetzel, R. Thermodynamics of Aβ(1-40) amyloid fibril elongation. *Biochemistry.* **44**(38), 12709–12718 (2005).
53. Buell, A. K., Dobson, C. M., and Welland, M. E. Measuring the kinetics of amyloid fibril elongation using quartz crystal microbalances. *Methods Mol. Biol.* **849**, 101–119 (2012).
54. Kellermayer, M. S., Karsai, A., Benke, M., Soós, K., and Penke, B. Stepwise dynamics of epitaxially growing single amyloid fibrils. *Proc. Natl. Acad. Sci.* **105**(1), 141–144 (2008).
55. Pinotsi, D., Buell, A. K., Galvagnion, C., Dobson, C. M., Schierle, G. S. K., and Kaminski, C. F. Direct observation of heterogeneous amyloid fibril growth kinetics via two-color super-resolution microscopy. *Nano Lett.* **14**(1), 339–345 (2013).
56. Ferkinghoff-Borg, J., Fonslet, J., Andersen, C. B., Krishna, S., Pigolotti, S., Yagi, H., Goto, Y., Otzen, D., and Jensen, M. H. Stop-and-go kinetics in amyloid fibrillation. *Phys. Rev. E.* **82**(1), 010901 (2010).
57. Wördehoff, M. M., Bannach, O., Shaykhalishahi, H., Kulawik, A., Schiefer, S., Willbold, D., Hoyer, W., and Birkmann E. Single fibril growth kinetics of α-synuclein. *J. Mol. Biol.* **427**(6), 1428–1435 (2015).

58. Knowles, T. P., Waudby, A., Devlin, G. L., Cohen, S. I. A., Aguzzi, A., Vendruscolo, M., Terentjev, E. M., Welland, M. E., and Dobson, C. M. An analytical solution to the kinetics of breakable filament assembly. *Science.* **326**(5959), 1533–1537 (2009).
59. Knowles, T. P. J., White, D. A., Abate, A. R., Agresti, J. J., Cohen, S. I. A., Sperling, R. A., Genst, E. J. D., Dobson, C. M., and Weitz, D. A. Observation of spatial propagation of amyloid assembly from single nuclei. *Proc. Natl. Acad. Sci.* **108**(36), 14746–14751 (2011).
60. Kodaka, M. Interpretation of concentration-dependence in aggregation kinetics. *Biophys. Chem.* **109**(2), 325–332 (2004).
61. Kodali, R. and Wetzel, R. Polymorphism in the intermediates and products of amyloid assembly. *Curr. Opin. Struct. Biol.* **17**(1), 48–57 (2007).
62. Powers, E. T. and Powers, D. L. Mechanisms of protein fibril formation: nucleated polymerization with competing off-pathway aggregation. *Biophys. J.* **94**(2), 379–391 (2008).
63. Van der Waals, J. D., Threfall, R., Adair, J. F., and Roth, F. *The Continuity of the Liquid and Gaseous States.* In: Physical memoirs, selected and translated from foreign sources, under the direction of the Physical Society of London. Taylor and Francis, London, 1890.
64. Becker, R. and Döring, W. The kinetic treatment of nuclear formation in supersaturated vapors. *Ann. Phys.* **24**(719), 752 (1935).
65. Frenkel, I. L. *Kinetic Theory of Liquids.* Dover Publications, New York; 1955.
66. Lothe, J. and Pound, G. M. Concentration of clusters in nucleation and the classical phase integral. *J. Chem. Phys.* **48**(4), 1849–1852 (1968).
67. Reiss, H., Katz, J. L., and Cohen, E. R. Translation–rotation paradox in the theory of nucleation. *J. Chem. Phys.* **48**(12), 5553–5560 (1968).
68. Kikuchi, R. The translation–rotation paradox in the nucleation theory. *J. Stat. Phys.* **1**(2), 351–375 (1969).
69. Shi, Y., Zhou, W., Lu, A. Y., Fang, W., Lee, Y. H., Hsu, A. L., Kim, S. M., Kim, K. K., Yang, H. Y., and Li, L. J. van der Waals epitaxy of MoS_2 layers using graphene as growth templates. *Nano Lett.* **12**(6), 2784–2791 (2012).
70. Abraham, F. F. *Homogeneous Nucleation: The Pretransition Theory of Vapor Condensation.* Academic Press, New York, 1974.
71. Debenedetti, P. G. *Metastable Liquids: Concepts and Principles.* Princeton University Press, Princeton; 1996.
72. Sear, R. P. Quantitative studies of crystal nucleation at constant supersaturation: experimental data and models. *Cryst. Eng. Comm.* **16**(29), 6506–6522 (2014).
73. Fletcher, N. H. J. Size effect in heterogeneous nucleation. *J. Phys. Chem.* **29**(3), 572–576 (1958).
74. Hienola, A. I., Winkler, P. M., Wagner, P. E., Vehkamäki, H., Lauri, A., Napari, I., and Kulmala, M. Estimation of line tension and contact angle from heterogeneous nucleation experimental data. *J. Phys. Chem.* **126**(9), 094705 (2007).
75. Tavare, N. S. *Industrial Crystallization: Process Simulation Analysis and Design.* New York: Plenum Press; 1995.
76. Oosawa, F. and Kasai, M. A theory of linear and helical aggregations of macromolecules. *J. Mol. Biol.* **4**(1), 10–21 (1962).

77. Wegner, A. Spontaneous fragmentation of actin filaments in physiological conditions. *Nature*. **296**(5854), 266–267 (1982).
78. Wang, Y. Petty, S., Trojanowski, A., Knee, K., Goulet, D., Mukerji, I., and King, J. Formation of amyloid fibrils in vitro from partially unfolded intermediates of human γc-crystallin. *Invest. Ophthalmol. Vis. Sci.* **51**(2), 672–678 (2010).
79. Binger, K. J., Pham, C. L. L., Wilson, L. M., Bailey, M. F., Lawrence, L. J., Schuck, P., and Howlett, G. J. Apolipoprotein C-II amyloid fibrils assemble via a reversible pathway that includes fibril breaking and rejoining. *J. Mol. Biol.* **376**(4), 1116–1129 (2008).
80. Cohen, S. I. A., Vendruscolo, M., Dobson, C. M., and Knowles, T. P. J. Nucleated polymerisation in the presence of pre-formed seed filaments. *Int. J. Mol. Sci.* **12**(9), 5844–5852 (2011).
81. Wegner, A. and Savko, P. Fragmentation of actin filaments. *Biochemistry.* **21**(8), 1909–1913 (1982).
82. Rochet, J. C. and Lansbury, P. T. Amyloid fibrillogenesis: hemes and variations. *Curr. Opin. Struct. Biol.* **10**(1), 60–68 (2000).
83. Ferrone, F. A., Hofrichter, J., Sunshine, H. R., and Eaton, W. A. Kinetic studies on photolysis-induced gelation of sickle cell hemoglobin suggest a new mechanism. *Biophys. J.* **32**(1), 361 (1980).
84. Ferrone, F. A., Hofrichter, J., and Eaton, W. A. Kinetics of sickle hemoglobin polymerization: I. Studies using temperature-jump and laser photolysis techniques. *J. Mol. Biol.* **183**(4), 591–610 (1985).
85. Ferrone, F. A., Hofrichter, J., and Eaton, W. A. Kinetics of sickle hemoglobin polymerization. II. A double nucleation mechanism. *J. Mol. Biol.* **183**(4), 611–631 (1985).
86. Padrick, S. B. and Miranker, A. D. Islet amyloid: phase partitioning and secondary nucleation are central to the mechanism of fibrillogenesis. *Biochemistry.* **41**(14), 4694–4703 (2002).
87. Ruschak, A. M. and Miranker, A. D. Fiber-dependent amyloid formation as catalysis of an existing reaction pathway. *Proc. Natl. Acad. Sci.* **104**(30), 12341–12346 (2007).
88. Meisl, G., Yang, X., Hellstrand, E., Frohm, B., Kirkegaard, J. B., Cohen, S. I. A., Dobson, C. M., Linse, S., and Knowles, T. P. J. Differences in nucleation behavior underlie the contrasting aggregation kinetics of the Aβ40 and Aβ42 peptides. *Proc. Natl. Acad. Sci.* **111**(26), 9384–9389 (2014).
89. Hellstrand, E., Boland, B., Walsh, D. M., and Linse, S. Amyloid β-protein aggregation produces highly reproducible kinetic data and occurs by a two-phase process. *ACS Chem. Neurosci.* **1**(1), 13–18 (2009).
90. Cohen, S. I. A., Linse, S., Luheshi, L. M., Hellstrand, E., White, D. A., Rajah, L., Otzen, D. E., Vendruscolo, M., Dobson, C. M., and Knowles, T. P. J. Proliferation of amyloid-β42 aggregates occurs through a secondary nucleation mechanism. *Proc. Natl. Acad. Sci.* **110**(24), 9758–9763 (2013).
91. Granas, A. and Dugundji, J. *Fixed Point Theory*. Springer Science & Business Media, Berlin; 2013.
92. Cohen, S. I. A., Vendruscolo, M., Dobson, C. M., and Knowles, T. P. J. Nucleated polymerization with secondary pathways. II. Determination of self-consistent solutions to growth processes described by non-linear master equations. *J. Chem. Phys.* **135**(6), 065106 (2011).

93. Ferrone, F. Analysis of protein aggregation kinetics. *Meth. Enzymol.* **309**, 256–274 (1999).
94. Sanders, J. E., Goldstein, B. S., and Leotta, D. F. Skin response to mechanical stress: adaptation rather than breakdown — a review of the literature. *J. Rehabil. Res. Dev.* **32**, 214–226 (1995).
95. Lee, C. F. Thermal breakage of a discrete one-dimensional string. *Phys. Rev. E.* **80**, 31134 (2009).
96. Loveday, S. M., Wang, X. L., Rao, M. A., Anema, S. G., and Singh, H. β-lactoglobulin nanofibrils: effect of temperature on fibril formation kinetics, fibril morphology and the rheological properties of fibril dispersions. *Food Hydrocoll.* **27**(1), 242–249 (2012).
97. Lee, C. F. Thermal breakage of a semiflexible polymer: breakage profile and rate. *J. Phys. Condens. Matter.* **27**(27), 275101 (2015).
98. Tessarz, P., Mogk, A., and Bukau, B. Substrate threading through the central pore of the Hsp104 chaperone as a common mechanism for protein disaggregation and prion propagation. *Mol. Microbiol.* **68**(1), 87–97 (2008).
99. Zhu, L., Zhang, X. J., Wang, L. Y., Zhou, J. M., and Perrett, S. Relationship between stability of folding intermediates and amyloid formation for the yeast prion Ure2p: a quantitative analysis of the effects of pH and buffer system. *J. Mol. Biol.* **328**(1), 235–254 (2003).
100. Hammer, N. D., Schmidt, J. C., and Chapman, M. R. The curli nucleator protein, CsgB, contains an amyloidogenic domain that directs CsgA polymerization. *Proc. Natl. Acad. Sci.* **104**(30), 12494–12499 (2007).
101. Škerget, K., Vilfan, A., Pompe-Novak, M., Turk, V., Waltho, J. P., Turk, D., and Žerovnik, E. The mechanism of amyloid-fibril formation by stefin B: temperature and protein concentration dependence of the rates. *Proteins.* **74**(2), 425–436 (2009).
102. Xue, W.-F., Homans, S. W. and Radford, S. E. Systematic analysis of nucleation-dependent polymerization reveals new insights into the mechanism of amyloid self-assembly. *Proc. Natl. Acad. Sci.* **105**(26), 8926–8931 (2008).
103. Ferguson, N., Berriman, J., Petrovich, M., Sharpe, T. D., Finch, J. T. and Fersht, A. R. Rapid amyloid fiber formation from the fast-folding WW domain FBP28. *Proc. Natl. Acad. Sci.* **100**(17), 9814–9819 (2003).
104. Uversky, V. N., Li, J. and Fink, A. L. Metal-triggered structural transformations, aggregation, and fibrillation of human α-synuclein a possible molecular link between Parkinson's disease and heavy metal exposure. *J. Biol. Chem.* **276**(47), 44284–44296 (2001).
105. Nielsen, L., Khurana, R., Coats, A., Frokjaer, S., Brange, J., Vyas, S., Uversky, V. N. and Fink, A. L. Effect of environmental factors on the kinetics of insulin fibril formation: elucidation of the molecular mechanism. *Biochemistry.* **40**(20), 6036–6046 (2001).
106. Ballauff, M. and Wolf, B. A. Degradation of chain molecules. 1. Exact solution of the kinetic equations. *Macromolecules.* **14**(3), 654–658 (1981).
107. Ballauff, M. and Wolf, B. A. Degradation of chain molecules. 2. Thermodynamically induced shear degradation of dissolved polystyrene. *Macromolecules.* **17**(2), 209–216 (1984).

108. Tanigawa, M., Suzuto, M., Fukudome, K., and Yamaoka, K. Changes in molecular weights and molecular weight distributions of differently stranded nucleic acids after sonication: gel permeation chromatography/low angle laser light scattering evaluation and computer simulation. *Macromolecules.* **29**(23), 7418–7425 (1996).
109. Guérin, G., Wang, H., Manners, I., and Winnik, M. A. Fragmentation of fiberlike structures: sonication studies of cylindrical block copolymer micelles and behavioral comparisons to biological fibrils. *J. Am. Chem. Soc.* **130**(44), 14763–14771 (2008).
110. Hill, T. L. Length dependence of rate constants for end-to-end association and dissociation of equilibrium linear aggregates. *Biophys. J.* **44**(2), 285 (1983).
111. Hong, L. and Yong, W.-A. Simple moment-closure model for the self-assembly of breakable amyloid filaments. *Biophys. J.* **104**(3), 533–540 (2013).
112. Schreck, J. S. and Yuan, J.-M. A kinetic study of amyloid formation: fibril growth and length distributions. *J. Phys. Chem. B*, **117**(21), 6574–6583 (2013).
113. Jaynes, E. T. Information theory and statistical mechanics. *Phys. Rev.* **106**(4), 620 (1957).
114. Jaynes, E. T. On the rationale of maximum-entropy methods. *Proc. IEEE.* **70**(9), 939–952 (1982).
115. Pressé, S., Ghosh, K., Lee, J., and Dill, K. A. Principles of maximum entropy and maximum caliber in statistical physics. *Rev. Mod. Phys.* **85**(3), 1115 (2013).
116. Michaels. T. C. T. and Knowles, T. P. J. Role of filament annealing in the kinetics and thermodynamics of nucleated polymerization. *J. Chem. Phys.* **140**(21), 214904 (2014).
117. Esler, W. P., Stimson, E. R., Jennings, J. M., Vinters, H. V., Ghilardi, J. R., Lee, J. P., Mantyh, P. W., and Maggio, J. E. Alzheimer's disease amyloid propagation by a template-dependent dock-lock mechanism. *Biochemistry.* **39**(21), 6288–6295 (2000).
118. Nguyen, P. H., Li, M. S., Stock, G., Straub, J. E., and Thirumalai, D. Monomer adds to preformed structured oligomers of Aβ-peptides by a two-stage dock–lock mechanism. *Proc. Nat. Acad. Sci.* **104**(1), 111–116 (2007).
119. Scheibel, T., Bloom, J., and Lindquist, S. L. The elongation of yeast prion fibers involves separable steps of association and conversion. *Proc. Natl. Acad. Sci. U. S. A.* **101**(8), 2287–2292 (2004).
120. Bodenstein, M. and Lutkemeyer, H. Quasi-steady state assumption. *Z. Phys. Chem.* **114**, 208–236 (1924).
121. Benson, S. W., *et al. Foundations of Chemical Kinetics.* McGraw-Hill, New York, 1960.
122. Michaelis, L. and Menten, M. L. Die kinetik der invertinwirkung. *Biochem. Z.* **49**(333–369), 352 (1913).
123. Buell, A. K., Galvagnion, C., Gaspar, R., Sparr, E., Vendruscolo, M., Knowles, T. P. J., Linse, S., and Dobson, C. M. Solution conditions determine the relative importance of nucleation and growth processes in α-synuclein aggregation. *Proc. Natl. Acad. Sci.* **111**(21), 7671–7676 (2014).
124. Sunde, M. and Blake, C. C. F. From the globular to the fibrous state: protein structure and structural conversion in amyloid formation. *Q. Rev. Biophys.* **31**(01), 1–39 (1998).
125. Prusiner, S. B. Novel proteinaceous infectious particles cause scrapie. *Science.* **216**(4542), 136–144 (1982).

126. Kelly, J. W. The alternative conformations of amyloidogenic proteins and their multi-step assembly pathways. *Curr. Opin. Struct. Biol.* 8(1), 101–106 (1998).
127. Xing, Y., Nakamura, A., Korenaga, T., Guo, Z., Yao, J., Fu, F., Matsushita, T., Kogishi, K., Hosokawa, M., Kametani, F., *et al.* Induction of protein conformational change in mouse senile amyloidosis. *J. Biol. Chem.* 277(36), 33164–33169 (2002).
128. Hong, L., Qi, X., and Zhang. Dissecting the kinetic process of amyloid fiber formation through asymptotic analysis. *J. Phys. Chem. B.* **116**(23), 6611–6617 (2011).
129. Peterson, S. A., Klabunde, T., Lashuel, H. A., Purkey, H., Sacchettini, J. C., and Kelly, J. W. Inhibiting transthyretin conformational changes that lead to amyloid fibril formation. *Proc. Natl. Acad. Sci.* **95**(22), 12956–12960 (1998).
130. Shammas, S. L., Garcia, G. A., Kumar, S., Kjaergaard, M., Horrocks, M. H., Shivji, N., Mandelkow, E., Knowles, T. P. J., Mandelkow, E., and Klenerman, D. A mechanistic model of tau amyloid aggregation based on direct observation of oligomers. *Nat. Commun.* **6**, 7025 (2015).
131. Lee, C. C., Nayak, A., Sethuraman, A., Belfort, G., and McRae, G. J. A three-stage kinetic model of amyloid fibrillation. *Biophys. J.* **92**(10), 3448–3458 (2007).
132. Andersen, C. B., Yagi, H., Manno, M., Martorana, V., Ban, T., Christiansen, G., Otzen, D. E., Goto, Y., and Rischel, C. Branching in amyloid fibril growth. *Biophys. J.* **96**(4), 1529–1536 (2009).
133. Pallitto, M. M. and Murphy, R. M. A mathematical model of the kinetics of β-amyloid fibril growth from the denatured state. *Biophys. J.* **81**(3), 1805–1822 (2001).
134. Dobson, C. M. Principles of protein folding, misfolding and aggregation. In *Seminars in Cell & Developmental Biology*, Vol. 15. Elsevier, New york; 2004. pp. 3–16.
135. Luheshi, L. M., Crowther, D. C., and Dobson, C. M. Protein misfolding and disease: from the test tube to the organism. *Curr. Opin. Chem. Biol.* **12**(1), 25–31 (2008).
136. Munishkina, L. A., Cooper, E. M., Uversky, V. N., and Fink, A. L. The effect of macromolecular crowding on protein aggregation and amyloid fibril formation. *J. Mol. Recgn.* **17**(5), 456–464 (2004).
137. Magno, A., Caflisch, A., and Pellarin, R. Crowding effects on amyloid aggregation kinetics. *J. Phys. Chem. Lett.* 1(20), 3027–3032 (2010).
138. Hung, A. Y., Haass, C., Nitsch, R. M., Qiu, W. Q., Citron, M., Wurtman, R. J., Growdon, J. H., and Selkoe, D. J. Activation of protein kinase c inhibits cellular production of the amyloid beta-protein. *J. Biol. Chem.* **268**(31), 22959–22962 (1993).
139. Wang, D. S., Dickson, D. W., and Malter, J. S. β-Amyloid degradation and Alzheimer's disease. *BioMed. Res. Intl.* **2006**, 58406 (2006).
140. Buxbaum, J. D., Thinakaran, G., Koliatsos, V., OCallahan, J., Slunt, H. H., Price, D. L., and Sisodia, S. S. Alzheimer amyloid protein precursor in the rat hippocampus: transport and processing through the perforant path. *J. Neurosci.* **18**(23), 9629–9637 (1998).
141. Gunawardena, S. and Goldstein, L. S. B. Disruption of axonal transport and neuronal viability by amyloid precursor protein mutations in drosophila. *Neuron.* **32**(3), 389–401 (2001).
142. Haass, C. and Selkoe, D. J. Soluble protein oligomers in neurodegeneration: lessons from the Alzheimer's amyloid β-peptide. *Nat. Rev. Mol. Cell Biol.* 8(2), 101–112 (2007).

143. Simone, A. D., Esposito, L., Pedone, C., and Vitagliano, L. Insights into stability and toxicity of amyloid-like oligomers by replica exchange molecular dynamics analyses. *Biophys. J.* **95**(4), 1965–1973 (2008).
144. Walsh, D. M. and Selkoe, D. J. Aβ oligomers — A decade of discovery. *J. Neurochem.* **101**(5), 1172–1184 (2007).
145. Fändrich, M. Oligomeric intermediates in amyloid formation: structure determination and mechanisms of toxicity. *J. Mol. Biol.* **421**(4), 427–440 (2012).
146. Stefani, M. The oligomer species: mechanistics and biochemistry. In *Amyloid Fibrils and Prefibrillar Aggregates: Molecular and Biological Properties*, Otzen D.E. (Ed.) John Wiley & Sons, New York, pp. 127–150, 2013.
147. Small, D. H., Mok, S. S., and Bornstein, J. C. Alzheimer's disease and Aβ toxicity: from top to bottom. *Nat. Rev. Neurosci.* **2**(8), 595–598 (2001).
148. Ono, K., Condron, M. M., and Teplow, D. B. Structure–neurotoxicity relationships of amyloid β-protein oligomers. *Proc. Natl. Acad. Sci.* **106**(35), 14745–14750 (2009).
149. Wong, P. T., Schauerte, J. A., Wisser, K. C., Ding, H., Lee, E. L., Steel, D. G., and Gafni, A. Amyloid-β membrane binding and permeabilization are distinct processes influenced separately by membrane charge and fluidity. *J. Mol. Biol.* **386**(1), 81–96 (2009).
150. Williams, T. L. and Serpell, L. C. Membrane and surface interactions of Alzheimers Aβ peptide—Insights into the mechanism of cytotoxicity. *FEBS J.* **278**(20), 3905–3917 (2011).
151. Capone, R., Quiroz, F. G., Prangkio, P., Saluja, I., Sauer, A. M., Bautista, M. R., Turner, R. S., Yang, J., and Mayer, M. Amyloid-β-induced ion flux in artificial lipid bilayers and neuronal cells: resolving a controversy. *Neurotox. Res.* **16**(1), 1–13 (2009).
152. Kawahara, M. Neurotoxicity of β-amyloid protein: oligomerization, channel formation and calcium dyshomeostasis. *Curr. Pharm. Des.* **16**(25), 2779–2789 (2010).
153. Engel, M. F. M., Khemtémourian, L., Kleijer, C. C., Meeldijk, H. J. D., Jacobs, J., Verkleij, A. J., de Kruijff, B., Killian, J. A., and Häppener, J. W. M. Membrane damage by human islet amyloid polypeptide through fibril growth at the membrane. *Proc. Natl. Acad. Sci.* **105**(16), 6033–6038 (2008).
154. Xue, W.-F., Hellewell, A. L., Gosal, W. S., Homans, S. W., Hewitt, E. W., and Radford, S. E. Fibril fragmentation enhances amyloid cytotoxicity. *J. Biol. Chem.* **284**(49), 34272–34282 (2009).
155. Demuro, A., Smith, M., and Parker, I. Single-channel Ca^{2+} imaging implicates Aβ1–42 amyloid pores in Alzheimers disease pathology. *J. Cell Biol.* **195**(3), 515–524 (2011).
156. Schauerte, J. A., Wong, P. T., Wisser, K. C., Ding, H., Steel, D. G., and Gafni, A. Simultaneous single-molecule fluorescence and conductivity studies reveal distinct classes of Aβ species on lipid bilayers. *Biochemistry.* **49**(14), 3031–3039 (2010).
157. Jang, H., Arce, F. T., Capone, R., Ramachandran, S., Lal, R., and Nussinov, R. Misfolded amyloid ion channels present mobile β-sheet subunits in contrast to conventional ion channels. *Biophys. J.* **97**(11), 3029–3037 (2009).
158. Quist, A., Doudevski, I., Lin, H., Azimova, R., Ng, D., Frangione, B., Kagan, B., Ghiso, J., and Lal, R. Amyloid ion channels: a common structural link for protein-misfolding disease. *Proc. Natl. Acad. Sci. U. S. A.* **102**(30), 10427–10432 (2005).

159. Lin, H., Bhatia, R., and Lal, R. Amyloid β protein forms ion channels: implications for Alzheimer's disease pathophysiology. *FASEB J.* **15**(13), 2433–2444 (2001).
160. Shafrir, Y., Durell, S., Arispe, N., and Guy, H. R. Models of membrane-bound Alzheimer's Abeta peptide assemblies. *Proteins.* **78**(16), 3473–3487 (2010).
161. Bolognesi, B., Kumita, J. R., Barros, T. P., Esbjorner, E. K., Luheshi, L. M., Crowther, D. C., Wilson, M. R., Dobson, C. M., Favrin, G., and Yerbury, J. J. ANS binding reveals common features of cytotoxic amyloid species. *ACS Chem. Biol.* **5**(8), 735–740 (2010).
162. Olzscha, H., Schermann, S. M., Woerner, A. C., Pinkert, S., Hecht, M. H., Tartaglia, G. G., Vendruscolo, M., Hayer-Hartl, M., Hartl, F. U., and Vabulas, R. B. Amyloid-like aggregates sequester numerous metastable proteins with essential cellular functions. *Cell.* **144**(1), 67–78 (2011).
163. Querfurth, H. W. and LaFerla, F. M. Mechanisms of disease. *N. Engl. J. Med.* **362**(4), 329–344 (2010).
164. Demuro, A., Mina, E., Kayed, R., Milton, S. C., Parker, I., and Glabe, C. G. Calcium dysregulation and membrane disruption as a ubiquitous neurotoxic mechanism of soluble amyloid oligomers. *J. Biol. Chem.* **280**(17), 17294–17300 (2005).
165. Holtzman, D. M., Goate, A., Kelly, J., and Sperling, R. Mapping the road forward in Alzheimers disease. *Sci. Trans. Med.* 3(114), 114ps48 (2011).
166. Selkoe, D. J. Resolving controversies on the path to Alzheimer's therapeutics. *Nat. Med.* **17**(9), 1060–1065 (2011).
167. Gong, Y., Chang, L., Viola, K. L., Lacor, P. N., Lambert, M. P., Finch, C. E., Krafft, G. A., and Klein, W. L. Alzheimer's disease-affected brain: presence of oligomeric Aβ ligands (addls) suggests a molecular basis for reversible memory loss. *Proc. Natl. Acad. Sci.* **100**(18), 10417–10422 (2003).
168. LaFerla, F. M. Calcium dyshomeostasis and intracellular signalling in Alzheimer's disease. *Nat. Rev. Neurosci.* **3**(11), 862–872 (2002).
169. Turner, P. R., OConnor, K., Tate, W. P., and Abraham, W. C. Roles of amyloid precursor protein and its fragments in regulating neural activity, plasticity and memory. *Prog. Neurobiol.* **70**(1), 1–32 (2003).
170. Hong, L., Huang, Y.-J., and Yong, W.-A. A kinetic model for cell damage caused by oligomer formation. *Biophys. J.* **109**, 1338–1346, 2015.
171. Gregory S. M., Cavenaugh, A., Journigan, V., Pokorny, A., and Almeida, P. F. F. A quantitative model for the all-or-none permeabilization of phospholipid vesicles by the antimicrobial peptide Cecropin A. *Biophys. J.* **94**(5), 1667–1680 (2008).
172. Gregory, S. M., Pokorny, A., and Almeida, P. P. F. Magainin 2 revisited: a test of the quantitative model for the all-or-none permeabilization of phospholipid vesicles. *Biophy. J.* **96**(1), 116–131 (2009).
173. Eyring, H. The activated complex in chemical reactions. *J. Chem. Phys.* **3**(2), 107–115 (1935).
174. Anslyn, E. V. and Dougherty, D. A. *Modern Physical Organic Chemistry*. University Science Books, John Wiley & Sons, New York; 2006.
175. Cabaleiro-Lago, C., Quinlan-Pluck, F., Lynch, I., Lindman, S., Minogue, A. M., Thulin, E., Walsh, D. M., Dawson, K. A., and Linse, S. Inhibition of amyloid β protein fibrillation by polymeric nanoparticles. *J. Am. Chem. Soc.* **130**(46), 15437–15443 (2008).

176. Yoo, S. I., Yang, M., Brender, J. R., Subramanian, V., Sun, K., Joo, N. E., Jeong, S.-H., Ramamoorthy, A., and Kotov, N. A. Inhibition of amyloid peptide fibrillation by inorganic nanoparticles: functional similarities with proteins. *Angew. Chem. Int. Ed.* **50**(22), 5110–5115 (2011).
177. Liao, Y.-H., Chang, Y.-J., Yoshiike, Y., Chang, Y.-C., and Chen, Y.-R. Negatively charged gold nanoparticles inhibit Alzheimer's amyloid-β fibrillization, induce fibril dissociation, and mitigate neurotoxicity. *Small.* **8**(23), 3631–3639 (2012).
178. Huang, C., Cheng, H., Hao, S., Zhou, H., Zhang, X., Gao, J., Sun, Q.-H., Hu, H., and Wang, C. Heat shock protein 70 inhibits α-synuclein fibril formation via interactions with diverse intermediates. *J. Mol. Biol.* **364**(3), 323–336 (2006).
179. Liberek, K., Lewandowska, A., and Zietkiewicz, S. Chaperones in control of protein disaggregation. *EMBO J.* **27**(2), 328–335 (2008).
180. Xu, L.-Q., Wu, S., Buell, A. K., Cohen, S. I. A., Chen, L.-J., Hu, W.-H., Cusack, S. A., Itzhaki, L. S., Zhang, H., Knowles, T. P. J., *et al.* Influence of specific HSP70 domains on fibril formation of the yeast prion protein Ure2. *Philos. Trans. R. Soc. B.* **368**(1617), 20110410 (2013).
181. Cohen, S. I. A., Arosio, P., Presto, P., Kurudenkandy, F. R., Biverstal, H., Dolfe, L., Dunning, C., Yang, X., Frohm, B., Vendruscolo, M., *et al.* A molecular chaperone breaks the catalytic cycle that generates toxic Aβoligomers. *Nat. Struct. Mol. Biol.* **22**(3), 207–213 (2015).
182. Landau, L. D. and Lifshitz, E. M. *Statistical Physics, Part 1. Course of Theoretical Physics*, Vol. 5. Butterworth Heinemann, London, 1980. p. 468.
183. Librizzi, F. and Rischel, C. The kinetic behavior of insulin fibrillation is determined by heterogeneous nucleation pathways. *Protein Sci.* **14**(12), 3129–3134 (2005).
184. Tan, P. and Hong, L. Modeling fibril fragmentation in real-time. *J. Chem. Phys.* **139**(8), 084904 (2013).
185. Fu, S.-Y., Yue, C.-Y., Hu, X., and Mai, Y.-M. Characterization of fiber length distribution of short-fiber reinforced thermoplastics. *J. Mater. Sci. Lett.* **20**(1), 31–33 (2001).
186. Pöschel, T., Brilliantov, N. V., Frömmel, C. Kinetics of prion growth. *Biophys. J.* **85**(6), 3460–3474 (2003).
187. Tarantola, A. *Inverse Problem Theory and Methods for Model Parameter Estimation.* SIAM, Philadelphia; 2005.
188. Michaels, T. C. T., Garcia, G. A., and Knowles, T. P. J. Asymptotic solutions of the oosawa model for the length distribution of biofilaments. *J. Chem. Phys.* **140**(19), 194906 (2014).
189. Lando, S. K. *Lectures on Generating Functions*, Vol. 23. American Mathematical Society, Providence; 2003.
190. Michaels, T. C. T., Yde, P., Willis, J. C. W., Jensen, M. H., Otzen, D., Dobson, C. M., Buell, A. K., and Knowles, T. P. J. The length distribution of frangible biofilaments. *J. Chem. Phys.* **143**(16), 164901 (2015).
191. Hong, L. Link the fluorescence and TEM studies on amyloid fiber formation — To reconstruct the fiber length distribution based on the knowledge of moments evolution. *arXiv preprint arXiv: 1212.4555,* 1–9 (2012).

192. Nilsson, M. R. Techniques to study amyloid fibril formation in vitro. *Methods*, **34**(1), 151–160 (2004).
193. Khurana, R., Coleman, C., Ionescu-Zanetti, C., Carter, S. A., Krishna, V., Grover, R. K., Roy, R., and Singh, S. Mechanism of thioflavin T binding to amyloid fibrils. *J. Struct. Biol.* **151**(3), 229–238 (2005).
194. Biancalana, M. and Koide, S. Molecular mechanism of thioflavin-T binding to amyloid fibrils. *Biochim. Biophys. Acta.* **1804**(7), 1405–1412 (2010).
195. Naiki, H., Higuchi, K., Hosokawa, M., and Takeda, T. Fluorometric determination of amyloid fibrils in vitro using the fluorescent dye, thioflavine T. *Analt. Biochem.* **177**(2), 244–249 (1989).
196. Naiki, H., Higuchi, K., Nakakuki, K., and Takeda, T. Kinetic analysis of amyloid fibril polymerization in vitro. *Lab. Invest.* **65**(1), 104–110 (1991).
197. Jackson, M. and Mantsch, H. H. The use and misuse of ftir spectroscopy in the determination of protein structure. *Crit. Rev. Biochem. Mol. Biol.* **30**(2), 95–120 (1995).
198. Pelton, J. T. and McLean, L. R. Spectroscopic methods for analysis of protein secondary structure. *Anal. Biochem.* **277**(2), 167–176 (2000).
199. Greenfield, N. J. Using circular dichroism spectra to estimate protein secondary structure. *Nat. Protoc.* **1**(6), 2876–2890 (2006).
200. Li, G., Zhou, P., Shao, Z., Xie, X., Chen, X., Wang, H., Chunyu, L., and Yu, T. The natural silk spinning process. *Eur. J. Biochem.* **268**(24), 6600–6606 (2001).
201. Berne, B. J. and Pecora, R. *Dynamic Light Scattering: With Applications to Chemistry, Biology, and Physics*. Courier Corporation, New York; 2000.
202. Aragon, S. R. and Pecora, R. Theory of dynamic light scattering from polydisperse systems. *J. Chem. Phys.* **64**(6), 2395–2404 (1976).
203. Berne, B. J. Interpretation of the light scattering from long rods. *J. Mol. Biol.* **89**(4), 755–758 (1974).
204. John, S. R. *Macmillan Encyclopedia of Physics*. Simon & Schuster, Macmillan, New York; 1996.
205. Kowalewski, T. and Holtzman, M. H. In situ atomic force microscopy study of Alzheimers β-amyloid peptide on different substrates: new insights into mechanism of β-sheet formation. *Proc. Natl. Acad. Sci.* **96**(7), 3688–3693 (1999).
206. Ward, R., Jennings, K., Jepras, R., Neville, W., Owen, D., Hawkins, J., Christie, G., Davis, J., George, A., Karran, E., *et al.* Fractionation and characterization of oligomeric, protofibrillar and fibrillar forms of β-amyloid peptide. *Biochem. J.* **348**, 137–144 (2000).
207. Khurana, R., Ionescu-Zanetti, C., Pope, M., Li, J., Nielson, L., Ramírez-Alvarado, M., Regan, L., Fink, A. L., and Carter, S. A. A general model for amyloid fibril assembly based on morphological studies using atomic force microscopy. *Biophys. J.* **85**(2), 1135–1144 (2003).
208. Ban, T., Hamada, D., Hasegawa, K., Naiki, H., and Goto, Y. Direct observation of amyloid fibril growth monitored by thioflavin t fluorescence. *J. Biol. Chem.* **278**(19), 16462–16465 (2003).
209. Ban, T., Hoshino. M., Takahashi, S., Hamada, D., Hasegawa, K., Naiki, H., and Goto, Y. Direct observation of Aβ amyloid fibril growth and inhibition. *J. Mol. Biol.* **344**(3), 757–767 (2004).

210. Ban, T., Yamaguchi, K., and Goto, Y. Direct observation of amyloid fibril growth, propagation, and adaptation. *Accnt. Chem. Res.* **39**(9), 663–670 (2006).

211. Meisl, G., Kirkegaard, J. B., Arosio, P., Michaels, T. C. T., Vendruscolo, M., Dobson, C. M., Linse, S., and Knowles, T. P. J. Molecular mechanisms of protein aggregation from global fitting of kinetic models. *Nat. Prot.* **11**(2), 252–272 (2016).

Kinetic Studies on Protein Aggregation With and Without the Presence of Crowders

5

John S. Schreck, John Bridstrup and Jian-Min Yuan

Drexel University, Department of Physics, Philadelphia, Pennsylvania 19104, USA

In this review, we begin by describing the importance of studying the processes of protein self-assembly for understanding the formation of natural protein polymers, disease-causing amyloids, and amyloids as a material. We then discuss various methods developed in treating protein aggregation, before focusing on the kinetic approach. We review the developments of the kinetic methods including the Becker–Döring (BD) model and the generalized Smoluchowski model as applied to protein aggregation. Motivated by the processes taking place in living organisms, we present a general discussion on the effects of molecular crowding on protein stability and reactions. Then we present recent results obtained on the effects of molecular crowding on protein aggregation using kinetic models.

1. Introduction

Most diseases are related to failures of protein materials.[1] Examples can be found in Hutchinson–Gilford progeria syndrome,[2] osteogenesis imperfacta,[3] sickle cell disease,[4] and cancers.[5] Hutchinson–Gilford progeria syndrome is associated with a type of premature aging, caused by mutations of lamin influencing the lamin microstructures on the cell nuclear membrane.[5] A point mutation of tropocollagen causes the reduction of the maximal strains and stress that collagen can support, resulting in osteogenesis imperfacta.[3] Sickle cell disease arises from a single mutation on the two beta subunits of hemoglobin.[4] Finally, breast cancer cells are on average softer than normal breast cells, because the actin polymers in cancer cells are shorter and more disordered than those in normal cells.[5] If we understand the mechanisms of these protein failures, correction and prevention of these failures can give us insights into how to treat and prevent diseases. One type of protein failure is associated with protein misfolding and protein aggregation to form protein amyloids.[6,7] Since neurodegenerative diseases, such as Alzheimer's, Parkinson's disease, and type 2 diabetes, are related to this type of failure; protein aggregation has recently become an active field of research.[6,7] On the other hand, the growth of naturally occurring functioning protein polymers also has been investigated for many years, for example, in the studies of actin polymers and microtubules. The actin polymer, for instance, is related to the cytoskeleton structure, stiffness, and shape of cells, but also is responsible for tumor migration properties resulting in metastasis of cancer cells.[5] Furthermore, because of their uniform structure and nanoscale size, protein amyloids can serve as templates for conducting nanowires when coated with gold or silver.[8] It is due to all of these reasons (neurodegenerative diseases, abnormal functioning of protein polymers, and amyloid as a material) that protein aggregation becomes an important field of investigation.

The processes of protein aggregation have been studied using a wide variety of methods. For small system sizes and short-time scales, all-atom molecular dynamics (MD) calculations with explicit solvent models provide the most detailed information on the structures of peptides/proteins, nuclei, and aggregates at the molecular level.[9–11] Using replica exchange MD and other metadynamics methods, MD calculations have already provided important structural information about monomeric spices (natively unfolded or compact), small aggregates, or fibrils and their interactions with molecular dyes or inhibitors.[9,10] Even knowledge on prenucleus formation, fibril elongation, or fibril growth at the early stages of aggregation has been obtained. However, since the interested length scale goes from atomic size to micrometers, and the interested time scale is from femtosecond to years, an all-atom MD calculation providing a complete picture of the entire aggregation process currently is far from reality.

To extend the size and time scales beyond those reachable by the atomistic MD models, coarse-graining models, such as discrete molecular dynamics (DMD)[12,13] or other models,[14,15] have been developed. Many models fall into this category and they come with different levels of length resolution.[9] A model could represent side chains only by particles, the entire amino acid residue by a few particles, or even the entire peptide monomer as one particle, and so on. The dynamics can be evolved according to a force field or by Monte Carlo methods. These models can include up to hundreds or more monomers in the system, and may take into consideration the interactions with the environment, such as membrane or surface interactions, and can also account for molecular-level structural differences, for example, between β-amyloid peptide-40 (Aβ-40) and Aβ-42, and so on. In general, these types of models fill the needs of intermediate length and time scales of protein aggregation, including nucleation, elongation, growth of fibrils, and even plague formation. There are also kinetic studies based on mass-action laws, which can reach the time and length scales relevant to aggregation. They belong to mean-field ensemble-averaged and continuous models. We present some details of their developments in the next section.

Methods based on statistical mechanics have also been developed for treating protein aggregation. For instance, both canonical ensemble[16] and grand canonical ensemble methods[17,18] have been used for this purpose. The advantage of statistical mechanical methods is the reduction of computation time and power required. Models based on statistical mechanics can also reach the longest time scales as well as large length scales. However, the disadvantage with this approach is that in general it can only give us equilibrium properties of aggregation.

Since protein environments inside cells are crowded, how this crowdedness affects protein stability and protein reactions, such as aggregation, is an important problem to investigate. It is particularly relevant because *in vivo* Aβ concentration is in the nanomolar range, but in *in vitro* experiments the critical concentration for aggregation is higher than micromolar. However, in reality people do get Alzheimer's disease. Is it because the crowdedness *in vivo* makes the effective concentration of Aβ much higher than nanomolar? It is to answer questions like this that in the second part of the article we focus on the effects of molecular crowding on protein aggregation.

The organization of this review article is as follows: in Section 2, we put the development of a kinetic approach to protein aggregation in perspective as it was shaped by the works of others. The kinetic approach is then reviewed in Section 3, along with the results of calculations compared to experimental data. As an introduction, the importance of the effects of molecular crowding on protein stability and aggregation is described and a brief history on the subject by the group presented in Section 4. Then applications of simple kinetic models to study the effects of molecular crowding on the equilibrium and kinetic properties of protein aggregation are presented in Section 5.

2. Developments of the Kinetic Approach

Kinetic approaches applied to the self-assembly of particles, such as water drops, colloids, or molecules, have a long history. They often involve a discrete set of deterministic mean-field equations following the mass-action laws. Among the early attempts using the kinetic approach for assembly are those by Smoluchowski[19] and Becker and Döring.[20] We discuss the BD model first and the Smoluchowski equation later in a generalized form.

Becker and Döring were given credit for introducing the natural idea that monomer addition and subtraction to a cluster or an aggregate are the main mechanism for the initiation, growth, and disintegration of an assembly.[20] This mechanism of monomer addition and subtraction, which can be described as:

$$M_n + M_1 \underset{k_{\text{off}}}{\overset{k_{\text{on}}}{\rightleftharpoons}} M_{n+1}, \tag{1}$$

for any aggregate of size $N \geq n_c$, where $a_n = k_{\text{on}}$ and $b_{n+1} = k_{\text{off}}$ are the monomer addition and subtraction rate constants, respectively, is an essential part of many later models. M_r refers to the aggregate of size r proteins. One important feature of the BD model is that analytic solutions exist for the equilibrium and steady states (see Section 3). Burton[21] reviewed a variety of ways this model has been used in discussing condensation and phase transitions. Wattis,[22] however, gave a more mathematical analysis of the properties of the model.

In their studies of the polymerization of proteins, such as the formation of F-actins from G-actins, Oosawa and coworkers[23,24] had included nucleation and its reverse reactions in the set of the mass-action equations. The kinetic reaction for the production or disassembly of nuclei can be described by the following expression.[24]

$$n_c M_1 \underset{k^*}{\overset{k_n}{\rightleftharpoons}} M_{n_c} \tag{2}$$

where k_n is the nucleus formation rate constant, and k_* is the nucleus disassembly rate constant. For the critical nuclei, only monomer addition applies, while concentrations of aggregates that are smaller than the critical nucleus are assumed to be zero. These assumptions are consistent with the classic theory of nucleation, where the critical nucleus is assumed to be at the top of a free energy barrier when it is plotted as a function of polymer size. Oosawa also obtained approximate analytic solutions to the number and mass concentrations of actin polymers. Their model is one of the first kinetic models (if not the very first) developed to study protein aggregations. Actin polymers, of course, are just one of naturally forming polymers playing important structural roles in the shape and in the migration of living cells. However, as discussed in the following section, Oosawa's model has also been incorporated in treatments of amyloid formation of proteins responsible for neurodegenerative diseases. Numerous reviews on the kinetics of amyloid formation are available.[25–29] An extensive review for works before 2013 can be found in Ref. 30.

In their treatment of self-aggregation of protein amyloid formed by aberrant biological processes, Knowles *et al.*[31] extended the Oosawa model by considering also the fragmentation of aggregates, as described by Smoluchowski kinetics.[19] Smoluchowski kinetics is described by the following expression:

$$M_r + M_s \underset{k_{\text{off}}}{\overset{k_{\text{on}}}{\rightleftharpoons}} M_{r+s}, \tag{3}$$

where r and s are greater or equal to the n_c, the size of the critical nucleus, and the merging and breakage kernels (e.g., the rate constants) are $a_{r,s}$ and b_{r+s}, respectively. Knowles *et al.*[31] describe the fragmentation process as secondary nucleation because it can provide nuclei and stable aggregates for further growth. An important contribution that they made are the analytic solutions to the number (P) and mass (M) concentrations of filaments based on fixed point analysis of the mapping of P and M. Their results have been used to fit experimental data for the formation of many protein amyloids and has also been used to discuss the scaling relation of the lag phase to protein concentration. For simplicity, they assume that all breakage rates are constant and independent of the size of the filaments. Knowles' model[31] has further been investigated by Hong *et al.*[32] in terms of the scaling relations and asymptotic behaviors.

It is under this backdrop that we have carried out our investigation of protein aggregation solving kinetic equations. To get more detailed information about the protein aggregation processes, for example, the evolution of the size distribution of the aggregates, we found that we need to improve upon Knowles' model by including in addition to the fragmentation of filaments of any size greater than n_c the corresponding forward reactions of Eq. 3. Furthermore, in general, the elongation and breakage rate constants should be size dependent. Our model[33] fits the experimental results of CgsB[34] and β-LAC fibrils[35] quite well, and could be used to fit a variety of experimental data of amyloid-forming systems. One example could be the data from Tycko *et al.*,[36] who recently measured the time evolution of the size distributions of the aggregates of β-amyloid peptides (Aβ) and fitted the experimental data using a less-descriptive kinetic model. More details of our treatment[33] are reviewed in Section 3.

One basic assumption of the aforementioned kinetic models, starting with the Oosawa model, is being challenged by the experimental results of Lee *et al.*[37] Based on the classical nucleation theory, the oligomers smaller than the nucleus have free energies less than that of the nucleus, which sits at the top of the free energy barrier, but higher than that of the free monomers. Thus, it is reasonable to assume that the population of these oligomers to be zero. Usually the oligomers cannot be detected by thioflavin T (ThT) fluorescence, only larger aggregates like fibrils. By using the fluorophore FlAsH and the mutant Cys–Cys-amyloid-β_{1-40} as amyloid protein, Lee *et al.*[37] verified the detection of nonfibrillar oligomers. Based on their observations, they suggest that the mechanism of nucleated conformational conversion for Aβ aggregation is when spherical oligomers form first and then slowly convert to amyloid fibrils. This mechanism is shown in Fig. 1(a)(ii). It is not

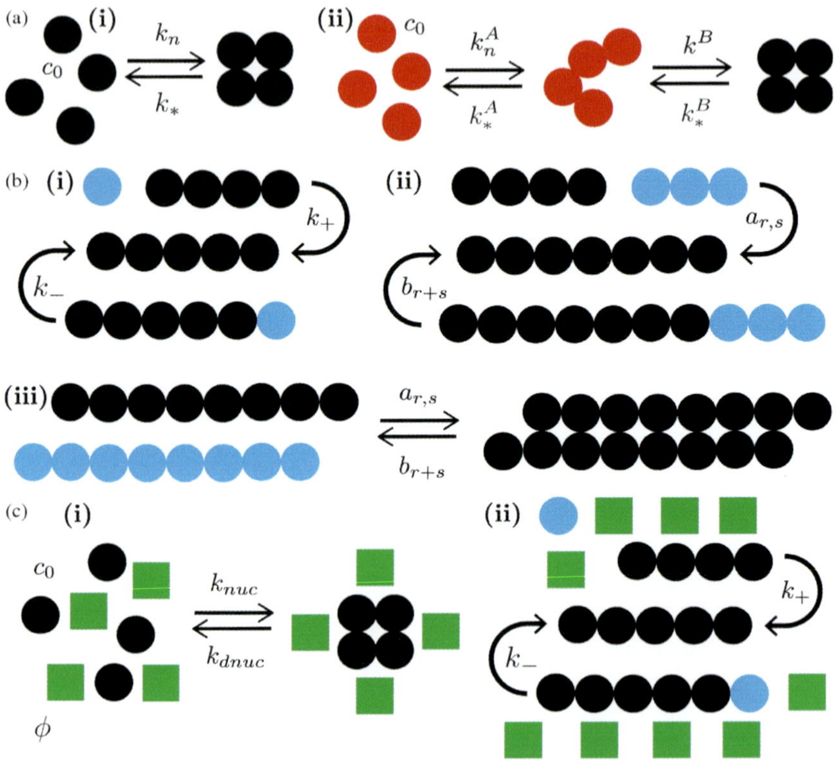

Figure 1: Various reactions possibly involved in protein aggregation are illustrated. In (a)(i), the nucleation step is illustrated. In (a)(ii), a two-step nucleation mechanism is illustrated. Proteins first join to form a small critical nucleus (state A), which then undergo a structural or conformation rearrangement (state B). In (b), several elongation mechanisms are illustrated; (i) monomer addition and monomer subtraction, (ii) end-to-end fibril merging of two aggregates of varying size, and (iii) lateral association of two fibrils of any size that may or may not bind in register. In (c)(i), the nucleation mechanism with crowders in solution at volume fraction ϕ is illustrated. (ii) Monomer addition and subtraction reactions in the presence of crowders. In (a) and (c), c_0 is the initial monomer concentration present in solution. In all figures, a circle represents a protein and a square represents a crowder.

clear how common this mechanism is among all amyloid proteins. In any case, this suggests that the protein aggregation process may be more complicated than previously thought. This mechanism is to a certain extent supported by the work of Saric *et al.*,[38] which claims that at physiological conditions amyloid protein goes through a two-step nucleation process that is mediated by nonspecific attractive interactions. Furthermore, the size of the critical nucleus is not constant but increases with protein concentration and inter-protein interactions. Additionally, Cohen *et al.*[39] also observed fibril-catalyzed secondary nucleation, that is, nuclei formed on the surface of existing fibrils, corresponding to the heterogeneous nucleation first observed by Ferrone *et al.*[40] in the study of sickle cell disease. Thus, these recent results cast doubt on the rigorous or literary application of the classical nucleation theory to protein aggregation.

The kinetic models give us different information from the MD or coarse-graining studies, which provide atomic- or molecular-level description of the aggregation processes, including structural and geometric information. As noted, the kinetic models belong to a mean-field ensemble-averaged description between the microscopic and macroscopic level, and are here to provide a means to fit experimental data. We anticipate more such discrepancies being revealed as more comparisons between kinetic and molecular level studies become available. It will lead to our eventual understanding of the protein aggregation processes. Of course, atomic level understanding can improve upon the kinetic models as well. A kinetic model that utilizes to some extent the two-step nucleation into consideration is described in Ref. 33

3. Our Kinetic Model

In this section, we introduce a model for protein aggregation where the formation of small critical nuclei containing n_c number of proteins are the first aggregates to form from an initial monomer concentration of amyloid proteins, c_0, and may grow into β-sheet dominated structures containing thousands of proteins. We take the nucleus as the oligomer capable of reaching the peak of the rate-limiting free energy barrier, and that aggregates containing fewer than n_c number of proteins are assumed to be thermodynamically unstable.[31,40] The assumption that small oligomeric aggregates can be ignored may not be true, as numerous experiments have shown that they may exist at concentrations below the critical fibril concentration.[37,41–45] Since experiments and simulations have shown that the pathways to fibrils could be sequence dependent,[12,46] we consider in our description of fibril formation a variety of fibril growth and fragmentation mechanisms. These include, but are not limited to, monomer addition and subtraction from the fibril ends. Moreover, it is entirely reasonable that for proteins and aggregates in solution, collisions between monomers and aggregates of varying shapes and sizes will occur, and some of these collisions could yield larger aggregates, while others could result in aggregates breaking apart.[47] We contend that the pathways for amyloid proteins to aggregate into fibrils would be better captured by a model that includes many aggregation steps, including dimer–dimer, trimer–hexamer, and other types of merging reactions. The former examples are possible pathways for Aβ(1–40) and Aβ(1–42) amyloid fibril formation, respectively.[48] In addition, the rates for aggregation or fragmentation steps could depend on the size and shape of the aggregates, and/or the conformation of monomers and aggregates, which may render some aggregation or fragmentation reactions more likely than others. For example, monomer addition is often considered as the most favorable pathway for protein aggregates to grow larger. Using classical coagulation theory, one can easily show that for proteins and aggregates diffusing in solvent, the monomer addition mechanism is the most probable growth pathway. However, this theory also says that other pathways, such as fibril merging or breaking, can contribute to the overall growth or shrinkage of the fibrils, respectively.

Before writing down our generalized kinetics model for protein aggregation, we first consider the Oosawa model and Knowles' model for nucleated polymerization of proteins into aggregates. In the Oosawa model, the pathways for protein aggregates to grow or shrink in size are monomer addition and monomer subtraction, for example, BD kinetics[20,49] as described by Eq. 1, combined with nucleus production as described by Eq. 2. By imposing the law of mass action, the mass flux, $\mathcal{J}_r(t)$, can be defined as the net rate of clusters of size r growing to size $r+1$ at a time t for clusters at least as large as the size of the critical nuclei as:

$$\mathcal{J}_r(t) = k_{\text{on}} c_r(t) c_1(t) - k_{\text{off}} c_{r+1}(t) \tag{4}$$

where $\{c_r(t)\}_{r=1}^{\infty}$ are the concentrations of all allowed clusters at any time t. Starting with an initial concentration of monomer protein, c_0, the mass-action equations governing the time evolution of aggregates of any size in the Oosawa model can be written as:

$$\frac{dc_r(t)}{dt} = \mathcal{J}_{r-1}(t) - \mathcal{J}_r(t) + \frac{1}{2}(k_n c_1(t)^{n_c} + k_* c_{n_c})\delta_{r,n_c} \tag{5}$$

$$\frac{dc_1(t)}{dt} = -\frac{d}{dt}\left[\sum_{i=n_c}^{\infty} i c_i(t)\right], \tag{6}$$

which describes every possible way to construct the aggregate of size r, and includes nuclei production and disassembly reactions.[24] The Kronecker delta δ_{r,n_c} is equal to one when $r = n_c$ and equal to zero otherwise. The equation in the second line describes the evolution of the monomer concentration and follows from mass conservation. As we have noted earlier, it is a common assumption to take $k_* = 0$ and regard the formation of critical clusters to be an irreversible process.

BD type kinetics is just one type of reversible reactions that can involve protein monomers and aggregates. Knowles' model for protein aggregation incorporates fibril fragmentation reactions, and is a combination of the Oosawa model with the backward reactions of Smoluchowski kinetics as given in Eq. 3.[31,50] In particular, the fragmentation reactions in their model describe the breaking of aggregates of any size into two aggregates each containing a varying number of proteins (greater than n_c). The resulting equations for the evolution of cluster concentrations containing r number of proteins, $c_r(t)$, in Knowles' model are:

$$\begin{aligned}\frac{dc_r(t)}{dt} &= \mathcal{J}_{r-1}(t) - \mathcal{J}_r(t) \\ &\quad - k_-(r-1)c_r(t) + 2k_- \sum_{i=r+1}^{\infty} c_i(t) \\ &\quad + k_n c_1(t)^{n_c}\delta_{r,n_c} \\ \frac{dc_1(t)}{dt} &= -\frac{d}{dt}\left[\sum_{i=n_c}^{\infty} ic_i(t)\right],\end{aligned} \tag{7}$$

where k_- is the size-independent aggregate fragmentation rate constant. The earlier equation holds for any $r \geq n_c + 1$, while $c_r = 0$ for $1 < r < n_c$. Knowles *et al.*[31] and others[32,33,,51] have showed that strong fragmentation can act as an effective secondary nucleation mechanism, whereby fragmenting aggregates may act as seeds that speed up elongation of fibrils during early times of aggregation.

In Knowles' model, the quantities that are typically the most useful for comparing theory with experiments involving amyloid protein aggregation are the zeroth and first moments of the fibril size distribution that describe the number of fibrils, $P(t)$, and the protein mass in fibrils $M(t)$. They can be defined as:

$$P(t) \equiv \sum_{i=n_c}^{\infty} c_r(t) \tag{8}$$

$$M(t) \equiv \sum_{i=n_c}^{\infty} rc_r(t), \tag{9}$$

where $M(t)$ has often been compared with experimental characterizations of amyloid formation as monitored, for example, by fluorescence utilizing dyes like ThT and

Congo red, which are known to bind strongly to fibrils containing significant β-sheet content,[52–54] but may not bind as strongly to other aggregates that may be present, such as small oligomers, that do not contain significant β-sheet content.[55,56] The average length of fibrils at any time may be expressed as:

$$L(t) = \frac{M(t)}{P(t)}. \tag{10}$$

The quantities $M(t)$, $L(t)$, and the size distributions of fibrils as a function of time may be readily measured in experiments.[35,57–59] Since these quantities will depend on the relative values of the rate constants in a kinetics model, predicting them accurately will require that a model correctly takes into account aggregation and fragmentation reactions that meaningfully contribute to the growth or shrinkage of fibrils, respectively. In general, inclusion of filament fragmentation processes in Eq. 7 fits experimental fluorescence data better than the Oosawa model.[31] However, despite its success in accurately fitting a variety of amyloid-forming fluorescence data, a main limitation of Knowles' model is the fact that the fragmentation reactions are not balanced by the corresponding reverse processes of merging of two aggregates of any size greater than n_c into a single aggregate. Consequently, the model describes an unphysical scenario where breakage reactions release an infinite amount of heat into the environment, as is noted in Ref. 60

Aggregate merging may include the following reactions: joining of two smaller disordered oligomers into a larger aggregate, end-to-end association of β-sheet dominated fibrils which preferentially grow in one direction, and lateral association of two fibrils.[61,62] The merging of two otherwise disjoint clusters has already been experimentally observed in other self-assembling systems of polypeptides,[63–80] but it is not yet clear to what extent reactions such as end-to-end association contribute to fibril elongation in amyloid protein systems. Several recent in silico studies have observed the phenomena in MD simulations of amyloid protein systems.[14,81,82] In particular, coarse-grained MD simulations of Aβ showed that small oligomers may join with other fibril seeds.[81] Additionally, coarse-grained simulations of α-synuclein aggregation observed oligomers, each containing about 30 proteins, merging into a larger composite aggregate. In this study, the fibrils were mainly observed to grow longer by monomer addition, but merging was observed when the ends of two fibrils were properly aligned and proximal to one another.[82] Other coarse-grained studies suggest that monomer addition is the predominant elongation mechanism, but merging of two filaments into a single filament also contributes to fibril elongation, including linear aggregates that do not necessarily have the same morphologies.[14]

For the remainder of this section, we focus on mass-action kinetics descriptions that have included aggregate merging terms, with a focus on the formulation that we presented in Ref. 33. It is a straightforward exercise to include the reverse of the breakage reactions in Knowles' model, and is accomplished by adopting Smoluchowski kinetics, Eq. 3, that takes into account both forward and backward reactions involving two clusters of any sizes. Here, we assumed that the kernels in Eq. 3 are symmetric, $a_{r,s} = a_{s,r}$ and $b_{r+s} = b_{s+r}$. As we have noted earlier, the kernels may be dependent on system properties such as the size and shape of the protein aggregates, as well as the conformation of the proteins in aggregates. We discuss the kernels in more detail as follows. For aggregates of any size, it follows from the law of mass-action that the mass flux from concentrations $c_r(t)$ and $c_s(t)$ going to $c_{r+s}(t)$ may be written as:

$$W_{r,s}(t) = a_{r,s}c_r(t)c_s(t) - b_{r+s}c_{r+s}(t) \tag{11}$$

where $c_r = 0$ if $r < n_c$ except for $r = 1$. If $n_c > 2$, $W_{1,1}(t) = 0$. Starting from an initial monomer protein concentration c_0, we may describe a set of coupled nonlinear differential equations for describing the temporal changes to $c_r(t)$ as described by Eq. 3 as:

$$\begin{aligned} \frac{dc_r(t)}{dt} &= \frac{1}{2}\sum_{s=1}^{r-1} W_{s,r-s}(t) - \sum_{s=1}^{\infty} W_{r,s}(t) + \frac{1}{2}k_n c_1(t)^{n_c}\delta_{r,n_c} \\ \frac{dc_1(t)}{dt} &= -\frac{d}{dt}\left[\sum_{i=n_c}^{\infty} rc_r(t)\right], \end{aligned} \tag{12}$$

where the factor of one-half in the first term compensates for the fact that it counts every way to construct the aggregate of size r twice. The third term is the Oosawa homogeneous nucleation rate. Just as in Eqs. 5 and 7, the equation in the second line describes the evolution of the monomer concentration and follows from mass conservation. Equation 12 can be solved numerically for systems that describe aggregates containing tens of thousands of proteins.[33] Depending on the form of the rate constants, some closed-form expressions may be obtained for $M(t)$ and $L(t)$.

Before investigating solutions to the set of equations given in Eq. 12 for different kernels, we first point out some expectations for the quantities $M(t)$, $P(t)$, and $L(t)$ that are often used to fit experimental data. The reactions for the joining of two aggregates will not be expected to play a major role in the formation of small critical clusters early on in assembly because the merging terms depend on fibrils being present in solution. Small aggregate formation will still be driven by primary homogeneous nucleation events, thus $M(t)$ for the Smoluchowski model and Knowles' model will predict the same behavior for short times. However, fibril merging may play a significant role at times beyond the lag phase when two clusters that are at least as large as the critical nucleus have formed, and are in large numbers in solution, and may come into contact with each other and join to become one cluster. This means that inclusion of aggregate-joining terms in a mass-action kinetics model may have a significant impact on $P(t)$, and therefore $L(t)$, at later times. However, the aggregate-joining terms may not affect $M(t)$ very strongly, if at all, which would explain why models that exclude aggregate merging reactions are still able to characterize the fibril mass time dependence as measured in experiments,[33,60] but not necessarily $L(t)$ for the same experimental systems.[33] The quantities $M(t)$ and $P(t)$ therefore may relax to their equilibrium or steady-state values at different time scales. Therefore, adjusting the rate constants in order to best fit $M(t)$ to experimental data may not be a sufficient procedure to parametrize the model so that it then correctly predicts the length distributions, or other quantities describing the aggregates. Thus, understanding the different possible fragmentation and aggregate merging reactions in amyloid systems is necessary to correctly predict quantities like $M(t)$ and $L(t)$, as well as the size distributions for fibrils, as time evolves.

We first consider the case when all forward reaction rates $a_{r,s}$ are assumed to be constant and equal to k_+, and all reverse reaction rates b_{r+s} are assumed to be constant and equal to k_-. We note that while concentration-independent backward rates may be a reasonable assumption, taking all forward rates to be equal to k_+ for reactions involving arbitrary cluster sizes may not represent a physically reasonable approximation for experiments in solution because larger clusters should diffuse more slowly than smaller ones. However, taking all forward reactions to be equal with rate constant k_+ is the simplest case, and helps to shed light on the dependencies of $c_r(t)$ on aggregation processes including merging unmodulated by the sizes of aggregates. We adopt it here for now and then generalize to a diffusion-limited example where the rates depend on aggregate size.

Choosing a constant forward rate constant also offers some practical solutions to the infinite set of differential equations in Eq. 12. In particular, the number of fit parameters can be kept to a minimum. Additionally, analytical solutions are obtainable for $M(t)$ and $L(t)$. In this model formulation, the quantity $M(t)$ can be shown to be independent of fibril merging processes.[33,60] Specifically, Ref. 60 used fixed-point mapping to obtain analytic results for the quantities $M(t)$ and $P(t)$, and showed $M(t)$ does not depend on k_+, while the average number of fibrils at a given time $P(t)$, and therefore $L(t)$, explicitly depend on k_+. Thus, Knowles' model and the Smoluchowski approach describe different behavior for $L(t)$, especially at long times when fibril merging may come to dominate. In Ref. 33 we used both models to fit $M(t)$ to ThT data for CgsB fibrils and showed that Knowles' model predicts that fibril lengths initially grow rapidly, but then breakage of longer filaments takes over and $L(t)$ eventually converges toward an aggregate size more typical of an oligomer (average size on the order of 10) rather than a fibril. By contrast, fitting $M(t)$ to the CgsB data as predicted by the Smoluchowski model showed that aggregates also initially grow rapidly in length, the fibrils are longer than those predicted by Knowles' model for the same data, and are much less likely to break up at large times, with average fibril lengths containing thousands of proteins, which is more realistic. Additionally, $L(t)$ as predicted by the Smoluchowski model for CgsB fibrils approached a constant equilibrium value much faster than in Knowles' model.

We next consider the more realistic case when the Smoluchowski model is equipped with size-dependent kernels. As noted earlier, one of the main advantages in choosing constant forward and backward rate constants is that the number of free parameters can be kept to a minimum, and analytical solutions are obtainable for $M(t)$ and $P(t)$. For the size-dependent kernels, one simply cannot allow all of the different rate constants to be adjustable parameters as fitting data becomes pathological. Simple statistical mechanics considerations regarding the stability of monomer proteins, as well as some approximations in describing the size and shape of the proteins in aggregates, can help to simplify the kernels while still keeping to a minimum the number of parameters and leaving open the possibility for analytical solutions. For example, Hill has shown for Smoluchowski-type reactions[83] that the equilibrium constant may be written as:

$$K_{r,s} = A\left(\frac{rs}{r+s}\right)^{n} \tag{13}$$

where A is a constant prefactor and n is a number that is related to the motional degrees of freedom of the aggregate, which includes translational, rotational, and vibrational contributions. By assuming the proteins are spherical in shape, are connected along a linear string, and that the reactions are diffusion-controlled bimolecular reactions, the merging and fragmentation rate constants $a_{r,s}$ and b_{r+s} may be written as:

$$a_{r+s} = \frac{k_{-}(r\ln s + s\ln r)}{rs(r+s)} \tag{14}$$

$$b_{r,s} = \frac{k_{+}(rs)^{n-1}(r\ln s + s\ln r)}{(r+s)^{n+1}} \tag{15}$$

where k_+ and k_- are taken to be constants. These expressions predict that monomer addition has the highest association rate, while merging of two clusters of the same size has the slowest merging association rate. Similarly, breakage at the middle of a fibril that results in two clusters of equal size is predicted to be the fastest shrinkage process, while monomer subtraction is predicted to be the slowest shrinkage process.

Hong *et al.*[52] have reported closed-form expressions for $P(t)$ and $M(t)$ for Knowles' extended Smoluchowski model that includes merging processes for the kernels given in Eq. 14. The authors found by adjusting the values of k_+, k_-, k_n, and monomer addition and subtraction rate constants k_{on} and k_{off}, respectively, that values of $n \approx 1 - 3$ best fit experimental data for a variety of amyloid aggregate-forming protein systems including Sup35 and CgsB.[52] Free fibrils in solution have $n \approx 4 - 6$, and so the range of values of $n \approx 1 - 3$ suggests that the fibrils lose some of their motional freedom that may be due to fibril bundling, cross-linking, extended network formation, and so on.[52]

In our Smoluchowski formulation given in Eq. 12,[33] we were inspired by the usage of Smoluchowski equations in colloidal science where the fractal dimension D_f is used in size-dependent kernels.[84,85] The fractal dimension for protein aggregation is a simple way to quantify the irregularity of formed fibrils where those with different morphologies may merge together, as well as multiple fibrils coming together to form disjointed clusters, or possibly extended networks when multiple clusters become linked together as in the cytoskeleton — F-actin fibrils.[84–87] For spherical aggregates $D_f = 3$, while for linear aggregates $D_f = 1$. We chose to keep n fixed at $n = 4$ and instead used the fractal dimension, D_f, as an adjustable parameter for fitting experimental data. The forward reaction rates for a diffusion-limited reaction that includes D_f now take the form,

$$a_{r,s} = k_P(r^{-1/D_f} + s^{-1/D_f}) \tag{16}$$

where k_P is a constant prefactor. We then express the breakage rate kernel b_{r+s} by combining Eq. 13 with Eq. 16 to arrive at:

$$b_{r+s} = k_M \left(\frac{rs}{r+s}\right)^n (r^{-1/D_f} + s^{-1/D_f}) \tag{17}$$

where k_M is now a composite, constant prefactor. In our formulation, the adjustable parameters are k_P, k_M, k_n, and D_f, which is only one more than in Knowles' model, but two fewer than the approach taken by Hong *et al.* In Ref. 33, we showed that k_P, k_n, k_M, and D_f can be adjusted to fit both $M(t)$ and $L(t)$ to experimental data for β-LAC fibrils with high accuracy. The results for $M(t)$, $P(t)$, and $L(t)$ are also plotted in Fig. 2, where the top and middle plots ($L(t)$ and $M(t)$, respectively) show

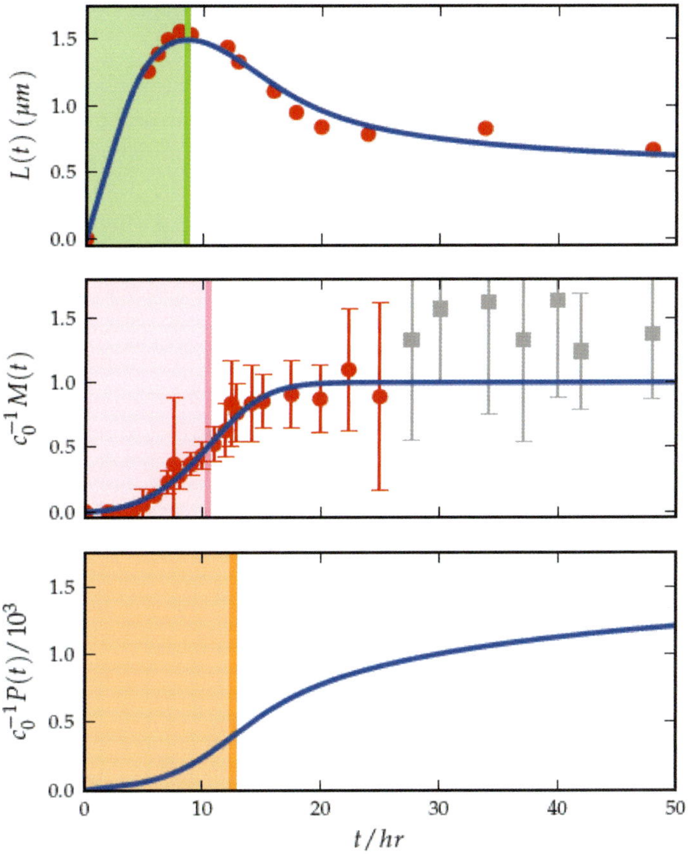

Figure 2: Top plot: The average length of fibrils, $L(t)$, is plotted (blue curve) along side AFM measurements (red dots) of β-LAC fibrils. The green region illustrates when the average length is growing longer. The white region illustrates when fibrils are shrinking in size before reaching a constant value at long times. Middle plot: The average mass in fibrils, $M(t)$, is plotted (blue curve) alongside the ThT data for the β-LAC fibrils (red dots). The pink region illustrates when the monomeric proteins are rapidly joining fibrils. The white region illustrates when growth slows down and reaching a point when all proteins are in aggregates (around 20h). Bottom plot: The predicted average number of fibrils, $P(t)$. The orange region illustrates when the growth of the number of fibrils present is increasingly growing. The white region shows when the growth in fibril numbers is slowing down. $P(t)$ eventually reaches a constant value at long times. These plots are based on those presented in Ref. 33.

the experimental data alongside the fits. The bottom plot shows the predicted profile for $P(t)$.

In Fig. 2, the three plots clearly show that there are several different regimes when elongation and shrinkage reactions dominate, and when they both contribute. For example, the top plot in Fig. 2 shows that aggregates initially grow very rapidly in length when nucleation and monomer addition are the principal mechanisms driving monomer proteins to join aggregates. Additionally, the number of aggregates are few as most proteins remain free in solution. However, the breakage and fibril merging terms eventually begin to contribute as $L(t)$ reaches a peak value of about 1.5 micrometers at the 8.5 h mark before starting to decline. The slope of $M(t)$ becomes increasingly less steep once half the monomer proteins have joined aggregates and occurs at 10.4 h. The quantity $P(t)$ does not reach its peak value until the system reaches equilibrium, however, the slowdown in the growth of new fibrils starts at 12.6 h. Stronger fragmentation would cause $P(t)$ to grow much more rapidly, while $L(t)$ would drop off more rapidly after reaching a peak value. Stronger fibril merging would cause $P(t)$ to grow more slowly while $L(t)$ would fall off less than that in the figure after reaching its peak value. The plot clearly shows that even at 50 h, both $L(t)$ and $P(t)$ still have not reached their equilibrium values, while $c_0^{-1}M(t)$ reaches 1 around 20 h when monomers are completely starved (note that c_0 is the total concentration of protein).

We found the fractal dimension for these fibrils to be $D_f \approx 1.34$, a value that is consistent with more open structures seen in the atomic force microscopy (AFM) characterizations. The AFM images of the β-LAC fibrils[84] are also similar to those seen in electron microscopy images of, for example $A\beta(1-40)$ and $A\beta(1-42)$, that show linear fibril structures in extended networks of varying protein density.[88] Furthermore, the rate constants in the Smoluchowski model that best fit $M(t)$ and $L(t)$ for the β-LAC fibrils also predict length distributions at different times that are consistent with the same quantities as measured by AFM for β-LAC fibrils.[84] Thus, our results show that the concept of fractal dimension applied to protein aggregation may be a useful concept to assist in characterizing fibril formation when using mass-action type models.

Here, we have only considered a few of the types of kernels that have been used in the protein aggregation literature, but there may be others that predict $M(t)$ and $P(t)$ that compare with experimental data well, and that may also admit closed-form expressions. As an example, a recent investigation by Lee has shown that the fragmentation rates for the breaking of semiflexible polymers into fragments of different sizes are all roughly equivalent and independent of cluster size.[89] This finding supports the usage of constant fragmentation rates, such as in Knowles' Smoluchowski model. Future investigations could find solutions to Eq. 12 where all of the fragmentation rates are held constant, but where the forward rates would remain size dependent if the merging reactions are diffusion-limited. In such a scenario, the backward rates could be combined with Eq. 13 to derive an expression for the kernel $a_{r,s}$ as:

$$a_{r,s} = A\left(\frac{rs}{r+s}\right)^n k_-, \tag{18}$$

where A is a prefactor and can be used alongside k_- and k_n (and perhaps n) as adjustable parameters when fitting experimental data. This kernel predicts that reactions involving smaller clusters combining with larger clusters will have higher association rates than clusters of equivalent size. Equation 18 also predicts that monomer addition is the fastest elongation process.

New extensions or applications of the aforementioned kinetics models may also consider including fit parameters that describe various features of the aggregation, such as fractal dimension and n, as they provide insight into features of the aggregates that are not directly taken into account in kinetics equations, such as the three-dimensional (3D) geometry of the proteins and of aggregates. Additionally, researchers may consider measuring $M(t)$ and $L(t)$ in experiments, which can then both be used to fit a Smoluchowski-type mass-action model and should provide more insight into the mechanism occurring in experiments. Furthermore, we have only considered models that describe the fibrillization of amyloid proteins through the lens of nucleated polymerization. More descriptive models that are capable of describing the conformational changes in the proteins could provide a clearer picture of the underlying mechanisms that drive monomers to aggregate into fibrils that are dominated by β-sheet content. In a previous work,[33] we discussed adding conformational dependences to nucleated polymerization reactions involving critical nuclei. Reactions that describe conformational changes may potentially be used in Smoluchowski-like models to describe the transition of unstructured oligomers into those that contain significant β-structure that may go onto form β-dominated fibrils. A model of this type could be combined with size-dependent kernels and compared with, for example, FlAsH fluorescence, whose agent is able to bind more readily to oligomer-sized structures with less β-content than ThT or Congo red florescence.[37] Still there are other approaches for describing the rate constants. One such way would be to probe the dominant processes using coarse-grained models and use the predicted rates in a mass-action model such as the Smoluchowski model, rather than fitting the model to experimental data. This approach could possibly be used to decipher which reactions contribute to nucleation, growth, and shrinkage of fibrils, as well as helping to understand what might be missing from these models.

4. Effects of Molecular Crowding on Protein Stability and Reactions

In a living organism, the environments that a protein molecule is faced with can be very crowded by other macromolecules and confined by compartment lipid membranes. The presence of these crowders and confinement may affect the stability as well as the reaction properties of a protein.[90,91] If we treat approximately the crowders as hard spheres, the driving force for these effects is entropic in nature. Inorganic materials are usually dominated by bond energies, but for biomolecules such as protein and DNA, entropy plays an important role because the hydrogen bonds and van der Waals interactions are weak compared to chemical bonds. Thus, in a small energy range, many conformations of similar energies exist. The reduction of the depletion zone or excluded volume, corresponding to overall higher entropy, is the driving force for the enhanced stability of the compact structures of proteins and for protein association. These effects of macromolecular crowding are well-studied subjects[92,93] and a few recent reviews can be found in Refs. 94–97

In this section, we outline a brief history of the group's activities in this field. Our interest in the field of crowding as well as confinement effects, which are similar, was first stimulated by the experimental work of Yen Wei's group at Drexel, in which they observed that the thermal stability of horseradish peroxidase and acid phosphatase, encapsulated in mesoporous silica, increases as the pore size of the silica matrix decreases. The pore size is controlled by a sol–gel process using glucose or fructose as a template, which can later be dissolved in water.[98] We have carried out simulations using an HP lattice model,[99] which shows confinement increases the melting point of the lattice peptide[98] and is consistent with the experimental results. We also studied the effects of confinement on the kinetics of refolding and unfolding. Wei's group later extended their study to cytochrome c and observed increased stability of encapsulated protein against unfolding induced by denaturants, such as urea.[100] On the theoretical side, we showed using the HP model plus Brownian dynamics that both the folding and unfolding times increase with the number of crowders in solution, but the equilibrium constant is shifted toward the folded state, indicating that crowding enhances stability.[100]

Yang and his group measured directly the effects of macromolecular crowding on the unfolding forces of ubiquitin using a single-molecule method based on AFM.[101] They pulled a polymer of eight ubiquitin proteins, prepared by protein engineering, using an AFM tip until a domain unfolded. Repeating the experiment several times, they obtained the average unfolding force. They observed more than a 10% increase in the unfolding force caused by the presence of dextran (40 kDa at 300 g/L and 0.25 m/s pulling rate). The results can be interpreted in terms of a model developed based on scaled particle theory (SPT).[101] Ping *et al.*[102] also carried out simulations using a 3D polymer model to calculate the depletion zone and the depletion force based on the Asakura–Oosawa theory,[103] and obtained results in qualitative agreement with the experiment. The results showed that the depletion force depends on both the volume fraction and size of the crowders.[102]

Gai and his group[104,105] have studied trpzip-m1 and some designed peptides with β-strand structures, including crowding experiment on trpzip. Following their work, Hoppe and Yuan[106] investigated the effects of crowders on the folding of β-sheet peptides by using an implicit crowding method where the peptides are represented by bead-lattice models. The density of states of the system was calculated using the Wang–Landau method,[107] from which we obtained the heat capacity and melting temperature of the system. The method was applied to trpzip-m1 and two designed peptides, one with three and the other with four β-strands.[106] For the wild-type peptide, tryzip-m1, both experimental results[105] and our calculations show that melting temperature increases as crowder concentration increases, that is, peptide stability increases with the volume fraction of the crowder, but not so for the designed peptides. This is because conformations can be found for these two systems that are more compact than the designed peptides.[104]

A surprising depletion-force-generated flow was serendipitously discovered by Hoppe.[108] In idealized 2D simulations, we assume that a hard ellipse with its center fixed at the origin is bombarded from all sides by crowder particles (hard disks), which are initially randomly positioned and directed with their velocities drawn from a Gaussian distribution. All hard disks and the rigid ellipse are enclosed within a lager 2D rigid square box. Additionally, all collisions between disks and between a disk and the interior and exterior boundaries are treated elastically rigorously and the simulation was done using DMD. After a transient period, the average density of the system is analyzed along the minor and major axes of the ellipse, which shows oscillatory behavior with the largest peak near the elliptic boundary. This behavior is well-known and is due to the average shell structure formed by the crowders near a rigid surface. The surprise is that when the average velocity vectors of disks are analyzed, they form a vortex flow in each of the four quadrants.[105] For example, in the first quadrant the vortex flow is counterclockwise and in 2nd quadrant the flow is clockwise, so that there is neither net flow nor net angular momentum in the system. There is a simple explanation for the generation of vortices from the initial random motions of the disks. It is well-known that the depletion force tends to drive particles from the most convex part of surface to a flatter region of the surface, because it will reduce the total area of the depletion zone, driven by the second law of thermodynamics. This moving tendency sets up the flow, but because the box is finite, a counterclockwise vortex is formed, for example, in the first quadrant.

The Asakura–Oosawa theory[103] predicts that for two hard spheres (e.g., proteins) in a solution with many crowders, the net force existing between these two hard spheres is attractive. This is because association of two hard spheres (proteins) tends to reduce the depletion zone of the system, thus crowders have more space to move into. This is the reason why macromolecular crowding favors protein assembly.[109] However, in general, as discussed in Sections 2 and 3, the self-aggregation processes of proteins are very complicated, involving nucleation, monomer addition and subtraction, elongation of filaments, fragmentation and association of aggregates, or nucleated conformational transformations, and even heterogeneous nucleation, and so on. In principle, to understand the effects of crowders on protein aggregation processes completely, we need to understand the crowding effects on each of the above steps, separately. Such an understanding may take a while to accomplish.

It is thus encouraging to see the recent publications on this subject by Minton, Hall and Hoppe.[110,111,118,119] Although greatly simplified kinetic and equilibrium models are employed in these studies, some of the concepts introduced can form a basis for more rigorous equilibrium and kinetic studies. We have carried out such studies, the details of which are the topic of the next section. In short, still simple, but often-used, equilibrium and kinetic models (i.e., BD or Oosawa models) are employed to study the effects of crowders on the equilibrium and kinetic properties of protein aggregation.

5. Equilibrium and Kinetic Studies on the Effects of Crowders on Protein Aggregation

Several recent studies have applied theoretical approaches to studying the effect of crowders on the equilibrium properties of protein aggregation, as well as other aspects of their dynamics.[110–119] Experimental studies have also been done.[120–123] For example, Rosin *et al.*[120] used ThT fluorescence to study the kinetics of actin filament formation in the presence of several crowding agents (dextran, TMAO, and urea) which showed that, for dextran and TMAO, the critical concentration of monomers is lowered as the volume fraction of crowders is increased and the lag phase of aggregation is significantly decreased. Hatters *et al.*[122] achieved similar results from a turbidity study of human ApoC-II in the presence of dextran. It is not so trivial, however, to say that crowders simply favor aggregation. It has been shown[124–129] that for certain systems the presence of crowders can have the opposite effect and inhibit crowding, and in general there are competing factors along the pathways to aggregation, this has come to be known as the dual effect of crowders. It is evident then that the subject of the effects of macromolecular crowders on aggregation is still open to new theories and new interpretations, but limited by the relatively small amount of experimental data available. In this section, we introduce our method for including the crowding effect in equilibrium and kinetic models and discuss some of the results and implications.

Considering only homogeneous nucleation and monomer addition and subtraction, as is illustrated in Fig. 1(c)(i–ii), we may make use of the equilibrium constant relationship

$$K_r = \frac{\bar{a}_r}{\bar{a}_1\bar{a}_{r-1}} = e^{\frac{-\Delta G_r}{k_B T}}, \tag{19}$$

where $\Delta G_r = G_r - G_{r-1} - G_1$ is the free energy difference between the associating monomer-polymer pair and the aggregate, and $\bar{a}_r$ is the chemical activity of an r-mer. Multiplying K_r by K_{r-1}, K_{r-2}, and so on, as follows,

$$\begin{aligned} &\frac{\bar{a}_r}{\bar{a}_1\bar{a}_{r-1}} * \frac{\bar{a}_{r-1}}{\bar{a}_1\bar{a}_{r-2}} \cdots \\ &= \frac{\bar{a}_r}{\bar{a}_1^r} \end{aligned} \tag{20}$$

we obtain

$$\frac{\bar{a}_r}{\bar{a}_1^r} = e^{-\frac{\Delta G_r^0}{k_B T}}, \tag{21}$$

where $\Delta G_r^0 = G_r - rG_1 = \mu_r - r\mu_1$ is the free energy difference between r independent monomers and an r-mer, for one mole. Plugging in $\bar{a}_r = \gamma_r c_{\text{eq},r}$ and rearranging, we have:

$$c_{\text{eq},r} = \frac{\gamma_1^r}{\gamma_r} e^{-\frac{\Delta G_r^0}{k_B T}} c_{\text{eq},1}^r. \tag{22}$$

Comparison with previous work[49] tells us that

$$\frac{\gamma_1^r}{\gamma_r} e^{-\frac{\Delta G_r^0}{k_B T}} = \frac{a_{r-1}\dots a_1}{b_r\dots b_2} = Q_r, \tag{23}$$

where Q_r is the partition function, a_r and b_r are the association and dissociation rate constants for an r-mer, respectively, and the superscript 0 indicates the value in a crowder-less solution. All that is left then is to determine the rate constants and specify the activity coefficients. SPT allows us to write down an expression for the activity coefficient of a sphero-cylindrical rod suspended in spherical crowders,[110,130]

$$\ln \gamma = \ln(1-\phi) + A_1 z + A_2 z^2 + A_3 z^3 \tag{24}$$

where

$$\begin{aligned} A_1 &= R^3 + 3R^2 + 3R + 1.5L(R^2 + 2R + 1) \\ A_2 &= 3R^3 + 4.5R^2 + 4.5L(R^2 + R) \\ A_3 &= 3R^3 + 4.5LR^2 \\ z &= \frac{\phi}{1-\phi}, \end{aligned} \tag{25}$$

where R is the ratio of the sphero-cylindrical and crowder radii, r_{sc}/r_c, and L is the ratio of cylindrical length to diameter. Following Minton,[110] we apply the approximation

$$\ln \gamma_r = \ln \gamma_{r-1} + \frac{d \ln \gamma}{dr} = \ln \gamma_{r-1} + \frac{d \ln \gamma}{dL}\frac{dL}{dr} \tag{26}$$

and solving recursively for γ_r we get

$$\gamma_r = \gamma_1 \alpha^{r-1} \tag{27}$$

where

$$\begin{aligned} \alpha &= \exp\left(\frac{d \ln \gamma}{dL}\frac{dL}{dr}\right) \\ &= \exp\left\{\frac{2}{3}\left(\frac{r_1}{r_{sc}}\right)^3 * [1.5(R^2 + R + 1)z \right. \\ &\quad \left. + 4.5(R^2 + R)z^2 + 4.5R^2 z^3]\right\}, \end{aligned} \tag{28}$$

in which r_1 is the radius of a protein monomer. We consider the case where the total mass concentration of protein in the system is conserved, that is

$$c_0 = \sum_{r=1}^{\infty} r Q_r^0 \left(\frac{\gamma_1}{\alpha}\right)^{r-1} c_{\mathrm{eq},1}^r, \tag{29}$$

where c_0 is the initial mass of proteins. It is likely, however, that in living cells monomer concentration is held roughly constant,[131] whereas in *in vitro* experiments the total mass is constant. Thus, this approximation is most useful for comparison with available experiments. If the rate constants a_r^0 and b_r^0 are independent of r (the size of the polymer), analytic results for the equilibrium concentration of monomers may be expressed as:[49]

$$c_0 = \sum_{r=1}^{\infty} r(\lambda x)^{r-1} x$$
$$\lambda = \frac{a\gamma_1}{b\alpha}$$
$$x = c_{\text{eq},1} \tag{30}$$

which converges to

$$c_0 = \frac{x}{(1-\lambda x)^2}. \tag{31}$$

and may then be inverted to find

$$c_{\text{eq},1} = \frac{2c_0}{1 + 2\lambda c_0 + \sqrt{1 + 4\lambda c_0}}. \tag{32}$$

This result has been shown to be in good agreement with experiments.[132]

The kinetics are more complicated and, in general, analytic results are not possible, although approximate analytic results have been given.[23,31,133] Additionally, nucleation should play an explicit role and should particularly affect the lag phase of aggregation.[133] We make use of the transition state theory (TST) to study the effect of chemical activity of the rate constants. In TST, an intermediate step between reactants and products, the transition state, is assumed to be in quasi-equilibrium with both reactants and products. The Eyring equation,[134]

$$a_{r-1} = \kappa \frac{k_B T}{h} e^{-\beta \Delta G_r^{\ddagger}}, \tag{33}$$

is derived based on this assumption, where $\Delta G_r^{\ddagger} = G_r^{\ddagger} - G_{r-1} - G_1$ and a_{r-1} is the association rate constant for an $(r-1)$-mer and a monomer forming an r-mer. $G_r^{\ddagger}$ is the free energy of the transition state and G_{r-1} and G_1 are the free energies of an $(r-1)$-mer and a monomer, respectively. For one mole we have

$$G_{r-1} = \mu_{r-1} = \mu_{r-1}^0 + k_\text{B} T \ln(\gamma_{r-1} c_{r-1}). \tag{34}$$

Another approximation is introduced based on the fact that the transition state should more closely resemble the associated state than that of the reactants. Thus,

$$\gamma_r^{\ddagger} \approx \gamma_r, \tag{35}$$

and Eq. (33) then becomes:

$$a_{r-1} = \frac{\gamma_1}{\alpha} a^0_{r-1}. \tag{36}$$

Based on the argument made about the geometry of the transition state, the dissociation rate constant should be unaffected. The full set of rate equations, including nucleation, are:

$$\frac{dc_r}{dt} = \begin{cases} n(k_{dnuc}c_{n_c} - k_{nuc}c_1^{n_c}) - \sum_{r=n_c}^{\infty} \mathcal{J}_r & \text{if } r = 1 \\ 0 & \text{if } 1 < r < n_c \\ k_{nuc}c_1^{n_c} - k_{dnuc}c_{n_c} - \mathcal{J}_{n_c} & \text{if } r = n_c \\ \mathcal{J}_{r-1} - \mathcal{J}_r & \text{if } r > n_c, \end{cases} \tag{37}$$

where k_{nuc} and k_{dnuc} are the assembly and breaking rate constants for nucleation, respectively, n_c is the critical nucleus size and $\mathcal{J}_r$ is the flux given by

$$\mathcal{J}_r(t) = a_r c_r c_1 - b_{r+1} c_{r+1}. \tag{38}$$

The effect of crowders on the nucleation rate constant is given by

$$k_{nuc} = \left(\frac{\gamma_1}{\alpha}\right)^{n_c - 1} k^0_{nuc}. \tag{39}$$

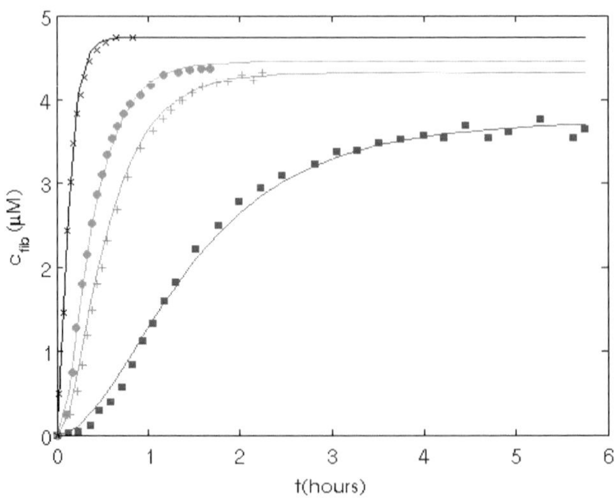

Figure 3: Fits for actin fibril formation in solutions of TMAO. The data are for $\phi = 0(+)$, $\phi = 0.0364(+)$, $\phi = 0.0721$, (•) and $\phi = 0.142$ (x).

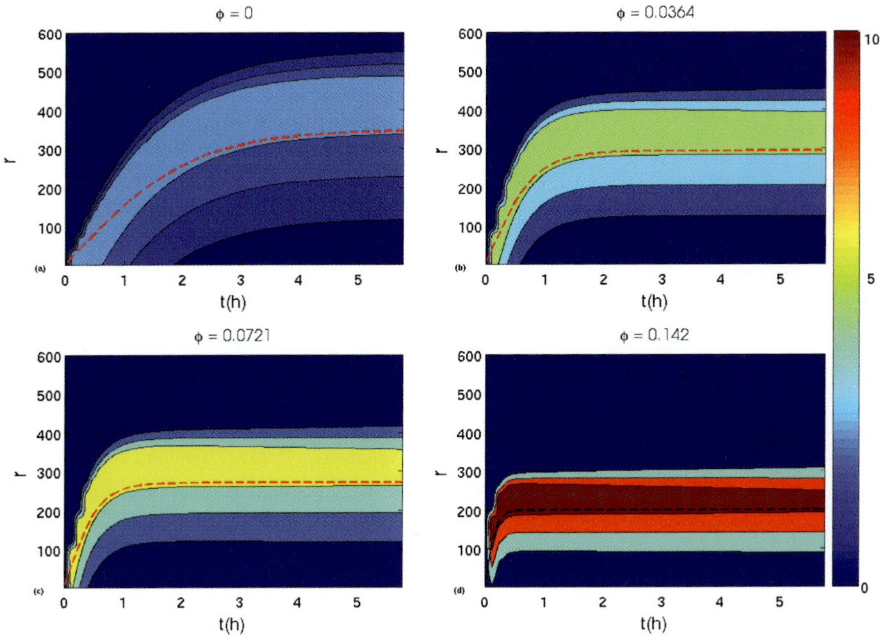

Figure 4: The polymer length distributions for $\phi = 0$ (blue), 0.0364 (magenta), 0.0721 (red), and 0.142 (black). The contours are in units of $10^{-5}\mu$M and the dashed line is the average length of polymers.

These equations then may be integrated numerically, with the effect of crowders on the rate constants explicitly included.

The results are in good agreement with the data available from Rosin *et al.*, where actin filament growth was monitored as a function of time in the presence of various concentrations of TMAO and dextran using ThT fluorescence. Fits of the model are shown in Fig. 3 for TMAO. An important aspect of this theory is that it allows for physically interpretable results, for instance, the radius of the crowders is shown to increase with increasing volume fraction. This is consistent with the fact that this radius is actually more of a correlation length for long range frictional interactions, and should increase as crowder concentration increases, as pointed out by Ioan *et al.*[135]

Figure 4 shows a contour plot of the length distribution of polymers as a function of time for the same system as in Fig. 3. An interesting result is that the increasingly large presence of crowders tends to force the polymer distribution into a larger number of small fibrils. This is due to the excluded volume effect the crowders have on the system, filaments simply cannot grow long since they are rigid and cannot occupy the same volume as the crowders. This result is also relevant given the current state of understanding of diseases, such as Alzheimer's, where it is believed that smaller oligomers are the primary cause of synaptic dysfunction and neuronal death.[136,137] For details regarding comparisons between the present model and and experimental results, we refer you to the original article.[132]

References

1. Buehler, M. J. and Yung, Y. C. Deformation and failure of protein materials in physiologically extreme conditions and disease. *Nat. Mater.* **8**(3), 175–188 (2009).
2. Dahl, K. N., Scaffidi, P., Islam, M. F., Yodh, A. G., Wilson, K. L., and T. Misteli, T. Distinct structural and mechanical properties of the nuclear lamina in Hutchinson–Gilford progeria syndrome. *Proc. Natl. Acad. Sci. USA*. **103**(27), 10271–10276 (2006).
3. Prockop, D. J. and Kivirikko, K. I. Collagens: molecular biology, diseases, and potentials for therapy. *Annu. Rev. Biochem.* **64**(1), 403–434 (1995).
4. Eaton, W. A. and Hofrichter, J. Sickle cell hemoglobin polymerization. *Adv. Protein Chem.* **4**, 63–279 (1990).
5. Chin, Y. R. and Toker, A. The actin-bundling protein palladin is an akt1-specific substrate that regulates breast cancer cell migration. *Mol. Cell.* **38**(3), 333–344 (2010).
6. Knowles, T. P., Vendruscolo, M., and Dobson, C. M. The amyloid state and its association with protein misfolding diseases. *Nat. Rev. Mol. Cell Biol.* **15**(6), 384–396 (2014).
7. Knowles, T. P. and Buehler, M. J. Nanomechanics of functional and pathological amyloid materials. *Nat. Nanotech.* **6**(8), 469–479 (2011).
8. Scheibel, T., Parthasarathy, R., Sawicki, G., Lin, X. M., Jaeger, H. and Lindquist, S. L. Conducting nanowires built by controlled self-assembly of amyloid fibers and selective metal deposition. *Proc. Natl. Acad. Sci. USA*. **100**(8), 4527–4532 (2003).
9. Morriss-Andrews, A. and Shea, J. E. Computational studies of protein aggregation: methods and applications. *Annu. Rev. Phys. Chem.* **66**, 643–666 (2015).
10. Zou, Y., Sun, Y., Zhu, Y., Ma, B., Nussinov, R., and Zhang, Q. Critical nucleus structure and aggregation mechanism of the c-terminal fragment of copper–zinc superoxide dismutase protein. *ACS Chem. Neurosci.* **7**(3), 286–296 (2016).
11. Carballo-Pacheco, M., Ismail, A. E., and Strodel, B. Oligomer formation of toxic and functional amyloid peptides studied with atomistic simulations. *J. Phys. Chem. B.* **119**(30), 9696–9705 (2015).
12. Urbanc, B., Betnel, M., Cruz, L., Bitan, G., and Teplow, D. B. Elucidation of amyloid beta-protein oligomerization mechanisms: discrete molecular dynamics study. *J. Am. Chem. Soc.* **132**(12), 4266–4280 (2010).
13. Wagoner, V. A., Cheon, M., Chang, I., and Hall, C. K. Impact of sequence on the molecular assembly of short amyloid peptides. *Proteins.* **82**(7), 1469–1483 (2014).
14. Barz, B. and Urbanc, B. Minimal model of self-assembly: emergence of diversity and complexity. *J. Phys. Chem. B.* **118**(14), 3761–3770 (2014).

15. Bieler, N. S., Knowles, T. P., Frenkel, D., and Vácha, R. Connecting macroscopic observables and microscopic assembly events in amyloid formation using coarse grained simulations. *PLoS Comput. Biol.* 8(10), e1002692 (2012).
16. Van Gestel J. and De Leeuw, S. A statistical–mechanical theory of fibril formation in dilute protein solutions. *Biophys. J.* **90**(9), 3134–3145 (2006).
17. Schreck, J. S. and Yuan, J. M. A statistical mechanical approach to protein aggregation. *J. Chem. Phys.* **135**(23), 235102 (2011).
18. Schreck, J. S. and Yuan, J. M. Statistical mechanical treatments of protein amyloid formation. *Int. J. Mol. Sci.* **14**(9), 17420–17452 (2013).
19. von Smoluchowski, M. Drei vorträge über diffusion, brownsche bewegung und koagulation von kolloidteilchen. *Physik. Z. B.* **17**, 557–571 (1916).
20. Becker, R. and Döring, W. Kinetische behandlung der keimbildung in übersättigten dämpfen. *Ann. Phys.* **24**, 719–752 (1935).
21. Burton, J. J. Nucleation theory. In *Statistical Mechanics, Part A: Equilibrium Techniques*, Ed. Bruce J. Berne. Waltham, MA, USA, pp. 195–234 (1975).
22. Wattis, J. A Becker–Döring model of competitive nucleation. *J. Phys. A: Math. Gen.* **32**(49), 8755 (1999).
23. Oosawa, F. and Kasai, M. Theory of linear and helical aggregations of macromolecules. *J. Mol. Biol.* **4**, 10–21 (1962).
24. Oosawa, F. and Asakura, S. *Thermodynamics of the Polymerization of Protein*. Academic Press, Cambridge, MA (1975).
25. Frieden, C. Protein aggregation processes: in search of the mechanism. *Protein Sci.* **16**(11), 2334–2344 (2007).
26. Murphy, R. M. Kinetics of amyloid formation and membrane interaction with amyloidogenic proteins. *Biochim. Biophys. Acta.* **1768**(8), 1923–1934 (2007).
27. Morris, A. M., Watzky, M. A., Agar, J. N., and Finke, R. G. Fitting neurological protein aggregation kinetic data via a 2-step, minimal/"ockham's razor" model: the Finke–Watzky mechanism of nucleation followed by autocatalytic surface growth. *Biochem.* **47**, 2413–2427 (2008).
28. Eisenberg, D. and Jucker, M. The amyloid state of proteins in human diseases. *Cell.* **148**(6), 1188–1203 (2012).
29. Cohen, S. I., Vendruscolo, M., Dobson, C. M., and Knowles, T. P. From macroscopic measurements to microscopic mechanisms of protein aggregation. *J. Mol. Biol.* **421**(2), 160–171 (2012).
30. Gillam, J. and MacPhee, C. Modelling amyloid fibril formation kinetics: mechanisms of nucleation and growth. *J. Phys. Condens. Matter.* **25**(37), 373101 (2013).
31. Knowles, T., Waudby, C., Devlin, G., Cohen, S., Aguzzi, A., Vendruscolo, M., Terentjev, E., Welland, M., and Dobson, C. An analytical solution to the kinetics of breakable filament assembly. *Science.* **326**(5959), 1533–1537 (2009).
32. Hong, L., Qi, X., and Zhang, Y. Dissecting the kinetic process of amyloid fiber formation through asymptotic analysis. *J. Phys. Chem. B.* **116**(23), 6611–6617 (2011).
33. Schreck, J. S. and Yuan, J. M. A kinetic study of amyloid formation: fibril growth and length distributions. *J. Phys. Chem. B.* **117**(21), 6574–6583 (2013).
34. Hammer, N. D., Schmidt, J. C., and Chapman, M. R. The curli nucleator protein, CsgB, contains an amyloidogenic domain that directs CsgA polymerization. *Proc Natl. Acad. Sci. USA.* **104**, 12494–12499 (2007).

35. Arosio, P., Beeg, M., Nicoud, L., and Morbidelli, M. Time evolution of amyloid fibril length distribution described by a population balance model. *Chem. Eng. Sci.* **78**, 21–32 (2012).
36. Qiang, W., Kelley, K., and Tycko, R. Polymorph-specific kinetics and thermodynamics of β-amyloid fibril growth. *J. Am. Chem. Soc.* **135**(18), 6860–6871 (2013).
37. Lee, J., Culyba, E., Powers, E., and Kelly, J. Amyloid-β forms fibrils by nucleated conformational conversion of oligomers. *Nat. Chem. Biol.* **70**(9), 602–609 (2011).
38. Šaric', A., Chebaro, Y. C., Knowles, T. P., and Frenkel, D. Crucial role of nonspecific interactions in amyloid nucleation. *Proc. Nat. Acad. Sci. USA.* **111**(50), 17869–17874 (2014).
39. Cohen, S. I., Linse, S., Luheshi, L. M., Hellstrand, E., White, D. A., Rajah, L., Otzen, D. E., Vendruscolo, M., Dobson, C. M., and Knowles, T. P. Proliferation of amyloid-β42 aggregates occurs through a secondary nucleation mechanism. *Proc. Nat. Acad. Sci. USA.* **110**(24), 9758–9763 (2013).
40. Ferrone, F., Hofrichter, J., and Eaton, W. Kinetics of sickle hemoglobin polymerization: I. studies using temperature-jump and laser photolysis techniques. *J. Mol. Biol.* **183**(4), 591–610 (1985).
41. Lomakin, A., Teplow, D., Kirschner, D., and Benedek, G. Kinetic theory of fibrillogenesis of amyloid β-protein. *Proc. Nat. Acad. Sci. USA.* **94**(15), 7942–7947 (1997).
42. Sabaté, R. and Estelrich, J. Evidence of the existence of micelles in the fibrillogenesis of β-amyloid peptide. *J. Phys. Chem. B.* **1090**(21), 11027–11032 (2005).
43. Ono, K., Condron, M., and Teplow, D. Structure–neurotoxicity relationships of amyloid β-protein oligomers. *Proc. Nat. Acad. Sci. USA.* **106**(35), 14745–14750 (2009).
44. Bernstein, S., Dupuis, N., Lazo, N., Wyttenbach, T., Condron, M., Bitan, G., Teplow, D., Shea, J., Ruotolo, B., Robinson, C., and Bowers, M. T. Amyloid-β protein oligomerization and the importance of tetramers and dodecamers in the aetiology of alzheimer's disease. *Nat. Chem.* **1**(4), 326–331 (2009).
45. Garzon-Rodriguez, W., Sepulveda-Becerra, M., Milton, S., and Glabe, C. Soluble amyloid Aβ-(1–40) exists as a stable dimer at low concentrations. *J. Biol. Chem.* **272**(34), 21037–21044 (1997).
46. Bitan, G., Kirkitadze, M. D., Lomakin, A., Vollers, S. S., Benedek, G. B., and Teplow, D. B. Amyloid β-protein (Aβ) assembly: Aβ(1–40) and Aβ(1–42) oligomerize through distinct pathways. *Proc. Natl. Acad. Sci. USA.* **1000**(1), 330–335 (2003).
47. Chandrasekhar, S. Stochastic problems in physics and astronomy. *Rev. Mod. Phys.* **150**(1), 1–89 (1943).
48. Petkova, A. T., Ishii, Y., Balbach, J. J., Antzutkin, O. N., Leapman, R. D., Delaglio, F., and Tycko, R. A structural model for alzheimer's β-amyloid fibrils based on experimental constraints from solid state NMR. *Proc. Natl. Acad. Sci. USA.* **99**, 16742–16747 (2002).
49. Wattis, J. An introduction to mathematical models of coagulation–fragmentation processes: a discrete deterministic mean-field approach. *Phys. D.* **222**(1), 1–20 (2006).
50. Masel, J., Jansen, V., and Nowak, M. Quantifying the kinetic parameters of prion replication. *Biophys. Chem.* **77**(2), 139–152 (1999).
51. Hong, L. and Yong, W. A. Simple moment-closure model for the self-assembly of breakable amyloid filaments. *Biophys. J.* **104**(3), 533–540 (2013).

52. Hartley, D., Walsh, D., Chianping, P., Diehl, T., Vasquez, S., Vassilev, P., Teplow, D., and Selkoe, D. Protofibrillar intermediates of amyloid β-protein induce acute electrophysiological changes and progressive neurotoxicity in cortical neurons. *J. Neurosci.* **19**(20), 8876–8884 (1999).
53. Nichols, M., Moss, M., Reed, D., Cratic-McDaniel, S., Hoh, J., and Rosenberry, T. Amyloid-β protofibrils differ from amyloid-β aggregates induced in dilute hexafluoroisopropanol in stability and morphology. *J. Biol. Chem.* **280**(4), 2471–2480 (2005).
54. Williams, A. D., Sega, M., Chen, M., Kheterpal, I., Geva, M., Berthelier, V., Kaleta, D. T., Cook, K. D., and Wetzel, R. Structural properties of aβ protofibrils stabilized by a small molecule. *Proc. Nat. Acad. Sci. USA.* **102**(20), 7115–7120 (2005).
55. Levine, H. Thioflavine T interaction with synthetic alzheimer's disease β-amyloid peptides: detection of amyloid aggregation in solution. *Protein Sci.* 20 (3), 404–410 (2008).
56. Bieschke, J., Zhang, Q., Powers, E., Lerner, R., and Kelly, J. Oxidative metabolites accelerate alzheimer's amyloidogenesis by a two-step mechanism, eliminating the requirement for nucleation. *Biochemistry.* **44**(13), 4977–4983 (2005).
57. Van Raaij, M., Van Gestel, J., Segers-Nolten, I., De Leeuw, S., and Subramaniam, V. Concentration dependence of β-synuclein fibril length assessed by quantitative atomic force microscopy and statistical–mechanical theory. *Biophys. J.* **95**(10), 4871–4878 (2008).
58. Xue, W., Homans, S., and Radford, S. Amyloid fibril length distribution quantified by atomic force microscopy single-particle image analysis. *Protein Eng. Des. Sel.* **22**(8), 489–496 (2009).
59. Adamcik, J., Jung, J., Flakowski, J., De Los Rios, P., Dietler, G., and Mezzenga, R. Understanding amyloid aggregation by statistical analysis of atomic force microscopy images. *Nat. Nano.* **5**(6), 423–428 (2010).
60. Michaels, T. C. and Knowles, T. P. Role of filament annealing in the kinetics and thermodynamics of nucleated polymerization. *J. Chem. Phys.* **140**(21), 214904 (2014).
61. Pallitto, M. M. and Murphy, R. M. A mathematical model of the kinetics of β-amyloid fibril growth from the denatured state. *Biophys. J.* **81**(3), 1805–1822 (2001).
62. Jain, S. and Udgaonkar, J. B. Evidence for stepwise formation of amyloid fibrils by the mouse prion protein. *J. Mol. Biol.* **382**(5), 1228–1241 (2008).
63. Hitchcock-DeGregori, S. E., Sampath, P., and Pollard, T. D. Tropomyosin inhibits the rate of actin polymerization by stabilizing actin filaments. *Biochemistry.* **27**(26), 9182–9185 (1988).
64. Kinosian, H. J., Selden, L. A., Estes, J. E., and Gershman, L. C. Actin filament annealing in the presence of ATP and phalloidin. *Biochemistry.* **32** (46), 12353–12357 (1993).
65. Sept, D., Xu, J., Pollard, T. D., and McCammon, J. A. Annealing accounts for the length of actin filaments formed by spontaneous polymerization. *Biophys. J.* **77**(6), 2911–2919 (1999).
66. Hatters, D. M., MacPhee, C. E., Lawrence, L. J., Sawyer, W. H., and Howlett, G. J. Human apolipoprotein C-II forms twisted amyloid ribbons and closed loops. *Biochemistry.* **39**(28), 8276–8283 (2000).

67. Conway, K. A., Lee, S. J., Rochet, J. C., Ding, T. T., Williamson, R. E., and Lansbury, P. T. Acceleration of oligomerization, not fibrillization, is a shared property of both α-synuclein mutations linked to early-onset parkinson's disease: implications for pathogenesis and therapy. *Proc. Nat. Acad. Sci. USA.* **97**(2), 571–576 (2000).
68. Andrianantoandro, E., Blanchoin, L., Sept, D., McCammon, J. A., and Pollard, T. D. Kinetic mechanism of end-to-end annealing of actin filaments. *J. Mol. Biol.* **312**(4), 721–730 (2001).
69. González, J. M., Vélez, M., Jiménez, M., Alfonso, C., Schuck, P., Mingorance, J., Vicente, M., Minton, A. P., and Rivas, G. Cooperative behavior of *Escherichia coli* cell-division protein ftsz assembly involves the preferential cyclization of long single-stranded fibrils. *Proc. Nat. Acad. Sci. USA.* **102**(6), 1895–1900 (2005).
70. Hatters, D. M., MacRaild, C. A., Daniels, R., Gosal, W. S., Thomson, N. H., Jones, J. A., Davis, J. J., MacPhee, C. E., Dobson, C. M., and Howlett, G. J. The circularization of amyloid fibrils formed by apolipoprotein C-II. *Biophys. J.* **85**(6), 3979–3990 (2003).
71. Yokoi, H., Kinoshita, T., and Zhang, S. Dynamic reassembly of peptide rada16 nanofiber scaffold. *Proc. Nat. Acad. Sci. USA.* **102**(24), 8414–8419 (2005).
72. Meehan, S., Knowles, T. P., Baldwin, A. J., Smith, J. F., Squires, A. M., Clements, P., Treweek, T. M., Ecroyd, H., Tartaglia, G. G., Vendruscolo, M. M., *et al.* Characterisation of amyloid fibril formation by small heat-shock chaperone proteins human αa-, αb-and r120g αb-crystallins. *J. Mol. Biol.* **372**(2), 470–484 (2007).
73. Thorn, D. C., Ecroyd, H., Sunde, M., Poon, S., and Carver, J. A. Amyloid fibril formation by bovine milk α_{s2}-casein occurs under physiological conditions yet is prevented by its natural counterpart, α_{s1}-casein. *Biochemistry.* **47**(12), 3926–3936 (2008).
74. Binger, K. J., Pham, C. L., Wilson, L. M., Bailey, M. F., Lawrence, L. J., Schuck, P., and Howlett, G. J. Apolipoprotein C-II amyloid fibrils assemble via a reversible pathway that includes fibril breaking and rejoining. *J. Mol. Biol.* **376**(4), 1116–1129 (2008).
75. Juárez, J., Taboada, P., and Mosquera, V. Existence of different structural intermediates on the fibrillation pathway of human serum albumin. *Biophys. J.* **96**(6), 2353–2370 (2009).
76. Ecroyd, H. and Carver, J. A. Crystallin proteins and amyloid fibrils. *Cell. Mol. Life Sci.* **66**(1), 62–81 (2009).
77. Jordens, S., Adamcik, J., Amar-Yuli, I., and Mezzenga, R., Disassembly and reassembly of amyloid fibrils in water–ethanol mixtures. *Biomacromolecules.* **12**(1), 187–193 (2010).
78. Teoh, C. L., Pham, C. L., Todorova, N., Hung, A., Lincoln, C. N., Lees, E., Lam, Y. H., Binger, K. J., Thomson, N. H., Radford, S. E., *et al.* A structural model for apolipoprotein C-II amyloid fibrils: experimental characterization and molecular dynamics simulations. *J. Mol. Biol.* **405**(5), 1246–1266 (2011).
79. Yang, S., Griffin, M. D., Binger, K. J., Schuck, P., and Howlett, G. J. An equilibrium model for linear and closed-loop amyloid fibril formation. *J. Mol. Biol.* **421**(2), 364–377 (2012).
80. Arosio, P., Owczarz, M., Wu, H., Butté, A., and Morbidelli, M. End-to-end self-assembly of RADA 16-I nanofibrils in aqueous solutions. *Biophys. J.* **102**(7), 1617–1626 (2012).

81. Xu, L., Chen, Y., and Wang, X. Assembly of amyloid β peptides in the presence of fibril seeds: one-pot coarse-grained molecular dynamics simulations. *J. Phys. Chem. B.* **118** (31), 9238–9246 (2014).
82. Ilie, I. M., den Otter, W. K., and Briels, W. J. A coarse grained protein model with internal degrees of freedom. application to α-synuclein aggregation. *J. Chem. Phys.* **144**(8), 085103 (2016).
83. Hill, T. L. Length dependence of rate constants for end-to-end association and dissociation of equilibrium linear aggregates. *Biophys. J.* **44**(2), 285–288 (1983).
84. Arosio, P., Rima, S., Lattuada, M., and Morbidelli, M. Population balance modeling of antibodies aggregation kinetics. *J. Phys. Chem. B.* **116**(24), 7066–7075 (2012).
85. Nicoud, L., Owczarz, M., Arosio, P., and Morbidelli, M. A multiscale view of therapeutic protein aggregation: a colloid science perspective. *J. Biotechnol.* **10**(3), 367–378 (2015).
86. Feder, J., Jøssang, T., and Rosenqvist, E. Scaling behavior and cluster fractal dimension determined by light scattering from aggregating proteins. *Phys. Rev. Lett.* **53**, 1403–1406 (1984).
87. Owczarz, M., Bolisetty, S., Mezzenga, R., and Arosio, P. Sol–gel transition of charged fibrils composed of a model amphiphilic peptide. *J. Colloid Interface Sci.* **437**, 244–251 (2015).
88. Dovidchenko, N. V., Glyakina, A. V., Selivanova, O. M., Grigorashvili, E. I., Suvorina, M. Y., Dzhus, U. F., Mikhailina, A. O., Shiliaev, N. G., Marchenkov, V. V., Surin, A. K., *et al.* One of the possible mechanisms of amyloid fibrils formation based on the sizes of primary and secondary folding nuclei of Aβ40 and Aβ42. *J. Struct. Biol.* **194**(3):404–414 (2016).
89. Lee, C. F. Thermal breakage of a semiflexible polymer: breakage profile and rate. *J. Phys. Condens. Matter.* **27**(27), 275101 (2015).
90. Ellis, R. J. Macromolecular crowding: obvious but underappreciated. *Trends Biochem. Sci.* **26**(10), 597–604 (2001).
91. Minton, A. P. The influence of macromolecular crowding and macromolecular confinement on biochemical reactions in physiological media. *J. Biol. Chem.* **276**(14), 10577–10580 (2001).
92. Minton, A. P. Models for excluded volume interaction between an unfolded protein and rigid macromolecular cosolutes: macromolecular crowding and protein stability revisited. *Biophys. J.* **88**(2), 971–985 (2005).
93. Soranno, A., Koenig, I., Borgia, M. B., Hofmann, H., Zosel, F., Nettels, D., and Schuler, B. Single-molecule spectroscopy reveals polymer effects of disordered proteins in crowded environments. *Proc. Nat. Acad. Sci. USA.* **111**(13), 4874–4879 (2014).
94. Zhou, H. X., Rivas, G., and Minton, A. P. Macromolecular crowding and confinement: biochemical, biophysical, and potential physiological consequences. *Annu. Rev. Biophys.* **37**, 375 (2008).
95. Kuznetsova, I. M., Turoverov, K. K., and Uversky, V. N. What macromolecular crowding can do to a protein. *Int. J. Mol. Sci.* **15**(12), 23090–23140 (2014).
96. Politou, A. and Temussi, P. A. Revisiting a dogma: the effect of volume exclusion in molecular crowding. *Curr. Opin. Struct.* Biol. **30**, 1–6 (2015).
97. Sapir, L. and Harries, D. Is the depletion force entropic? Molecular crowding beyond steric interactions. *Curr. Opin. Colloid Interface Sci.* **20**(1), 3–10 (2015).

98. Ping, G., Yuan, J., Vallieres, M., Dong, H., Sun, Z., Wei, Y., Li, F., and Lin, S. Effects of confinement on protein folding and protein stability. *J. Chem. Phys.* **118**(17), 8042–8048 (2003).
99. Chan, H. S. and Dill, K. A. Transition states and folding dynamics of proteins and heteropolymers. *J. Chem. Phys.* **100**(12), 9238–9257 (1994).
100. Ping, G., Yuan, J. M., Sun, Z., and Wei, Y. Studies of effects of macromolecular crowding and confinement on protein folding and protein stability. *J. Mol. Recogn.* 17(5), 433–440 (2004).
101. Yuan, J. M., Chyan, C. L., Zhou, H. X., Chung, T. Y., Peng, H., Ping, G., and Yang, G. The effects of macromolecular crowding on the mechanical stability of protein molecules. *Protein Sci.* **17**(12), 2156–2166 (2008).
102. Ping, G., Yang, G., and Yuan, J. M. Depletion force from macromolecular crowding enhances mechanical stability of protein molecules. *Polymer.* **47**(7), 2564–2570 (2006).
103. Asakura, S. and Oosawa, F. Interaction between particles suspended in solutions of macromolecules. *J. Polym. Sci.* **33**(126), 183–192 (1958).
104. Xu, Y., Bunagan, M. R., Tang, J., and Gai, F. Probing the kinetic cooperativity of β-sheet folding perpendicular to the strand direction. *Biochemistry.* **47**(7), 2064–2070 (2008).
105. Mukherjee, S., Waegele, M. M., Chowdhury, P., Guo, L., and Gai, F. Effect of macromolecular crowding on protein folding dynamics at the secondary structure level. *J. Mol. Biol.* **393**(1), 227–236 (2009).
106. Hoppe, T. and Yuan, J. M. Protein folding with implicit crowders: a study of conformational states using the wang — landau method. *J. Phys. Chem. B.* **115**(9), 2006–2013 (2011).
107. Wang, F. and Landau, D. Determining the density of states for classical statistical models: a random walk algorithm to produce a flat histogram. *Phys. Rev. E.* **640**(5), 056101 (2001).
108. Hoppe, T. and Yuan, J. M. Entropic flows, crowding effects, and stability of asymmetric proteins. *Phys. Rev. E.* **80**(1), 011404 (2009).
109. Minton, A. P. Implications of macromolecular crowding for protein assembly. *Curr. Opin. Struct. Biol.* **10**(1), 34–39 (2000).
110. Minton, A. P. The effect of time-dependent macromolecular crowding on the kinetics of protein aggregation: a simple model for the onset of age-related neurodegenerative disease. *Front. Phys.* **2**, 48 (2014).
111. Hoppe, T. and Minton, A. P. An equilibrium model for the combined effect of macromolecular crowding and surface adsorption on the formation of linear protein fibrils. *Biophys. J.* **108**(4), 957–966 (2015).
112. Tjong, H. and Zhou, H. X. Prediction of protein solubility from calculation of transfer free energy. *Biophys. J.* **95**, 2601–2609 (2008).
113. Qin, S., Cai, L., and Zhou, H. X. A method for computing association rate constants of atomistically represented proteins under macromolecular crowding. *Phys. Biol.* **9**(6), 066008 (2012).
114. Goins, H. S. A. and M. Waxham, M. Macromolecular crowding and size effects on probe microviscosity. *Biophys. J.* **95**, 5362–5373 (2008).
115. Kim, Y. and Mittal, J. Crowding induced entropy–enthalpy compensation in protein association equilibria. *Biophys. J.* **95**, 5362–5373 (2008).

116. Kozer, G. H. N., Kuttner, Y. Y., and Schreiber, G. Protein–protein association in polymer solutions: from dilute to semidilute to concentrated. *Biophys. J.* **92**, 2139–2149 (2007).
117. Latshaw, D. C., Cheon, M., and Hall, C. K. Effects of macromolecular crowding on amyloid β (16–22) aggregation using coarse-grained simulations. *J. Phys. Chem. B.* **1180**(47), 13513–13526 (2014).
118. Hall, D. and Minton, A. Effects of inert volume-excluding macromolecules on protein fiber formation. I. equilibrium models. *Biophys. Chem.* **98**(1–2), 93–104 (2002).
119. Hall, D. and Minton, D. Effects of inert volume-excluding macromolecules on protein fiber formation. II. kinetic models for nucleated fiber growth. *Biophys. Chem.* **107**(3), 299–316 (2004).
120. Rosin, C., Schummel, P. H., and Winter, R. Cosolvent and crowding effects on the polymerization kinetics of actin. *Phys. Chem.* **17**(13), 8330–8337 (2015).
121. Munishkina, L. A., Cooper, E. M., Uversky, V. N., and Fink, A. L. The effect of macromolecular crowding on protein aggregation and amyloid fibril formation. *J. Mol. Recognit.* **17**(5), 456–464 (2004).
122. Hatters, D. M., Minton, A. P., and Howlett, G. J. Macromolecular crowding accelerates amyloid formation by human apolipoprotein C-II. *J. Biol. Chem.* **277**(10), 7824–7830 (2002).
123. Chebotareva, N. A., Filippov, D. O., and Kurganov, B. I. Effect of crowding on several stages of protein aggregation in test systems in the presence of α-crystallin. *Int. J. Biol. Macromol.* **80**, 358–365 (2015).
124. Mittal, S. and Singh, L. Macromolecular crowding decelerates aggregation of a β-rich protein, bovine carbonic anhydrase: a case study. *J. Biochem.* **156**(5), 273–282 (2014).
125. Kuznetsova, I. M., Zaslavsky, B. Y., Breydo, L., Turoverov, K. K., and Uversky, V. N. Beyond the excluded volume effects: mechanistic complexity of the crowded milieu. *Molecules.* **20**(1), 1377–1409 (2015).
126. Breydo, L., Reddy, K. D., Piai, A., Felli, I. C., Pierattelli, R., and Uversky, V. N. The crowd you're in with: effects of different types of crowding agents on protein aggregation. *Biochim. Biophys. Acta.* **1844**(2), 346–357 (2014).
127. Phillip, Y., Harel, M., Khait, R., Qin, S., Zhou, H. X., and Schreiber, G. Contrasting factors on the kinetic path to protein complex formation diminish the effects of crowding agents. *Biophys. J.* **103**(5), 1011–1019 (2012).
128. Cabaleiro-Lago, C., Quinlan-Pluck, F., Lynch, I., Dawson, K. A., and Linse, S. Dual effect of amino modified polystyrene nanoparticles on amyloid β protein fibrillation. *ACS Chem. Neurosci.* **1**(4), 279–287 (2010).
129. Co, N. T., Hu, C. K., and Li, M. S. Dual effect of crowders on fibrillation kinetics of polypeptide chains revealed by lattice models. *J. Chem. Phys.* **138**(18), 185101 (2013).
130. Cotter, M. A. Hard-rod fluid: scaled particle theory revisited. *Phys. Rev. A.* **10**, 625 (1974).
131. Haass, C., Schlossmacher, M. G., Hung, A. Y., Vigo-Pelfrey, C., Mellon, A., Ostaszewski, B. L., Lieberburg, I., Koo, E. H., Schenk, D., and Teplow, D. B. Amyloid beta-peptide is produced by cultured cells during normal metabolism. *Nature.* **359**(6393), 322–325 (1992).
132. Bridstrup, J. and Yuan, J. M. Effects of crowders on the equilibrium and kinetic properties of protein aggregation. *Chem. Phys. Lett.* **659**, 252–257 (2016).

133. Hong, L. and J. Lei, J. Statistical mechanical model for helix-sheet-coil transitions in homopolypeptides. *Phys. Rev. E.* **78**, 051904 (2008).
134. Eyring, H. The activated complex in chemical reactions. *J. Chem. Phys.* 3(2), 107–115 (1935).
135. Ioan, C. E., Aberle, T. and Burchard, W. Light scattering and viscosity behavior of dextran in semidilute solution. *Macromolecules*. **34**, 326–336 (2001).
136. Haass, C. and Selkoe, D. J. Soluble protein oligomers in neurodegeneration: lessons from the alzheimer's amyloid bold β-peptide. *Nat. Rev. Mol. Cell Biol.* **8**, 101–112 (2007).
137. Benilova, I., Karran, E., and De Strooper, B. The toxic Aβ oligomer and alzheimer's disease: an emperor in need of clothes. *Nat. Neurosci.* **15**, 349–357 (2012).

Thermodynamics and Kinetics of Protein Folding and Aggregation

6

Min-Yeh Tsai, Jian-Min Yuan† and Sheng Hsien Lin‡*

**Department of Chemistry, Rice University, Houston, TX 77005*
†Department of Physics, Drexel University, Philadelphia, PA 19104
‡Department of Applied Chemistry, National Chiao Tung University, Hsinchu, Taiwan 30010

Thermodynamic and kinetic properties of proteins affect their folding and aggregation processes, which are responsible for a range of biological functions in cells. This chapter aims to introduce simple models of protein folding and aggregation, the nucleation–growth kinetics of protein aggregation, and its study in thermodynamic interpretation of aggregation mechanism. We first review the two-state thermodynamics of protein folding and a generalized two-state scheme, i.e., Ising model. Several Ising-based protein models are discussed, including a generalized Ising model for studying mean-field theory of protein folding and a modified Wako-Saitô-Muñoz-Eaton (M-WSME) model for describing foldon behavior of protein. Also, we review some mechanisms of protein aggregation and a statistical mechanical approach in addressing aggregation equilibrium properties. Finally, we report our recent work on the kinetics of protein aggregation using a kinetic Ising model and discuss diverse aggregation scenarios through a kinetic survey over four different amyloid proteins.

1. Introduction

Proteins are linear biopolymers that are able to spontaneously fold into a highly organized, usually compact and stable conformation, referred to as the structure of a native protein. The one-dimensional sequence of a protein, encoded at the genetic level, amazingly contains all the information needed to specify its three-dimensional

(3D) structure. Since a polypeptide chain is composed of amino acids with charged, polar, or nonpolar side-chain residues, folding usually involves a range of weak non-covalent interactions (H-bonds, electrostatics, van der Waals forces, etc.) that all together result in a vast cost in conformational entropy of the chain upon formation of the native structure. On top of that, those interactions along with complicated solvent effects further raise the difficulty in understanding the folding mechanism. Natural proteins, however, have evolved to perceive such a folding complexity and thus optimize the interactions between side-chain residues. As a result, they can correctly fold within the biological time scale and carry out a variety of functions involved in every biological process in the cell, such as enzyme activity, signal transduction, transcription regulation, etc. On the other hand, the structure of a native protein can be disrupted upon increasing temperature or by addition of chemical denaturants, such as urea and guanidine hydrochloride. Such a process of the protein denaturation is responsible for not only structural changes of proteins but also the loss of their enzyme activities. The folding–unfolding of proteins is in fact reversible, which means protein conformational changes are dictated by the laws of thermodynamics. This statement can be further rephrased as that a native protein adopts the conformation that minimizes its free energy.[1] A native protein, however, must retain its conformational flexibility in order to function, and therefore, is only marginally stable in its physiological environment. Since proteins are structurally complex and their folding dynamics involve large ranges in both length and time scales, the problem of spontaneous folding continues to challenge the protein community.

Although proteins are synthesized to have specific structural folds, they perform a variety of functions in the cell, as mentioned earlier. Proteins, however, rarely work alone but interact with others (e.g., protein, nucleic acids) to carry out their biological tasks. Protein–protein interactions are essential to the functions of all cells. Aberrant interactions will cause the formation of toxic protein aggregates and may eventually lead to a fibrous structure as a result. The fibril aggregates, also referred to as amyloid fibrils, made of peptides or proteins *in vivo*, can often cause neurodegenerative disorders, such as Alzheimer's disease, Parkinson's disease, Huntington's disease, as well as prion-related diseases.[2] On the other hand, naturally formed polymers of proteins have also been implicated in functional uses by organisms ranging from bacteria to mammals. Almost all proteins under certain conditions, such as high temperature, form aggregates. In fact, the aggregate state of a protein may even be more stable than its native state, suggesting that amyloid-like aggregation is an universal property inherent in the nature of all polypeptide chains; a property that has both functional (e.g., amyloid materials) and pathological characteristics at nanoscale.[3] In high-level organisms that can learn, in particular, amyloids formed by prion proteins were found to be involved in the formation of long-term memory.[4,5] A recent computational study on a prion-like protein supported that its transition from soluble coiled-coil oligomers into insoluble prion-like fibers was triggered by force exerted by actin filaments of cytoskeleton in eukaryotic cells.[6] The mechanical insight into the formation of fibers would help localize active synapse area for forming long-term memory.

Amyloid fibrils are rich in cross-β structure,[7] including parallel and in-register β-sheets.[8,9] Such an assembly of protein aggregates is highly insoluble, compared with its constitutional protein components. The process of protein-to-fibril transformation therefore involves a significant structural change from one phase (soluble) to another one (insoluble) — a process called phase transition. To understand such a process, we can view it from two aspects: folding as well as aggregation. Understanding of how a single protein folds is one aspect, and understanding of how proteins aggregate is the other. The former involves primarily intramolecular interactions, while the latter includes intermolecular interactions, as illustrated in Fig. 1.

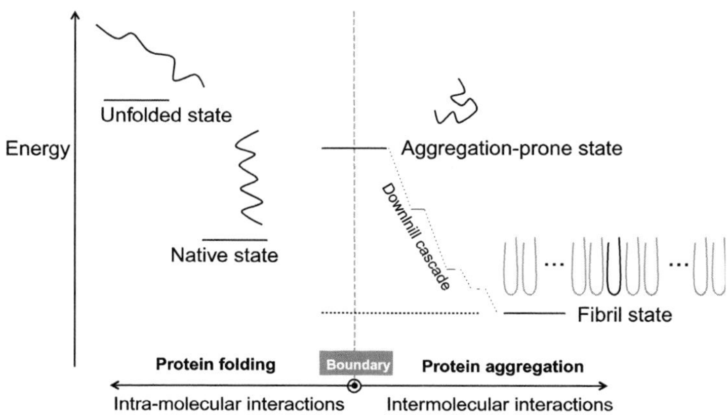

Figure 1: Schematic diagram of protein folding and aggregation is shown. The x and y axes denote, respectively, types of interactions and global energy state. The vertical dashed line drawn at the center indicates the boundary that separates protein folding and aggregation regimes. The left-hand side shows the folding regime where intramolecular interactions are dominant. The right-hand side presents the aggregation regime where proteins interact with each other via intermolecular interactions. Important protein conformational states are characterized (from left to right): unfolded state, native state, aggregation-prone state, and fibril state. If a protein gets misfolded, it can undergo a conformational transition into an aggregation-prone state, which is susceptible to form aggregates due to intermolecular contacts. In this situation, proteins would end up forming a global fibril state as a result.

In this chapter, we will review our study on protein folding, along with a review of the related literature. Meanwhile, we will also present our recent studies on protein aggregation. The organization of the article is as follows. In Sec. 2, we first review two-state thermodynamic descriptions of protein folding. Then, we review the use of simplified statistical mechanical models (i.e., Ising-like models) in describing protein secondary structure (e.g., α-helix, β-hairpin) as well as hierarchically more complex foldons. We also mention about a mean-field theory of protein folding on the basis of a generalized Ising model. In Sec. 3, we switch our topic into protein aggregation. We first give a brief review on the studies of protein aggregation mechanisms, followed by a short comment on the theoretical treatment of protein conformational changes involving the process of aggregation. In Sec. 4, we report our recent study on protein aggregation kinetics using a kinetic Ising model. We study thermodynamics and kinetics of the model and discuss a variety of aggregation scenarios based on a kinetic survey over four different amyloid proteins: (1) prion protein Sup35, (2) bovine insulin, (3) β_2-microglobulin, and (4) amyloid-β_{42} peptide. Finally, a brief conclusion is given in Sec. 5.

2. Simple Model in Protein Folding

2.1. *Two-state model*

Since the pioneer work of Anson and Mirsky[10,11] on the reversible nature of protein denaturation, scientists had realized that protein folding was governed by the laws of thermodynamics. The discovery of the reversible phenomenon was beautifully demonstrated in Anfinsen's Nobel Prize–winning work on the refolding of denatured ribonuclease. His thermodynamic hypothesis — *a native protein adopts the conformation that minimizes its free energy* — laid the basis for all subsequent thermodynamic studies of protein folding.[1] Thermodynamics deals with equilibrium properties such as enthalpy H and free energy G; they are all state functions. A state function refers to a property of protein system that depends only on its current equilibrium state, irrespective of the path by which the system arrives at the particular state. Although proteins exhibit a range of conformations, such as completely unfolded random coil, partially folded intermediates, and well-organized native structure, proteins generally demonstrate a two-state feature in their folding–unfolding process.[12] For example, the two-state behavior of cytochrome *c* (cyt *c*) was examined by calorimetry.[13] However, early kinetic studies in protein folding were limited to the time scale explored. A systematic study of protein folding kinetics was not feasible until the advent of new experimental techniques, such as high-resolution nuclear magnetic resonance (NMR) spectroscopy combined with hydrogen exchange (HX), time-resolved fluorescence techniques, mass spectroscopy, etc.[14–16] These novel techniques have provided better time resolution so as to reveal fast-folding events that were not previously available. In a systematic folding study, Plaxco *et al.* found that the folding stability and rates of proteins in fact correlated with their topology in the native state, suggesting that a native-centric simple model based on protein native topology could be sufficient in predicting proteins' thermodynamic and kinetic properties.[17] Theoretical studies, on the other hand, also provided explanations to proteins' simple folding kinetics.[18,19] Apparently, proteins are such a complex system with heterogeneous interactions when unfolded that their folding is similar to phase transitions of a finite system. The similarity therefore prompted us to develop a theoretical model that can describe the simple folding kinetics, along with its equilibrium properties.

The verification for the two-state folding thermodynamics is experimentally driven. A variety of experimental techniques, which are based on different physical and/or chemical principles, have been used to probe the process of protein folding/unfolding. Sometimes, the measured thermodynamic properties of folding, such as ΔG, ΔII, and ΔS, are consistent from different experimental techniques, and sometimes, they are not. Cytochrome *c* (cyt *c*), e.g., has been extensively studied using a variety of spectroscopy techniques,[20–22] illustrated in Fig. 2. Differential scanning calorimetry (DSC) provides the highest thermal stability (high folding–transition temperature), suggesting that the ΔH measured is a global property and its derived ΔG and ΔS should be interpreted on a global scale. Other spectroscopy techniques or probes, such as absorption spectroscopy, circular dichroism (CD), and fluorescence spectroscopy, however, measure protein's local properties pertaining to specific structural features, thereby giving rise to less thermal stability. Similar probe-dependent issue was also discussed on the determination of the folding-transition temperature of β-hairpin-forming peptides.[23]

2.2. *Ising-like model*

Although microscopic approaches to the thermodynamics and the kinetics of protein folding have been quite successful,[24–26] corresponding atomistic simulations in general are still limited to timescales shorter than biologically relevant ones, except for some fast-folding proteins.[27] Experimental studies have shown that many proteins are two-state folders. It is therefore reasonable to use a two-state–based model to form a theoretical framework in the generalization of protein folding studies. Ising-like models belong to such kinds. In simplified statistical mechanical models, one assumes the reduction of molecular degrees of freedom into coarse-grained

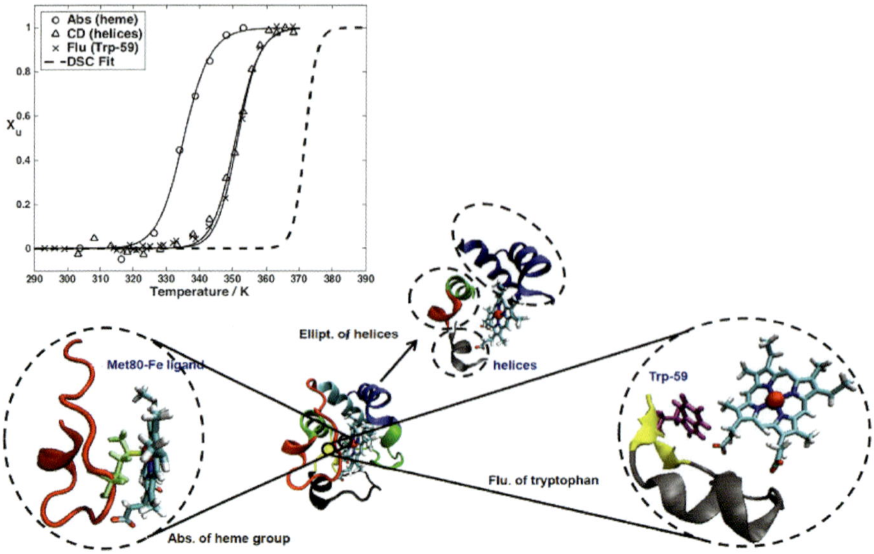

Figure 2: Thermal unfolding of cytochrome *c*(cyt *c*), measured using different types of spectroscopy techniques, is presented. Solid lines are numerical fits to the two-state model. The dashed line refers to a theoretical extrapolated result (at pH 7) of calorimetric measurements over a range of pH values (Figure reproduced and modified from Tsai's PhD thesis[22]). The cartoon diagram schematically exhibits different structural features (Met80-Fe ligand, helices, and Tryptophan-59) of cyt *c* that are detected using different experimental techniques (absorption spectroscopy [Abs], circular dichroism [CD], fluorescence spectroscopy [Flu], and differential scanning calorimetry [DSC], respectively).

ones in such a way that for each unit in a protein two states are sufficient to represent physically local configuration space explored during the folding–unfolding process.

The concept of coarse-grained structural unit can be treated using statistical mechanical models with varying levels of descriptions. The helix–coil theory, for example, was an early popular proposal to describe the formation of helical structures.[28,29] The mathematical technique used in the theory is the "transfer matrix method," which had been extensively applied to study the thermodynamics of a variety of helical variants[30] as well as other secondary structure motifs, e.g., β-sheet. Experimental studies using the helix–coil theory in general were quite successful because spectroscopy techniques such as CD in nature are particularly suitable in detecting protein helicity.[31,32] Understanding toward the dynamics of β-hairpin formation, however, was not as successful as that of helix formation. It was until late 1990s when Muñoz and coworkers combined fluorescence method and a new simple model (inspired by Wako and Saitô's earlier work,[33,34] officially named Wako-Saitô-Muñoz-Eaton (WSME) model[35] and has been analytically solved[36,37]) originally tailored to the topology of β-hairpin (with long-range interactions in sequence but spatially close) that one started to understand the thermodynamics, kinetics, and folding landscape of hairpin formation.[38,39] Later on, the model was further developed to deal with globular proteins with both α-helix and β-hairpin as well as tertiary interactions.[40] In the models mentioned earlier, the atomistic structure a protein or peptide is coarse grained at the level of residues or peptide bonds.

In the coarse-grained descriptions of protein, it is natural to consider a protein as a topological collection of peptide or residue units. On the one hand, one can significantly reduce a large amount of degrees of freedom of the main chain, in particular, in solvent. On the other hand, one can still keep the complexity in the description of protein configurational space to a certain degree. In some situations, however, one can further describe a single structural folding unit in terms of particular structural motifs because folding exhibits some repeating sequences of structural motifs in the native fold via tertiary interactions. The motifs can be α-helices, β-sheets, α/β, or hierarchically more complex foldons.[14,41,42] For example, cyt *c* was shown to contain five foldons. Each of them includes a collection of residues that all together demonstrate similar thermodynamic behavior due to cooperativity effect among them. Apparently, there is a need to fill the gap between the coarse-grained descriptions at the residue level and the foldon level. Some attempts have been made in finding the minimal number of peptide units required to form a foldon in protein using the energy landscape analysis along with some topological and sequence information.[43,44] To give a microscopic insight into the concept of foldon, we proposed an Ising model–based statistical mechanical model, modified from the WSME model (see the previous paragraph), where the protein Hamiltonian takes into account nonadditive many-body interactions between residues far apart in sequence. The modified WSME (M-WSME) model introduces empirical temperature parameters for hypothetical foldons in a peptide/protein to allow site-dependent properties of individual peptide units involved in a single foldon.[45] The site dependence reflects a cooperative property of foldon (see Fig. 3). Such an empirical treatment therefore offers a better experimental connection to probe-dependent techniques, independently verified by a molecular dynamics simulation study on a β-hairpin peptide.[46] We compare those exact site-dependent folding–unfolding curves with their mean-field result. We found that the mean-field result significantly alleviates thermodynamic site dependence (see Fig. 3, bottom), suggesting the effect of mean-field on local moieties of protein is the physical basis underlying the identification of foldon.

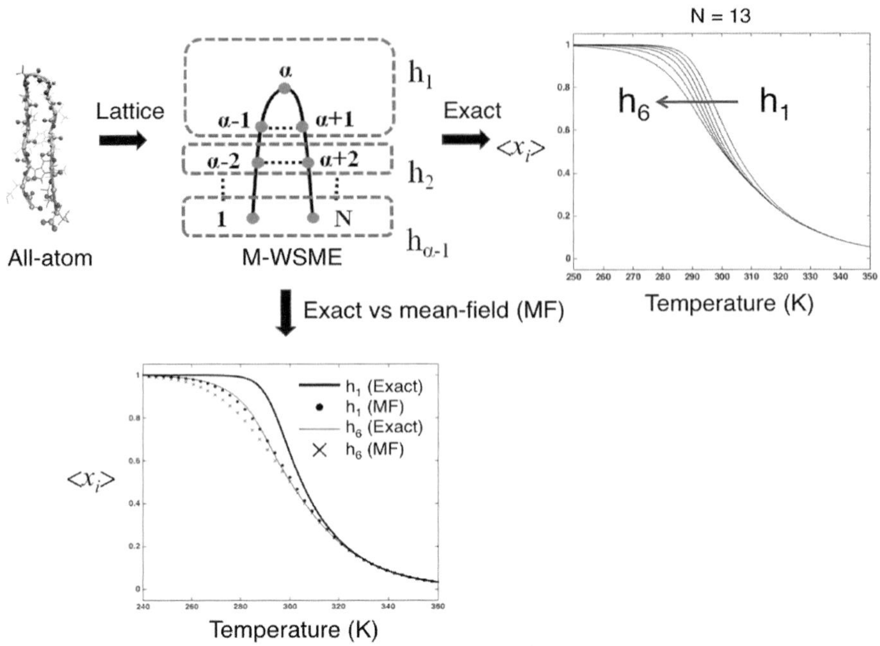

Figure 3: Hypothetical foldon(s) and site-dependent properties of a peptide/protein with a β-hairpin topology are shown. (Left) Schematic diagram of a β-hairpin is expressed in ribbon style with side-chain residues in licorice and backbone atoms in balls. (Middle) The β-hairpin is mapped into a lattice representation for statistical mechanical treatment using the M-WSME model ($N = 2\alpha - 1$, N is the number of units). (Right) Site-dependent folding–unfolding curves ($<x_i>$) are shown ($i = h_1 \sim h_6$). h_1 denotes the turn region while h_6 represents the terminal region. (Bottom) Folding–unfolding curves $<x_i>$ of h_1 (in black) and h_6 (in grey) are compared. Their exact solution (solid lines) exhibits distinct folding–unfolding behavior while their mean-field result (dots and crosses) shows similar behavior.

2.3. *Generalized Ising model*

Very often, theoretical studies of protein folding are devoted to either thermodynamics or kinetics. It is relatively challenging to connect both thermodynamics and kinetics within the same theoretical framework. Since experimental studies of folding–unfolding generally exhibit sigmoid behavior (e.g., induced by temperature, urea, GdnHCl, acids), it has been theoretically recognized as a quasi first-order phase transition using a spin glass analogy[24] or phase transition of finite systems.[47,48] Based on this, Liang *et al.* established a theoretical formalism that is compatible with the theory of phase transitions and meanwhile is applicable to both thermodynamics and kinetics. Starting from the scheme of the generalized Ising model (beyond nearest-neighbor interactions and/or with multi-state spins), the thermodynamics of protein folding–unfolding was described at the mean-field level.[49] As a result, the pairwise interactions in the conventional Ising model can be reduced as a one-body interaction object. The equations of state therefore can be solved in a self-consistent manner. In practical use, they assumed that the interactions involved in protein folding were primarily local in range (i.e., nearest neighbor) and uniform. The corresponding equations of state were derived in terms of an order parameter (the fraction of the folded state) that is related to experimentally measured quantity. In fact, any empirically inspired Ising-like folding Hamiltonian can be reformulated in terms of the generalized mean-field Ising model. We propose, e.g., a zipper-like model, inspired by the foldon construction of the native structure of cyt *c*,[41] to describe the classical "on-pathway" folding kinetics of the protein.[50] Moreover, a mean-field version of the M-WSME model is used to investigate the site-dependent properties of β-hairpin, as shown in Fig. 3. As has been commented in Sec. 2.2, mean-field treatment of localized interactions in protein folding has an important implication for foldon construction.

On the other hand, they applied the generalized Ising model to study the kinetics of protein folding–unfolding.[47,49] The kinetics involves time evolution of a collection of spin variables, which has been well formulated into a quantum master equation in terms of stochastic probability.[51] Although exact solution is provided for some particular cases,[51] the interactions specified are limited to short-range ones. Folding, however, is rather complicated and often involves a combination of short- and long-range interactions. To deal with the complex interaction scenario, Liang and coworkers derived a mean-field kinetic rate equation,[49] which allows for kinetic analysis for a variety of complicated models on the mean-field level (see Sec. 4.2 for its extension in protein aggregation).

3. Protein Aggregation

3.1. *Aggregation mechanism: a brief review*

The understanding toward the mechanism of protein aggregation depends on reliable kinetic measurements of fibril growth and insightful theoretical modeling. Quantitative kinetic measurement of aggregation is extremely difficult since the aggregation process involves certain stochastic fluctuation in forming nuclei and subsequent fibril seeds. It generally requires careful fine tuning in experimental condition and large amount of repetitive measurements in order to gain reliable results.[52] With insight gained from individual experimental methods and sometimes a combination of them,[53,54] researchers have proposed many theoretical models, which can be roughly divided into two categories: (1) nucleated polymerization (NP)-based mechanisms[55–57] and (2) nucleated conformational conversion (NCC) mechanisms[53,58,59] — a relatively new view. The NP-based model emphasizes that the formation of protein critical nuclei is the rate-limiting step of aggregation while the NCC model states that the key step associates with a conversion process by which the nuclei formed undergo a slow conformational transition into an active form of structure that allows subsequent fibril growth. Despite of the two different views described earlier, a general understanding toward kinetic pathways of aggregation has merged, summarized as nucleation, elongation, as well as some secondary pathways,[57,60,61] e.g., fragmentation,[52,62,63] lateral branching,[60,64,65] which all together seem to raise the difficulty in developing a unified kinetic pathway for proteins.

Theoretical treatments for the aforementioned models are primarily based on the well-known law of mass action. Researchers can refer to elementary processes of fibril formation and write down coupled differential equations accordingly that are either analytically or numerically tractable. Analytical solution[66–68] and exhaustive numerical calculations[59,69,70] for the models allow one to gain information in great detail, such as the time evolution of the concentration of filaments with any length and information about the length-dependent distribution. Although these works have greatly enhanced our knowledge about fibril formation, the approach used has to inevitably deal with a huge number of filament species with a wide distribution of length, i.e., more than thousands of filament species at least. The rationalization of a wide range of experimental results would depend highly on some additional assumptions used, due to a large number of parameters used. In fact, Auer and Kashchiev have showed that typical aggregation kinetic curves can be well described by only two parameters based on the theory of overall crystallization, suggesting the possibility in providing insight into the physical nature of the overall process of protein aggregation using only few basic properties.[71] With regard to protein aggregation mechanisms, Auer *et al.* used nucleation theory to analyze the thermodynamic conditions for one-step or two-step fibril nucleation.[72] Also, a simulation study was conducted to obtain aggregation mechanism in relation with temperature, concentration, as well as corresponding phase diagram.[73] Thus, it is of importance to understand protein aggregation from thermodynamic point of view. Recently, thermodynamic properties such as proteins' β-sheet propensity (see a review by Wu and Shea[74]) and the monomer-to-fibril (MtF) stability[75–78] started to receive much attention, suggesting the importance of thermodynamic descriptions in protein aggregation kinetics.[79] However, theoretical work regarding to this issue is still lacking.

In contrast to the mass-action approaches, we attempt to examine protein aggregation kinetics based on a simple model that allows the identification of phase transitions. The approach shall provide thermodynamic insight into the processes of nucleation-controlled aggregation. In Sec. 3.2, we review the use of a specialized Ising model, the Zimm-Brag (ZB) model, and its generalized version, Potts model, to study the thermodynamics of protein conformational changes and aggregation.

3.2. *Simple models of protein conformational changes and aggregation*

The ZB model or the like are revolutionary in the sense that they apply the powerful statistical mechanical techniques to the studies of protein conformational changes. Specifically, in the ZB model, one applies the spin-1/2 Ising model to study the properties of the two-state helix–coil transitions in proteins. However, there are several other secondary properties of proteins, besides helix and coil, which are important, such as, sheet and turn, etc., not to mention the more detailed classification of different types of helices, sheets, or turns. Especially, attempts have been made to include the sheet structures in the treatment.

To generalize the statistical mechanical treatment of the ZB model to include the sheet structure, Schreck and Yuan have applied recently the Potts model, so that transformations among sheet, helix, and coil can all be investigated in one theory.[80] Since the Potts model is a generalization of the two-state Ising model to a q-state problem, where q can be 2, 3, 4, or higher integer, it is not surprising that a Potts model is used in this generalization. However, since long-range interactions in the sequence space (but spatially close) are involved in the sheet or turn structures of proteins, how to carry out the generalization is nontrivial. One tool makes the ZB theory powerful is the use of 2×2 transfer matrices, it is beneficial to keep this technique in the theory. If L is the longest range of interactions between residues, careful thinking suggests that one can align L residues in a column and the whole sequence becomes a row of columns. The trick treats long-range interactions as nearest-neighbor interactions. According to this diagram it produces, we can write down the transfer matrices. The repeating unit in the ZB theory is, however, now replaced by a product of a set of matrices. The number of matrices increases with L. The dimension of the matrices therefore becomes $q^L \times q^L$, i.e., the dimension of the matrices depends on L and q.

As in the ZB theory, the partition function can now be expressed in terms of products of the set of transfer matrices discussed earlier. By numerically diagonalizing the transfer matrices, we can calculate exactly the thermodynamic properties of protein systems, e.g., phase transitions associated with the conformational changes. For details, we refer the readers to the original paper,[80] in which, models of helix–sheet–coil transformations were studied.

It is very interesting to extend the statistical mechanical treatment, in the spirit of the ZB model, to the studies of protein aggregation systems. This was first done by van Gestel and de Leeuw,[81] in which they constructed a quasi-grand canonical ensemble for two-state protein systems, where proteins are dominated by either helix (or sheet) and coil characteristics. In this treatment of protein aggregation, fundamental units are taken to be the protein monomers.

We have, on the other hand, taken a grand canonical ensemble approach to protein aggregation, by considering an aggregate phase in equilibrium with a solution phase. Since the number of protein monomers in the aggregate phase can fluctuate, it is natural to take the grand canonical ensemble approach. The equilibrium is defined by the equality of chemical potentials of protein monomers in the solution and in the aggregates. Furthermore, instead of using the two-state Ising models, we use a q-state Potts model to represent the characteristics of protein monomers. As in van Gestel's models, we define the effective Hamiltonian (free energy in reality) in terms of inter-protein energies between different states (helix, sheet, and coil) of monomers, and interfacial energies between protein monomers and solvent. However, we also include inter-filament interactions for multifilament fibrils. The grand canonical partition function can again be expressed as a product of transfer matrices. The dimension of the transfer matrices is equal to $q^{L_y} \times q^{L_y}$ for two-dimensional (2D) fibril structure and $q^{2L_y} \times q^{2L_y}$ for 3D fibril structures, where we assume the protofibrils contain L_y filaments.

Again by diagonalizing transfer matrices we can calculate numerically the equilibrium properties of the protein aggregation systems; for example, population distributions of aggregates of different size, and average number of filaments, average filament length, average number of sheet proteins in filaments. This method have been applied to Aβ(1–40) and Curli fibrils systems and results obtained are in good agreement with experimental data. For more details, the reader is referred to Refs. 82 and 83.

4. Protein Aggregation using a Kinetic Ising Model

In this chapter, we report our recent study on the kinetics of protein aggregation. We adopted a kinetic Ising model to describe phenomenologically the nucleation–growth kinetics of protein aggregation.

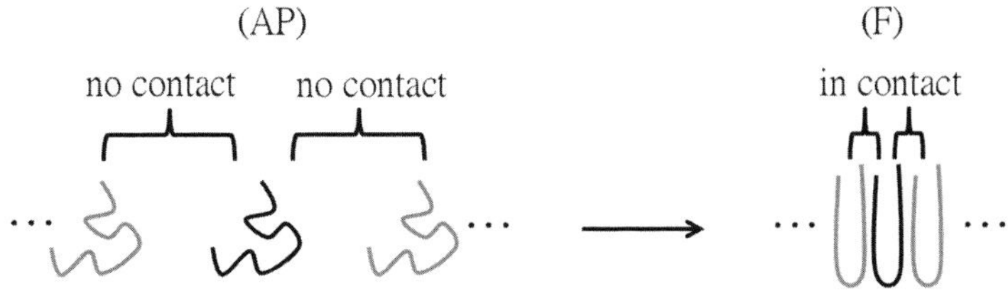

Figure 4: The two-state scheme of protein aggregation is schematically shown. The aggregation-prone (AP) state refers to a physical state of the monomeric protein (in *dark*) in solution, which is partially unfolded. The fibril state (F) refers to the other physical state of the protein (in *dark*) that forms β-sheet conformation in linear fibrils. The arrow indicates the transformation from the AP to the F state. (Figure adapted from Tsai *et al.*[90])

4.1. *Thermodynamics*

In the molecular description of protein aggregation, a protein undergoes a conformational transition from a partially unfolded (nonnative) state[84,85] or acid-denatured state[86,87] to a fibril-competent state that has the same β-strand/hairpin conformation in its monomeric form as the one present in fibrils. The transition takes place when monomeric proteins in solution self-assemble into a fibril aggregate through intermolecular interactions, for example, hydrophobic, electrostatic, and/or H-bond interactions. The partially unfolded state and the fibril-competent state are referred as the aggregation-prone[a] (AP) state[88,89] and the fibril (F) state, respectively. In other words, the aggregation process *per se* at single molecular level can be coarse grained into a two-state interacting scheme (see Fig. 4).

The Hamiltonian reads,

$$\hat{H}_s = -\sum_{i=1}^{N} \varepsilon_i \sigma_i - \sum_{i}^{N} \sum_{j \neq i}^{N} \mathcal{J}_{ij} \sigma_i \sigma_j \tag{1}$$

where σ_i is a binary variable; $\sigma_i = 1$ and $\sigma_i = -1$ denotes, respectively, the F state and the AP state of the ith protein. ε_i represents the site energy of the ith unit as a property of protein itself, which implicitly includes the solvent effect; $\mathcal{J}_{ij}$ describes the interaction energy between protein i and j units as a property of aggregation, which implicitly accounts for all the interactions required to form aggregates, e.g., hydrophobic, electrostatic, and H-bond interactions. These aggregation-related interactions are responsible for the cooperative effect in the Hamiltonian. The

[a]As a side comment, the theoretically proposed conformation N^* of a peptide in a lattice[89] model can also be identified as an AP state.[88]

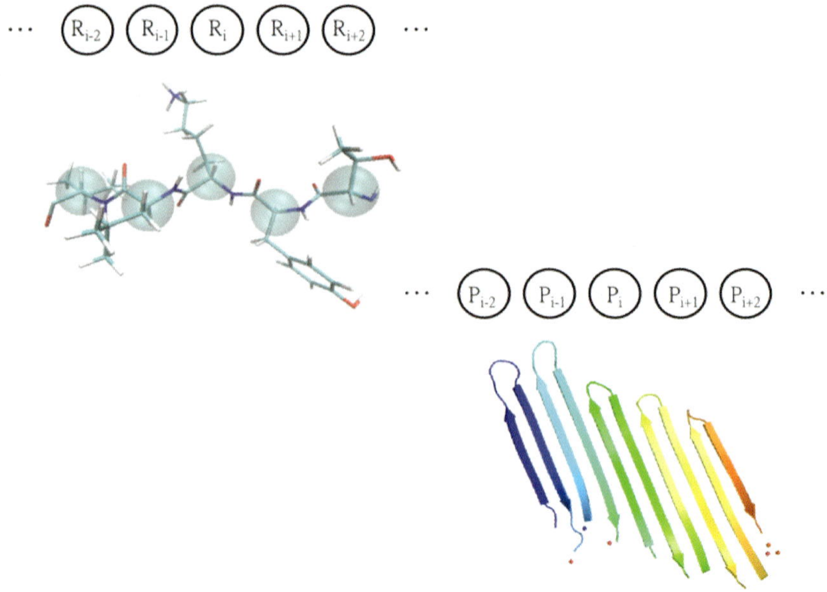

Figure 5: One-dimensional Ising model is used as a topological analogy in both protein folding and aggregation. Linear topology of the model in relation with a polypeptide chain (up) and a linear fibril (bottom) are shown. For a polypeptide chain, each unit refers to an amino acid residue/peptide bond while for a linear fibril each unit represents a protein/polypeptide.

Hamiltonian used in Eq. 1 is in fact a reminiscent of the conventional Ising model for protein folding (see Sec. 2.3). Here, same theoretical framework is applied to study protein aggregation. One needs to caution, however, that the two-state scheme used in the aggregation context is made at the protein level, not the residue level (see Fig. 5).

A typical configuration of protein aggregates in the Ising representation is represented by a collection of spin units $\{\sigma_1, \ldots \sigma_i, \ldots \sigma_N\}$. To connect the model to the physical quantities observed from aggregation experiments, we use the order parameter $\langle \sigma_i \rangle$ to describe the progress of protein aggregation. $\langle \ldots \rangle$ refers to the Boltzmann average. With that in mind, we start to deal with the thermodynamics of the model at the mean-field level. The $\langle \sigma_i \rangle$ at equilibrium is expressed as (see Appendix A.1)

$$\langle \sigma_i \rangle_{\rm eq} = \tanh\left(\frac{\langle E_i \rangle_{\rm eq}}{k_B T}\right) \tag{2}$$

where $\langle E_i \rangle_{\text{eq}}$ denotes the mean-field energy of unit i,

$$\langle E_i \rangle_{\text{eq}} = \varepsilon_i + \sum_{j \neq i}^{N} \mathcal{J}_{ij} \langle \sigma_j \rangle_{\text{eq}} \,. \tag{3}$$

From Eqs. (2) and (3), we see that given the units treated in the mean-field the equations of state in terms of $\langle \sigma_i \rangle_{\text{eq}}$ and $\langle \sigma_j \rangle_{\text{eq}}$ $(j \neq i)$ can be solved by computing a set of self-consistent equations as a result. In other words, the mean-field approximation allows for the reduction of a many-body interaction problem into a one-body problem, thereby simplifying the calculation.

In practical use, if we consider a one-dimensional model with nearest-neighbor interactions and meanwhile introduce homogeneity in the system. Eq. 3 becomes:

$$\begin{aligned} \langle E_i \rangle_{\text{eq}} &= \varepsilon_i + \mathcal{J}_{i,i-1} \langle \sigma_{i-1} \rangle_{\text{eq}} + \mathcal{J}_{i,i+1} \langle \sigma_{i+1} \rangle_{\text{eq}} \\ &\cong \varepsilon + 2\mathcal{J} \langle \sigma_i \rangle_{\text{eq}} \end{aligned} \tag{4}$$

Substituting Eq. 4 into Eq. 2 and dropping the subscript i, we can obtain:

$$\langle \sigma \rangle_{\text{eq}} = \tanh\left(\frac{\varepsilon + 2\mathcal{J} \langle \sigma \rangle_{\text{eq}}}{k_B T}\right), \tag{5}$$

which can be solved in a self-consistent manner. For practical applications, it is more convenient to use the mole fraction of aggregates $x_{\text{eq}} = (1 + \langle \sigma \rangle_{\text{eq}})/2$ to represent the progress of aggregation. Equation 5 therefore can be further transformed into the following equation:[47]

$$x_{\text{eq}} = \left[1 + \exp\left(-\frac{2\varepsilon + 4\mathcal{J}(2x_{\text{eq}} - 1)}{k_B T}\right)\right]^{-1}, \tag{6}$$

which describes x_{eq} (ranges from 0 to 1) in terms of ε and $\mathcal{J}$. Note that Eq. 6 refers to the equation of state of protein aggregation and that it reduces asymptotically to the simple two-state thermodynamic model (see Sec. 2.1) when $\mathcal{J} \approx 0$. In what follows we will provide another theoretical route, using a macroscopic approach, that leads to a similar result but their comparison shall deepen our understanding of the relationship between microscopic and macroscopic approaches.

In parallel with the treatment described earlier, Liang *et al.* considered same two-state theoretical construct in the context of solution thermodynamics/binary mixture and showed that the well-known regular solution model is thermodynamically equivalent to the nearest-neighbor Ising model on the mean-field level.[49] As a starting point for the treatment, we considered the free energy of mixing of the present aggregation system in terms of x_{eq} and applied the condition that minimizes the free energy in order to achieve equilibrium. The molar free energy of such a system on the basis of the regular solution model (also known as the Bragg–Williams approximation in binary mixture) is expressed as

$$G_{\mathrm{m}} = (1 - x_{\mathrm{eq}})\mu^0_{\mathrm{AP}} + x_{\mathrm{eq}}\mu^0_{\mathrm{F}} + k_B T[(1 - x_{\mathrm{eq}})\ln(1 - x_{\mathrm{eq}}) + x_{\mathrm{eq}}\ln x_{\mathrm{eq}}] + \Delta\chi(1 - x_{\mathrm{eq}})x_{\mathrm{eq}} \tag{7}$$

where G_m is the molar free energy and that μ^0_{AP} and μ^0_{F} refer to the chemical potentials of the AP state and the F state, respectively. $\Delta\chi$ represents the net AP–F interaction energy. At equilibrium, the free energy is minimized such that

$$\frac{\partial G_{\mathrm{m}}}{\partial x_{\mathrm{eq}}} = 0 = \mu^0_{\mathrm{F}} - \mu^0_{\mathrm{AP}} + k_{\mathrm{B}}T\ln\frac{x_{\mathrm{eq}}}{1 - x_{\mathrm{eq}}} + \Delta\chi(1 - 2x_{\mathrm{eq}}). \tag{8}$$

More concretely, Eq. 8 can be rearranged to yield the following equation of state:

$$x_{\mathrm{eq}} = \left[1 + \exp\left(-\frac{(\mu^0_{\mathrm{AP}} - \mu^0_{\mathrm{F}}) + \Delta\chi(2x_{\mathrm{eq}} - 1)}{k_B T}\right)\right]^{-1}. \tag{9}$$

By comparing Eqs. 9 and 6, we obtain:

$$\mu^0_{\mathrm{AP}} - \mu^0_{\mathrm{F}} = \Delta\mu = 2\varepsilon \tag{10}$$

and

$$\Delta\chi = 4\mathcal{J}. \tag{11}$$

These two relations are used to verify the correspondence between the Ising model and regular solution model at the mean-field level. These parameters are equilibrium properties and in principle can be inferred by fitting to experimental data. In other words, microscopic parameters such as the properties of protein and aggregation (ε and $\mathcal{J}$) in the Ising representation can be evaluated via thermodynamic parameters of binary mixture ($\Delta\mu$ and $\Delta\chi$).

4.2. *Kinetics*

The dynamics of a system can be described by the time evolution of a density operator, $\hat{\rho}(t)$, which satisfies the Liouville equation (the fundamental equation of statistical mechanics),

$$\frac{\partial \hat{\rho}}{\partial t} = -\frac{i}{\hbar}[\hat{H}, \hat{\rho}]. \tag{12}$$

This equation is a quantum analog of the classical version that describes the time evolution of the phase space distribution function. When $\hat{\rho}(t)$ is expressed in certain basis set, it becomes a density matrix, which describes a statistical ensemble of mixed quantum states and contains all information regarding to the dynamics of the system, interacting with external heat bath (see Appendix A.2). For example, we will be focusing on the time evolution of $\hat{\rho}^{(s)}(t)$ (s refers to "system" and heat bath variables have been averaged out), whose diagonal element $\hat{\rho}^{(s)}_{nn}(t)$ refers to the probability of a particular quantum state n at time t, also denoted by $P(\{\sigma_i\}, t)$ or $P(\sigma_1, \ldots, \sigma_i, \ldots, t)$. Off-diagonal elements, however, are associated with the coherence of the system and are not of interest here. Using the Markov approximation (no memory effect, purely stochastic), it follows that Eq. 12 can be further reduced to a master equation as shown below:

$$\begin{aligned} \frac{d}{dt}P(\sigma_1, \ldots, \sigma_i, \ldots; t) = &-\sum_i W(\sigma_i \to -\sigma_i)P(\sigma_1, \ldots, \sigma_i, \ldots; t) \\ &+\sum_i W(-\sigma_i \to \sigma_i)P(\sigma_1, \ldots, -\sigma_i, \ldots; t) \end{aligned} \tag{13}$$

Equation 13 treats the time evolution of a system's dynamics in terms of probability and that elementary processes of the system follow the principle of detailed balance, meaning that the system will equilibrate eventually, such that

$$W(\sigma_i \to -\sigma_i)P(\{\sigma_i\}, t) = W(-\sigma_i \to \sigma_i)P(\{-\sigma_i\}, t) \tag{14}$$

Note that W refers to the transition rate of flipping a spin and we assume that only one spin at a time is allowed to flip. The exact form of W is expressed by the Fermi golden rule (see Appendix A.2). The transitions are induced by external perturbation, here referring to thermal stimulation from heat bath.

Starting from quantum kinetics, our goal here is to be able to trace the progress of aggregation (using order parameter) as a function of time. The time dependence of such an order parameter $\langle \sigma_i \rangle$ can be readily understood and therefore defined by:

$$\langle \sigma_i(t) \rangle = \sum_{\{\sigma_1, \ldots, \sigma_i, \ldots\}} \sigma_i P(\sigma_1, \ldots, \sigma_i, \ldots; t), \tag{15}$$

and its rate form at the mean-field level is expressed as (see Appendix A.3)

$$\tau_i \frac{d}{dt} \langle \sigma_i \rangle = -\langle \sigma_i \rangle + \tanh\left(\frac{\langle E_i \rangle}{k_B T} \right) \tag{16}$$

where $\langle \sigma_i \rangle$ and $\langle E_i \rangle$ are same physical quantities used in Eqs. (2)–(4) except they are time dependent here. τ_i refers to a kinetic time scale of spin flipping (see Appendix A.3 for its quantum mechanical expression), an analogy to a monomeric protein transforming into a β-sheet conformation seen in a fibril.

It can be readily seen from Eq. 16 that the solution at equilibrium can be obtained by setting

$$\frac{d}{dt} \langle \sigma_i \rangle = 0. \tag{17}$$

Equation 2 is thus obtained. For practical usage, we again adopt nearest-neighbor interactions and assume that the system is uniform as has been used in Eq. 4. It follows from Eq. 16 that

$$\tau \frac{d}{dt} \langle \sigma \rangle = -\langle \sigma \rangle + \tanh\left(\frac{\varepsilon + 2\mathcal{J} \langle \sigma \rangle}{k_B T} \right) \tag{18}$$

or

$$\tau \frac{dx}{dt} = \frac{1}{2} - x + \frac{1}{2} \tanh\left(\frac{\varepsilon + 2\mathcal{J}(2x-1)}{k_B T} \right), \tag{19}$$

if $x = (1 + \langle \sigma \rangle)/2$ is used. The progress of aggregation x plays a role as a physical quantity connecting to experimental observables (e.g., variation in fluorescence intensity) with a normalization procedure being carried out properly.

In the description of protein aggregation kinetics, nucleation is the kinetic hallmark, characterized by an observation of lag time in kinetic traces, followed by a fast growth phase. These kinetic features can be studied using Eq. 19 as shown in Fig. 6.

4.3. *Applications*

First, we will demonstrate that the present model can be applied to investigate the aggregation of proteins under various experimental conditions, e.g., temperature, the initial protein concentration, seeding. We study the kinetic processes of aggregation for four different amyloid proteins: (1) prion protein Sup35, (2) bovine insulin, (3) β_2-microglobulin, and (4) amyloid-β_{42} peptide. All the fit curves are shown in Fig. 7, and the thermodynamic properties estimated for these proteins are summarized in Table 1. Note that for all numerical fittings we report fit parameters mainly

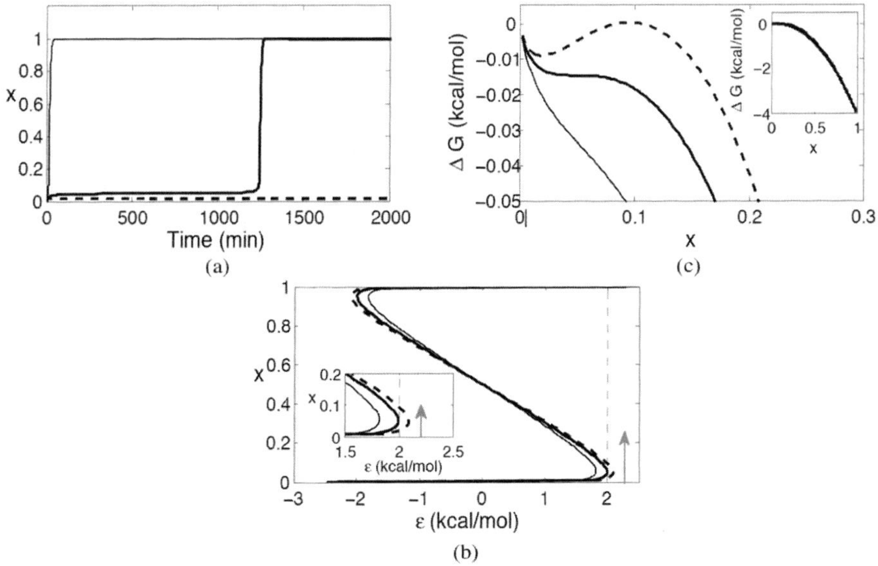

Figure 6: The nucleation–growth thermodynamics and kinetics of the present model are shown. (a) Three examples of kinetic curves showing no lag time (thin line), finite lag time (thick line), and infinite lag time (dashed line). (b) Corresponding thermodynamic analysis of (a) using Eq. 6 shows hysteresis loop on the phase diagram ($\varepsilon = 2\,\text{kcal/mol}$ is a global value used as indicated by a dashed line; the arrow indicates the direction of kinetic relaxation). *Inset*: an enlarged part of the loop near $\varepsilon = 2\,\text{kcal/mol}$. (c) Corresponding free energy profile of (a) using Eq. 7. *Inset*: zoom-out view. Reader is referred to the original reference for detailed figure caption. (Figure adapted from Tsai *et al.*[90])

in terms of ε and $\mathcal{J}$; however, they can be interpreted as thermodynamic quantities in $\Delta\mu$ and $\Delta\chi$, respectively, via the relations: $\Delta\mu = 2\varepsilon$ and $\Delta\chi = 4\mathcal{J}$ (see Sec. 4.1 and Eqs. 10 and 11). In the following, we discuss the aggregation kinetics of individual amyloid proteins.

4.3.1. *Prion protein Sup35*

The first example is a yeast prion protein called Sup35 (254 a.a.), which at physiological pH 7.5 under proper agitation conditions would self-assemble into β-sheet-enriched amyloid fibrils.[91] The experimental results[b] of aggregation kinetics as a function of temperature are shown along with the fittings using our model (see Fig. 7a). As can be seen from the figure, the effect of temperature on aggregation

[b] All the experimental data used in this work were extracted using the software, Engauge Digitizer 4.1, if not otherwise specified.

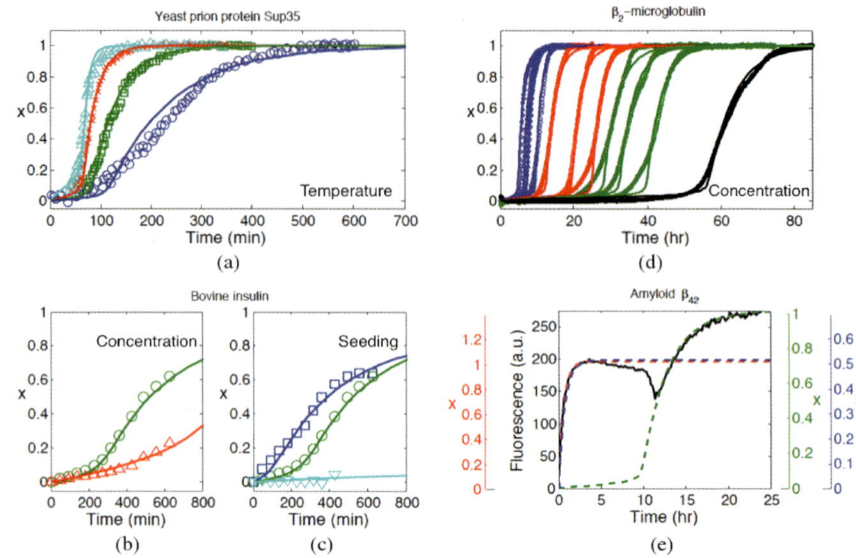

Figure 7: Numerical fitting/analysis of a variety of protein aggregation systems using the mean-field kinetic equation (Eq. 19) is shown. (a) Yeast prion protein Sup35 for temperature effect: 293 K (blue circle), 303 K (green square), 313 K (red cross), and 323 K (cyan triangle), measured by Sabate *et al.*[91] Solid lines refer to fit curves. (Figure adapted from Tsai *et al.*[90]) (b) Bovine insulin for protein concentration effect: 49 μM (red upper triangle) and 98 μM (green circle) and (c) seeding effect: 0 μM (cyan lower triangle), 0.21 μM (green circle), and 0.8 μM (blue square), both measured by Knowles *et al.*[68] These two plots show that the data for [Seed] = 0.8, 0.21, and 0 μM can be well reproduced by setting $x_0 = 0.15$, 0.1, and 0, respectively ($\varepsilon = 2, \mathcal{J} = 1.6$ kcal/mol were used as constants for that [Prot.] = 98 μM). Note that for both (b) and (c), other parameters such as temperature and τ were set to be constants and therefore served as global parameters. (d) β_2-Microglobulin for concentration effect, measured by Xue *et al.*[52] From left to right, 243.5 $\sim$ 61 μM (blue), 48.5 $\sim$ 17 μM (red), 16.7 $\sim$ 12.3 μM (green), and 8.4 μM (black), each curve was obtained on an average of 12 replicates from the original data. (e) Amyloid-β_{42} for detection of oligomeric intermediates, measured by Lee *et al.*[58] The kinetic trace was fitted separately by different stages: the early stage (blue and red dashed lines) and the latter stage (green dashed line). For these fit curves, the corresponding y axis is color coded and shown additionally on the sides. Note that except (d) and (e), the experimental data used in plots were digitized from the original papers published.

is to accelerate the aggregation process. We notice, however, that from the experimental data the temperature dependence on the change of lag time is not significant. Instead, the dependence correlates well with the half time of aggregation (at which the progress of aggregation reaches 50% completion), suggesting an intricate effect is involved at nucleation stage (in contrast to the case of β_2-microglobulin,

Table 1: Proteins and their aggregation properties are tabulated.

Protein[a]	Condition[b]	Exp.[c]	Type[d]	Time[e]	$\Delta\mu$ (kcal/mol)[f]	$\Delta\chi$ (kcal/mol)[g]
Sup35 (254)	pH 7.5	T	I/II	Half	$1.2 \sim 13.2$	$2.9 \sim 15.4$
Insulin (51)	60°C, pH 2.0	C+S	I/II	Half	$1 \sim 4$	$2.5 \sim 6.4$
β_2m (99)	37°C, pH 2.0	C	II-1	Lag	$9.6 \sim 9.8$	12.3 (10.5)
Aβ_{42} (42)	37°C, pH 7.4	-	II-2	Lag	10	12.6 (11.2)

[a]Number shown in parenthesis is the total number of amino acids residues in the protein. [b]The experimental condition carried out in the literature. [c]Type of experiments: T for varying temperature, C for varying protein concentration, S for varying seeding concentration. [d]Type of kinetic profiles: Type I, Type II, and subtypes (see Table 6 for details). [e]Type of macroscopic characteristic time for aggregation: half time and lag time. [f]β-Sheet/amyloid propensity. [g]Aggregation cooperativity. Experimetal value of β_2m and A$\beta_{42}$are shown in parenthesis, given by Refs. 77 and 92, respectively.

see Sec. 4.3.3). We assume that ε and $\mathcal{J}$ are both temperature-dependent (up to the linear approximation). In addition, the change of the growth slope (i.e., elongation rate) upon increasing temperature also suggests a significant temperature dependence on the kinetic time scale; therefore, τ can be assumed to be temperature dependent as well. We found that the aforementioned assumptions are required to globally fit the experimental data at different temperatures, and the fit parameters are listed in Table 2. The thermodynamic properties at 293, 303, 313, 323 K were obtained: $\Delta\mu = 2\varepsilon = 1.2, 5.2, 9.2, 13.2$ kcal/mol and $\Delta\chi = 4\mathcal{J} = 2.9, 7.3, 11.5, 15.4$ kcal/mol, respectively, while the kinetic time scale is predicted to be 130, 50, 25, and 10 min. Furthermore, the kinetic analysis using the Eyring equation provides the information about the thermodynamics of a potential activated complex, which can be calculated according to the temperature dependence of the kinetic time scales (see Appendix A.4). The results showed that $\Delta H^\dagger = 15.2$ kcal/mol, $\Delta S^\dagger = -0.025$ kcal/mol K, and that $\Delta G^\dagger$ 298K $= 22.7$ kcal/mol according to the equation, $\Delta G^\dagger = \Delta H^\dagger - T\Delta S^\dagger$. The prediction for $\Delta S^\dagger$ is negative suggests a decrease in entropy upon transforming from protein monomers (in the AP state) into the activated complex of the aggregates. This result implies that the conformation of such an activated state might have an ordered or at least partially ordered structure, compared with that of the AP state.

4.3.2. *Bovine insulin*

Taking on another example, we investigated bovine insulin protein (51 a.a.). The optimal fibril growth conditions for this protein were found to be 60°C and pH 2.0,

Table 2: Fit parameters for yeast prion protein Sup35.

	293 K	303 K	313 K	323 K
ε (kcal/mol)	0.6	2.6	4.6	6.6
$\mathcal{J}$ (kcal/mol)	0.73	1.83	2.87	3.85
τ (min)	130	50	25	10

Initial condition x_0 is globally set to be 0. Figure modified from Tsai *et al.*[90]

Table 3: Fit parameters for Bovine insulin.

Conc.\Seed	0 μM	0.21 μM	0.8 μM
98 μM		$\varepsilon = 2, \mathcal{J} = 1.6$	
	$x_0 = 0$	$x_0 = 0.1$	$x_0 = 0.15$
49 μM	—	$x_0 = 0.1$ $\varepsilon = 0.5, \mathcal{J} = 0.63$	—

and kinetic measurements of insulin aggregation was conducted using adsorption spectroscopy.[68] The experimental data for both the concentration-dependent and seeding effects are shown in Fig. 7b and c respectively, along with numerical fits (solid lines). Two different protein concentrations (98 and 49 μM) are used in Fig. 7b with a fixed amount of seeds (0.21 μM), while Fig. 7c demonstrates the seeding effect (0.8, 0.21, 0 μM) with a fixed protein concentration (98 μM). Such a cross-reference approach for independent variables allows a global examination of both the concentration and seeding effects. The fit parameters are detailed in Table 3. According to the fit results, we found that both ε and $\mathcal{J}$ are concentration dependent and that theoretical curves, in general, fit well to experimental data using same initial condition $x_0 = 0.1$ for taking into account the fixed seed concentration. Seeding effects, on the other hand, are simulated by varying the initial condition x_0 alone. Note that similar to prion protein Sup35, the dependence of lag time on concentration in this case is also not significant.

Both ε and $\mathcal{J}$ are in unit of kcal/mol. $T = 333$ K (60°C) and $\tau = 350$ min ($4.76 \times 10^{-5}\,s^{-1}$) are assumed to be global parameters.

4.3.3. *β_2-Microglobulin*

Regarding to the kinetics of concentration-dependent aggregation, Xue *et al.* carried out a comprehensive experimental study across a wide range of protein concentrations (from 8 to 244 μM) on β_2-microglobulin (99 a.a.) at 37°C and acidic condition (pH = 2).[52] The kinetic measurements[c] using the amyloid-specific dye, thioflavin T (ThT), are shown in Fig. 7d. First of all, the lag time observed from the experiment shows significant concentration dependence, which is different from the previous two cases for Sup35 and insulin. The difference between them suggests a distinct aggregation mechanism that is employed by Sup35 and insulin. Second, we found a significant anti-correlation between aggregation lag time and the slope of the fast growth phase; an increase in lag time comes with a decrease in the slope. As shown in the figure, the numerical fits globally capture the concentration effect on the kinetic trend as well as the correlation effect described earlier. As the concentration decreases, however, one may notice that the quality of the fit during the fast growth transition region is not as good as those at early and late stages. This comparison implies a slight deficiency of the present model in quantitative descriptions of the concentration effect. This imperfection in fitting, on the other hand, may be in part due to the choice for dye molecules (i.e., the use of ThT molecules), for which the molecular mechanism of its interactions with fibrils and/or oligomers have not been made clear.[58] Despite of the shortcomings, we note that the numerical results presented here are still the best fits available from the literature for a wide range of concentrations using only one single model.

Next, we look into the relationship between fit parameters (tabulated in Table 4) and concentration effect. The results show that an optimal $\mathcal{J} = 3.08$ kcal/mol (or $\Delta\chi = 12.3$ kcal/mol) displays as a global parameter while the change in concentration from 243.5 to 8.4 μM depends on the variation in ε from 4.88 to 4.78 kcal/mol ($\Delta\mu = 9.76$ to 9.65 kcal/mol), respectively, along with an increase in kinetic time scale τ (becoming slower). Such a short range of concentration variation in generating well-behaved global kinetic profile suggests the extreme sensitivity of the parameter, reflected in the delicate nature of nucleation process. It turns out that the fitted values of $\Delta\mu$ and $\Delta\chi$ are very close to the experimental value, 10.5 kcal/mol,

Table 4: Fit parameters for β_2-microglobulin.

Conc. (μM)	High	Medium	Low	Extremely Low
ε (kcal/mol)	4.88 ~ 4.80	4.82 ~ 4.76	4.79 ~ 4.76	4.78
$\mathcal{J}$ (kcal/mol)	3.08	3.08	3.08	3.08
τ (h)	1.5	2.5	4	7

$T = 310$ K (37°C); initial condition x_0 is globally set to be 0.

[c] We would like to thank Dr. Wei-Feng Xue and Prof. Sheena Radford for kindly providing their original data on β_2-microglobulin and allowing us to reproduce the experimental kinetic traces shown in Fig. 7d.

measured by Goto's group in a series study of amyloid stability of the protein.[77,94] Herein, we shall discuss the correlation between the two theoretical quantities and the experimentally measured quantity. First, we note that the experimental value was reported as the thermodynamic stability of fibrils with reference to the association of additional monomers onto existing seeds of fragmented fibrils. The quantity measured in fact is physically related to interactions between proteins, effectively taken into account in $\Delta\chi$. Thus, we suggest that $\Delta\chi$ is linked to the experimentally measured fibril stability. Finally, we found that the time constant τ has a significant dependence on concentration.

4.3.4. *Amyloid-β_{42} peptide*

As we have mentioned that some oligomeric intermediates exist during the nucleation process, the structure and properties of them can be investigated via experimental techniques or inferred from theoretical modeling. An oligomeric intermediate for $A\beta_{42}$, for example, has been identified experimentally using a novel chromophore (as a dye) called FlAsH, which can detect specifically the presence of some oligomers during the nucleation process.[58] The morphology of the detected oligomers was inferred to as disordered micelles, which would undergo a conformational transition into nuclei in the subsequent elongation of fibrils, supporting the nucleation conformational conversion (NCC) mechanism.[53] We propose that such a two-stage scheme can be treated individually by two independent time scales, one characterizing the formation of oligomeric intermediates and the other representing the formation of ordered fibrils where conformational transition from the oligomers into active fibril nuclei is the rate-limiting step. The experimental data,[d] measured at 37°C and pH 7.4, is shown in Fig. 7e. The whole aggregation process monitored is featured by the presence of a clear intermediate state (early phase), followed by a fibril elongation (late phase). We fit the kinetic trace using our model from two aspects. First, we carry out a numerical fit, only partially to the late phase that features the kinetic lag time prior to forming ordered fibril aggregates (*dashed-green* curve). The fit result also highlights the final kinetic plateau, which refers to the end of fibril aggregation. The result showed that $\Delta\chi$' is ~12.6 kcal/mol (with a prime sign for the late phase), consistent with the experimental value reported for this protein, 11.2 kcal/mol.[78,92] The value for $\Delta\mu$' was estimated to be ~10 kcal/mol, and $\tau = 3$ h. Note that $\Delta\mu = 2\varepsilon$ and $\Delta\chi = 4\mathcal{J}$ are used. In the description of the early phase, on the other hand, we examine it based on the assumption made

[d] We would like to thank Dr. Jiyong Lee and Prof. Jeffery W. Kelly, and coworkers for sending us their data on $A\beta_{42}$ and allowing us to reproduce the kinetic data shown in Fig. 7e.

for the structural properties of oligomeric intermediates. For example, there are primarily two dominant aggregation scenarios from literature.[70,74,95] One refers to a situation where the intermediates involved during nucleation are associated with ordered structures (termed Scen. 1), while for the other scenario the oligomeric intermediates should have characteristics of disordered micelles (termed Scen. 2). We performed separate numerical fits to the early kinetic trace, denoted by *dashed-red* and *dashed-blue* curves, respectively. The two fittings are of nearly same quality. The fits, nevertheless, differ in the use of the scale for the order parameters, x, depending on the structural property of the final oligomeric aggregates assumed. The formation of ordered aggregates, for example, is associated with a fully ordered transition for which x ranges from 0 to 1; the formation of disordered micelles, however, corresponds to an ordered-to-disordered transition, in which x ranges from 0 to 0.5. The kinetic time scale for the early phase from fitting in general is $5 \sim 6$ times faster than that of the late phase. The result suggests that the late phase involves a slow nucleated process, consistent with the rate-limiting NCC mechanism. Regarding to the early phase, the faster time scale corresponds to the formation of initial aggregates and that $\Delta\chi$ in general exhibits an appreciably small value, suggesting their formation undergoes primarily a non-cooperative process of aggregation. The $\Delta\mu$ for Scen. 1 was estimated to be 3 kcal/mol, suggesting a weaker aggregation propensity compared to the fibril-competent value (~10 kcal/mol); however, this mild aggregation propensity remains sufficient in forming an ordered oligomeric species. In contrast, Scen. 2 gave a marginal value of 0.01 kca/mol for $\Delta\mu$, indicating a weaker aggregation propensity for disordered oligomeric species. The significant difference in the aggregation propensity implies the importance of the structural property of intermediates in the thermodynamic interpretation of the early nucleation process (see Table 5).

Table 5: Fit parameters for amyloid-β_{42}.

	Early phase (nucleation)			Late phase (elongation)		
	ε	$\mathcal{J}$	τ	ε'	$\mathcal{J}'$	τ'
Scen. 1	1.5	0.1	0.6	5	3.145	3
Scen. 2	0.005	0.1	0.5			

ε and $\mathcal{J}$ are in unit of kcal/mol and τ, in unit of hours. $T = 310\,\text{K}$ (37°C). Initial condition x_0 is globally set to be 0. Scen. 1 refers to the mechanism via forming ordered intermediates while Scen. 2 indicates the mechanism via forming disordered intermediates. Note that $\Delta\mu = 2\varepsilon$ and $\Delta\chi = 4\mathcal{J}$.

4.4. *Aggregation scenarios*

We conclude this section with a few remarks, which may be helpful on way of interpreting protein aggregation kinetics using Eq. 19. Despite significant experimental work done on a range of proteins, it is still difficult to uniquely classify proteins into a specific aggregation pathway, probably due to the lack of systematic studies in multiple experimental variables. However, it is possible to use the existing data to give a flavor of how the present model can be used to classify different scenarios of protein aggregation. In mechanistic descriptions of aggregation pathways, a key role in revealing the mechanism lies in the properties of the intermediates involved during the aggregation process. A unified aggregation scenario based on the intermediate's properties has merged from both experimental[70,95] and coarse-grained model[74] studies. All these results suggest that the formation of amyloid fibrils can be achieved primarily through two different pathways, depending on the intrinsic property of proteins, such as β-strand/sheet propensity (or alternatively, aggregation or amyloidogenic propensity). For example, in the early stage of aggregation, proteins with high β-sheet propensity tend to form ordered aggregates, normally termed as nucleus, while those with low β-sheet propensity form disordered aggregates with amorphous properties, such as micelles. The following sections will discuss some related issues.

4.4.1. *Thermodynamic Interpretation of parameters*

Our model has a total of three parameters: two thermodynamic parameters and one kinetic time constant. The two thermodynamic parameters can be understood either from a microscopic point of view using the Ising model Hamiltonian (see Eq. 1) or from their correspondence to binary mixture on a macroscopic picture (see Eqs. 7–11). Using the Ising model representation, the thermodynamics of protein aggregation is determined solely by parameter pair $(\varepsilon, \mathcal{J})$. In protein aggregation energetics, ε is an intrinsic energy of a protein, which in the context of protein aggregation refers to the amyloid propensity of the protein. In contrast, $\mathcal{J}$ describes the cooperative effect upon aggregation, which requires interactions with neighboring proteins. Our investigations on four different kinds of proteins under a variety of experimental conditions provide a general understanding of the parameters in relation to environmental variables.

(1) Both ε and $\mathcal{J}$ have significant temperature dependence.
(2) Both ε and $\mathcal{J}$ have mild or little dependence in protein concentration.
(3) τ has significant temperature dependence but its dependence in protein concentration is not clear.

4.4.2. *Kinetic profiles of aggregation*

To gain a deeper thermodynamic understanding of a variety of experimental kinetic profiles, we look into those kinetic curves with Eq. 19 using reduced units, $T^* = k_B T/2\mathcal{J}$, $\varepsilon^* = \varepsilon/2\mathcal{J}$ and $t^* = t/\tau$. By reducing $\mathcal{J}$, the kinetics is determined by the relative contribution of the thermal energy ($k_B T$) and the amyloid propensity (ε) of proteins with respect to the cooperative interaction between neighboring proteins ($\mathcal{J}$). The t^* indicates time elapse in unit of the time scale τ. Fig. 8a shows, from left to right, the kinetic profiles for a systematic survey of values of T^* and ε^*. The leftmost

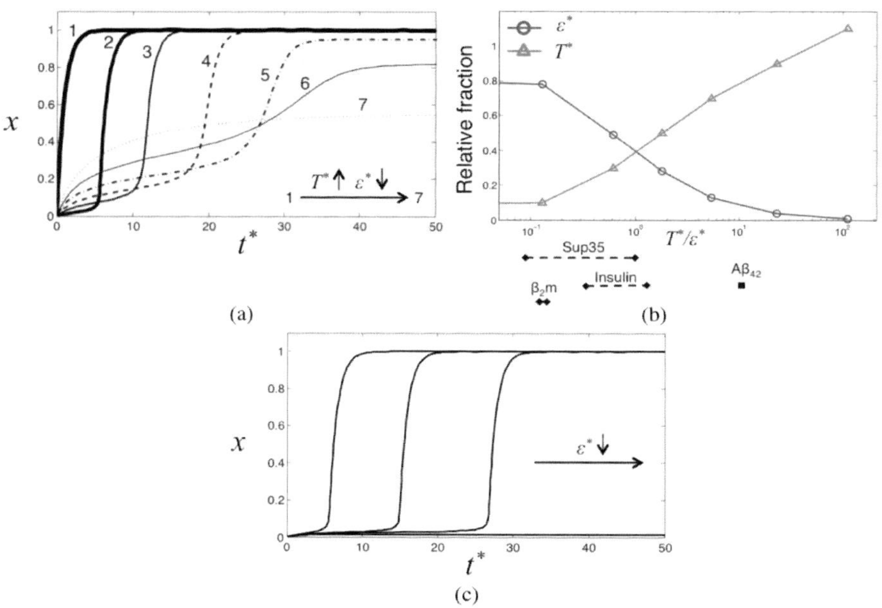

Figure 8: Kinetic profiles of the mean-field kinetic Ising model are shown. (a) The kinetic profiles are demonstrated in terms of reduced units $T^* = k_B T/2\mathcal{J}$, $\varepsilon^* = \varepsilon/2\mathcal{J}$, and $t^* = t/\tau$. The kinetic curves, from left to right, denote a gradual change in the values of the pair (T^*, ε^*): (0, 1.02), (0.1, 0.78), (0.3, 0.49), (0.5, 0.28), (0.7, 0.13), (0.9, 0.04), and (1.1, 0.01), respectively. These values were chosen to ensure a complete kinetic behavior being observed as well as an easier visual inspection of all the curves being overlaid. (b) The transition of aggregation driving force, quantified in the relative value of ε^*(in circles) and T^* (in triangles) as function of T^*/ε^*. The value of each data point, connected from left to right, corresponds sequentially to the profiles numbered in (a): kinetic curve 2 ~ 7 (curve 1 is not shown due to a poor scaling effect along the axis). The additional dashed bars with diamond heads shown below refer to a schematic range of predicted T^*/ε^* for the protein. (c) The effect of ε^* on the kinetic behavior, which modulates the lag time. From left to right, the curves denote $\varepsilon^* = 0.78$, 0.77, 0.768, and 0.76, respectively, while keeping $T^* = 0.1$ as a constant.

curve (labeled 1) refers to a situation where thermal energy $k_B T$ is negligible with respect to $\mathcal{J}$ and ε. The kinetic relaxation is primarily driven by the magnitude in ε (i.e., $\varepsilon \gg k_B T$) and featured by a monotonically increasing curve, indicating a transformation from a monomeric phase ($x \sim 0$) into a fibril phase ($x \sim 1$). Nucleation in this case is highly suppressed; observation of lag time is therefore not significant. The rightmost curve (labeled 7), however, signifies another extreme condition where thermal effect is dominant ($k_B T \gg \varepsilon$) in the kinetics. The kinetic curve would eventually relax into a thermally disordered phase ($x \sim 0.5$) since high thermal energy causes entropically favorable states. As we move from the leftmost to the rightmost curve (from curve 1 to 7), a long-lived intermediate, featured by its varying value of x (e.g., $x \sim 0.2$ for curve 5), appears prior to the growth stage. The trend observed from the kinetic modeling suggests that the structural properties (i.e., a degree of disorder) of the intermediate are variable quantities, depending on the relative magnitude of thermal energy and amyloid propensity in driving the aggregation process. Note that we do not intend to distinguish all the configurations compatible with a given value of x, as the current model addresses only the kinetics of order–disorder transition on a macroscopic scale.

As the contribution of thermal energy $k_B T$ or amyloid propensity ε increases or decreases, we can see significant change in resulting kinetic profile, as described earlier. The paired values of (T^*, ε^*) used in the kinetic curves (curve 1 to 7) are plotted against the ratio T^*/ε^*. Fig. 8b presents such increasing and decreasing trend in $k_B T$ and ε, respectively. Note that the names of the protein studied are listed below in an area where we use a bar length for each protein to indicate the range of the ratio along the axis. Also, we found that ε^* is responsible for the change in lag time. Figure 8c shows a sequence of kinetic profiles with a decreasing ε^* (from left to right). In other words, lag time will increase when $\mathcal{J}$ increases (or effectively ε decreases). To sum up, using the present model the kinetic profile of aggregation is determined by the relative contribution of $k_B T$ and ε as aggregation driving force. In addition, lag time can be modulated by the relative magnitude of ε and $\mathcal{J}$.

With the results obtained above, we proceed to apply them into the classification of protein aggregation. Three different categories are classified accordingly: (1) monomer-to-fibril (MtF) transition, (2) monomer-to-fibril with lag time (MtFL), and (3) monomer-to-disordered (MtD) transition. The first type, MtF, refers to a scenario where protein monomers transform from the AP state into the fibril F state without going through a significant lag time phase. In this case, amyloid propensity ε is the transition-driving force. In contrast, the second type MtFL indicates another situation where protein monomers in the AP state undergo a significant waiting time (i.e., lag time) before transforming into the fibril state F. The appearance of

Table 6: Types of kinetic profiles for aggregation kinetics.

Type	Driving force[a]	Lag time	Class.[b]	Intermediate
I	ε	No	MtF	No
II–1	$\varepsilon > k_B T$	Yes	$MtFL_1$	Ordered aggregates
II–2	$\varepsilon \sim k_B T$ (or $< k_B T$)	Yes	$MtFL_2$	Disordered aggregates
III	$k_B T$	No	MtD	No

[a]ε is related to amyloid propensity by $2\varepsilon = \Delta\mu$ and represents the driving force for aggregation while it is offset by the thermal energy $k_B T$. [b]Classification: monomer-to-fibril (MtF), monomer-to-fibril with lag time (MtFL), and monomer-to-disordered (MtD)

lag time phase is related to a long-lived intermediate phase, a macroscopic mixture of both AP and F states. In this case, both ε and $k_B T$ play a role in driving the aggregation process. Depending on the degree of thermal contribution $k_B T$, it can be further divided into two subsets: (1) MtFL with only a mild thermal contribution ($\varepsilon > k_B T$) and (2) MtFL with a comparable or strong thermal contribution ($\varepsilon \sim k_B T$ or $< k_B T$). The third type MtD highlights a scenario where thermal contribution plays a dominant role, which overwhelms the contribution from the amyloid propensity ε. Protein monomers are therefore transformed continuously into a macroscopic disordered nonfibril state without forming any appreciable intermediates. Table 6 summarizes the classification of different aggregation scenarios inferred from the present model.

4.4.3. *Classification of protein studied*

Herein, we have shown that the present kinetic model offers a physical connection to a variety of aggregation scenarios. First, protein's high β-sheet/amyloid propensity refers to a situation where aggregation is primarily driven by thermodynamic potential ε. Extremely high ε corresponds to a limiting case, denoted by Type I (MtF) in Table 6. Nucleation in this case is suppressed such that no significant lag time can be observed. However, half time, a more general property can be measured instead. Note that we make a clear distinction between lag time and half time. The physical meaning of lag time emphasizes the nucleated nature of aggregation process, while half time merely refers to a time at which the progress of aggregation reaches 50% completion. On the other hand, as the thermal energy $k_B T$ gets significant, large β-sheet/amyloid propensity would be offset by thermal contribution. As a result, lag time would emerge and that this case is termed Type II (MtFL). Depending on the relative contribution of ε and $k_B T$, the property of the accompanying intermediate would be different. If $\varepsilon > k_B T$, the presence of intermediate would not be significant, and the structural property of the intermediate is an ordered aggregate (Type II-1), for example, see curve 2 in Fig. 8a. β_2-m belongs to this kind. If $\varepsilon \sim k_B T$ or $\varepsilon < k_B T$, the property of the intermediate would be disordered (Type II-2), for example, see curve 5 and curve 6 in Fig. 8a. Aβ_{42} belongs to this category. Sup 35 and insulin, however, show a wide range in the ratio T^*/ε^* in the experimental condition studied, which to a certain degree cannot be exclusively classified into any specific category. We therefore assign them into a mixed group, Type I/Type II, where half time, not lag time, represents the characteristic timescale on a macroscopic scale.

Note that the classification of proteins reported in this section should not, in any case, be interpreted as a unique and exclusive outcome of the protein studied. In fact, we shall stress that kinetic profiles of aggregation can vary from condition to condition. A protein being identified as Type II-1 under a specified experimental condition, for example, can transform into another type, say Type II-2, under different experimental conditions and so forth. Kinetic profile analysis, therefore, should be carried out with caution and should also take into account experimental context.

5. Conclusions

In this chapter, we attempt to provide a general and unifying picture in bridging protein folding and aggregation. While being distinct in kinetic timescale and therefore seemingly irrelevant macroscopically, folding and aggregation usually involve conformational changes of individual proteins on a microscopic basis; subsequent oligomerization due to partially unfolded states or misfolded states can lead to the formation of aggregates. Protein aggregation is thus clearly intertwined with possible conformational states during the folding process. The two-state nature of folding allows us to employ a generalized two-state scheme (i.e., Ising-like model) in order to explore local, sub-local, and global properties of protein folding. These properties are associated with thermodynamic behaviors of foldon. Our studies show that protein localized interactions in the mean-field context is likely the physical basis of foldon construction in protein. The linear analogy of protein to one-dimensional fibril aggregates allows us to investigate protein aggregation within the same theoretical framework. Equilibrium properties of protein aggregation can be solved using a q-state Potts model with a grand canonical ensemble approach. Regarding to the kinetics of aggregation, kinetic Ising model exhibits nucleation–growth kinetics, which reveals diverse kinetic profiles employed by different types of amyloid-forming proteins (e.g., prion, β_2m, insulin, and Aβ_{42}) under different experimental conditions. The theoretical work reported in this chapter provides insight into the thermodynamic interpretation of protein aggregation kinetics. Future work will be directed toward the kinetic study on the multistate generalization.

Appendix A

A.1. *Thermal averaged quantity*

The conventional Ising Hamiltonian can be written in the generalized form:

$$\hat{H}_s = -\sum_{i=1}^{N} \sigma_i E_i(\{\sigma'\}) \tag{A.1}$$

where $E_i(\{\sigma'\})$ can be any function of $\{\sigma\} \equiv \{\sigma_1, \ldots \sigma_i, \ldots \sigma_N\}$ and in our case,

$$E_i = E_i(\{\sigma'\}) = \varepsilon_i + \sum_{j \neq i}^{N} \sigma_j \mathcal{J}_{ij}. \tag{A.2}$$

Putting Eq. A.2 into Eq. A.1, we get Eq. 1.

To evaluate thermodynamic quantities, it is essential to perform the calculation of the partition function,

$$\mathcal{Q} = \sum_{\{\sigma\}} \exp\left(-\frac{\hat{H}_s\{\sigma\}}{k_B T}\right) \tag{A.3}$$

However, the analytical treatment using Eq. A.3 in our case is not tractable due to complex many-body interactions. The mean-field approximation is well known for simplifying the calculation of the partition function, which can reduce effectively a many-body problem into a one-body problem. Therefore, from Eq. A.2, we define at equilibrium that:

$$\langle E_i \rangle_{\text{eq}} = \langle E_i(\{\sigma'\}) \rangle_{\text{eq}} = \varepsilon_i + \sum_{j \neq i}^{N} \langle \sigma_j \rangle_{\text{eq}} \mathcal{J}_{ij} \tag{A.4}$$

where $\langle E_i \rangle_{\text{eq}}$ is evaluated under the mean-field approximation with $\{\langle \sigma' \rangle_{\text{eq}}\} \equiv \{\langle \sigma_1 \rangle_{\text{eq}}, \ldots, \langle \sigma_i \rangle_{\text{eq}}, \ldots \langle \sigma_N \rangle_{\text{eq}}\}$.

It follows from Eq. A.3 that:

$$\mathcal{Q}(T) = \sum_{\{\sigma\}} \exp\left(\frac{\sum_{i=1}^{N} \sigma_i \langle E_i \rangle_{\text{eq}}}{k_B T}\right) = \prod_{i=1}^{N} q_i(T) \tag{A.5}$$

where $q_i(T)$ denotes the partition function of the ith unit,

$$q_i\ (T) = \sum_{\sigma_i} \exp\left(\frac{\sigma_i \langle E_i\rangle_{\text{eq}}}{k_B T}\right) = \exp\left(\frac{\langle E_i\rangle_{\text{eq}}}{k_B T}\right) + \exp\left(\frac{-\langle E_i\rangle_{\text{eq}}}{k_B T}\right). \tag{A.6}$$

Equation A.5 indicates that the system partition function can be written as a product of the partition function for individual units, thereby avoiding the many-body complexity. Thus, we can easily calculate the thermodynamic quantity,

$$\langle\sigma_i\rangle_{\text{eq}} = \sum_{\{\sigma\}} \sigma_i P(\{\sigma\}) = \frac{\sum_{\{\sigma\}} \sigma_i \cdot e^{\frac{\sigma_1\langle E_1\rangle_{\text{eq}}+\cdots\sigma_i\langle E_i\rangle_{\text{eq}}+\cdots\sigma_N\langle E_N\rangle_{\text{eq}}}{k_B T}}}{Q(T)}$$

$$= \frac{q_1 q_2 \ldots \left(e^{\langle E_i\rangle_{\text{eq}}/k_B T} - e^{-\langle E_i\rangle_{\text{eq}}/k_B T}\right)\ldots q_N}{q_1 q_2 \ldots \left(e^{\langle E_i\rangle_{\text{eq}}/k_B T} + e^{-\langle E_i\rangle_{\text{eq}}/k_B T}\right)\ldots q_N} = \tanh\left(\frac{\langle E_i\rangle_{\text{eq}}}{k_B T}\right) \tag{A.7}$$

A.2. *Kinetic Ising model and master equation*

The total Hamiltonian for a system coupled to a heat bath can be written as

$$\hat{H} = \hat{H}_{\text{s}} + \hat{H}_{\text{b}} + \hat{H}' \tag{A.8}$$

where $\hat{H}_{\text{s}}$ and $\hat{H}_{\text{b}}$ represent the Hamiltonian of the system (see Eq. 1) and the bath, respectively, and $\hat{H}'$ denotes the interaction Hamiltonian between the system and bath, assumed to be linear with respect to the system variables and given by:

$$\hat{H}' = \sum_i (\hat{\sigma}_i^+ + \hat{\sigma}_i^-)\hat{F} \tag{A.9}$$

where $\hat{F}$ denotes the interacting bath operator and that

$$\hat{\sigma}_i^+ = |1_i\rangle\langle -1_i| \quad \text{and} \quad \hat{\sigma}_i^- = |-1_i\rangle\langle 1_i|. \tag{A.10}$$

The interaction Hamiltonian used ensures that only one spin at a time is allowed to flip due to the interaction with the bath. For example, the matrix element,

$$\left\langle \{\sigma\}, \alpha \left| \sum_i \left(\hat{\sigma}_i^+ + \hat{\sigma}_i^+\right) \hat{F} \right| \{\sigma'\}, \beta \right\rangle, \tag{A.11}$$

would be nonvanishing only when $\{\sigma'\}$ differs from $\{\sigma\}$ by merely one σ_j' such that $\{\sigma_1', \ldots, \sigma_j', \ldots\} = \{\sigma_1, \ldots, -\sigma_j, \ldots\}$. Note that α and β in Eq. A.11 refer to the bath variables.

The dynamics of the system can be described by the time evolution of a density matrix, $\hat{\rho}(t)$, which evolves according to the Liouville equation,[96]

$$\frac{\partial \hat{\rho}}{\partial t} = -\frac{i}{\hbar}[\hat{H}, \hat{\rho}]. \tag{A.12}$$

Note that we are only interested in the time evolution of the diagonal elements of the density matrix $\hat{\rho}^{(s)}(t)$ (the heat bath variables have been averaged out), whose diagonal element, $\hat{\rho}^{(s)}_{nn}(t)$, represents the probability that the system in the state n at time t, also denoted by $P(\{\sigma\}, t)$ or $P(\sigma_1, \ldots, \sigma_i, \ldots; t)$. Using the Markov approximation, it follows from Eq. A.12 that:

$$\begin{aligned}\frac{d}{dt}P(\sigma_1, \ldots, \sigma_i, \ldots; t) = &-\sum_i W(\sigma_i \to -\sigma_i)P(\sigma_1, \ldots, \sigma_i, \ldots; t) \\ &+\sum_i W(-\sigma_i \to \sigma_i)P(\sigma_1, \ldots, -\sigma_i, \ldots; t)\end{aligned} \tag{A.13}$$

The transition rates are determined by the Fermi golden rule as follows:

$$W(\sigma_i \to -\sigma_i) = \frac{2\pi}{\hbar}\sum_{\alpha,\beta}^{\alpha\neq\beta} \rho^{(B)}_{\beta\beta}|\langle\alpha|\hat{F}_i|\beta\rangle|^2\delta(E_\alpha - E_\beta + 2E_i\sigma_i) \tag{A.14}$$

and

$$W(-\sigma_i \to \sigma_i) = \frac{2\pi}{\hbar}\sum_{\alpha,\beta}^{\alpha\neq\beta} \rho^{(B)}_{\beta\beta}|\langle\alpha|\hat{F}_i|\beta\rangle|^2\delta(E_\alpha - E_\beta - 2E_i\sigma_i) \tag{A.15}$$

where $\rho^{(\mathrm{B})}_{\beta\beta}$ denotes the Boltzmann distribution function for the β state of the heat bath, and $\delta(E_\alpha - E_\beta + 2E_i\sigma_i)$ and $\delta(E_\alpha - E_\beta - 2E_i\sigma_i)$ refer to the energetic resonance between the system and the bath expressed by the Dirac delta functions.

The aforementioned descriptions outline the theoretical framework of the kinetic Ising model. Equation A.13 is a master equation with transition rates given, which satisfy the principle of detailed balance. Although Glauber had written down the exact solutions to a kinetic Ising model,[51] our objective here is to apply a kinetic Ising model to the study of protein aggregation at the mean-field level.

A.3. *Time-evolution kinetic equation*

If we multiply both sides of Eq. A.13 by σ_k and sum over all configurations, $\{\sigma\}$, it follows that:

Left-hand side,

$$\frac{d}{dt}\sum_{\{\sigma\}}\sigma_k P(\{\sigma\},t) = \frac{d}{dt}\langle\sigma_k\rangle \tag{A.16}$$

Right-hand side,

$$\begin{aligned}
&= -\sum_{\{\sigma\}}\sigma_k W(\sigma_k \to -\sigma_k)P(\{\sigma_k\},t)\\
&\quad +\sum_{\{\sigma\}}\sigma_k W(-\sigma_k \to \sigma_k)P(\{-\sigma_k\},t) \quad [i=k]\\
&\quad -\sum_{\{\sigma\}}\sigma_k \sum_{i\neq k} W(\sigma_i \to -\sigma_i)P(\{\sigma_i\},t)\\
&\quad +\sum_{\{\sigma\}}\sigma_k \sum_{i\neq k} W(-\sigma_i \to \sigma_i)P(\{-\sigma_i\},t) \quad [i\neq k]
\end{aligned} \tag{A.17}$$

where the notation $P(\{-\sigma_i\},t) \equiv P(\sigma_1,\ldots,-\sigma_i,\ldots;t)$ and $P(\{\sigma_i\},t) \equiv P(\sigma_1,\ldots,\sigma_i,\ldots;t)$ are used for simplicity. Note that in Eq. A.17 one can exchange σ_i for $-\sigma_i$ for both $i = k$ and $i \neq k$ while keeping the individual sums unchanged due to the detailed balance among local transitions, e.g., $W(-\sigma_i \to \sigma_i)P(\{-\sigma_i\},t) = W(\sigma_i \to -\sigma_i)P(\{\sigma_i\},t)$. One can make such a substitution for the second and the fourth terms; it turns out that the third and the fourth terms would be cancelled. It follows from Eqs. A.16 and A.17 that

$$\begin{aligned}
\frac{d}{dt}\langle\sigma_k\rangle &= -2\sum_{\{\sigma\}}\sigma_k W(\sigma_k \to -\sigma_k)P(\{\sigma_k\},t)\\
&= -2\langle\sigma_k W(\sigma_k \to -\sigma_k)\rangle
\end{aligned} \tag{A.18}$$

Next, we shall focus our attention on the form of the transition rate, $W(\sigma_k \to -\sigma_k)$. In fact, a direct use of Eqs. A.14 and A.15 into Eq. A.18 is not analytically feasible. Glauber designed a form for $W(\sigma_k \to -\sigma_k)$, which aimed at describing the coupling between nearest neighbor units; he accordingly obtained the analytical solution to the kinetic equation in terms of the Bessel functions. Herein, we adopt a theoretically equivalent but relatively tricky way to deal with the equation. First, we split W into two parts that exhibit, respectively, mathematically even and odd properties in σ_i, as shown in Eq. A.19.

$$2W(\sigma_k \to -\sigma_k) = [W(\sigma_k \to -\sigma_k) + W(-\sigma_k \to \sigma_k)]_{\text{even}} \\ + [W(\sigma_k \to -\sigma_k) - W(-\sigma_k \to \sigma_k)]_{\text{odd}} \quad \text{(A.19)}$$

According to Eqs. A.14 and A.15, the sum in the "even" parentheses in Eq. A.19 will make no difference if we exchange σ_k for $-\sigma_k$, however, the sum in the "odd" parentheses, if same exchange is made, will result in the following symmetry difference, due to $\sigma_k = \pm 1$,

$$[W(\sigma_k \to -\sigma_k) - W(-\sigma_k \to \sigma_k)]_{\text{odd}} \\ = \sigma_k [W(\sigma_k \to -\sigma_k) - W(-\sigma_k \to \sigma_k)]_{\text{odd}}. \quad \text{(A.20)}$$

It follows from Eqs. (A.19) and (A.20) that we can rewrite Eq. A.18 as

$$-\frac{d}{dt}\langle\sigma_k\rangle = \langle [W(\sigma_k \to -\sigma_k) + W(-\sigma_k \to \sigma_k)]\sigma_k \rangle \\ + \langle [W(\sigma_k \to -\sigma_k) - W(-\sigma_k \to \sigma_k)] \rangle \quad \text{(A.21)}$$

where $\sigma_k^2 = 1$ is used. In the mean-field approximation, since correlation effects among spin units are ignored, the previous equation becomes:

$$-\frac{d}{dt}\langle\sigma_k\rangle = [\bar{W}(\sigma_k \to -\sigma_k) + \bar{W}(-\sigma_k \to \sigma_k)]\langle\sigma_k\rangle \\ + [\bar{W}(\sigma_k \to -\sigma_k) - \bar{W}(-\sigma_k \to \sigma_k)] \\ = [\bar{W}(\sigma_k \to -\sigma_k) + \bar{W}(-\sigma_k \to \sigma_k)] \\ \cdot \left\{ \langle\sigma_k\rangle + \frac{[\bar{W}(\sigma_k \to -\sigma_k) - \bar{W}(-\sigma_k \to \sigma_k)]}{[\bar{W}(\sigma_k \to -\sigma_k) + \bar{W}(-\sigma_k \to \sigma_k)]} \right\} \quad \text{(A.22)}$$

where $\bar{W}$'s refer to the mean-field reaction rates, which have similar expression with Eqs. A.14 and A.15 except that E_i is replaced by $\langle E_i \rangle$ in this case. Finally, we assume the canonical distribution for the heat bath, that is,

$$\bar{W}(-\sigma_k \to \sigma_k) = e^{\frac{2\langle E_i \rangle}{k_B T}} \bar{W}(\sigma_k \to -\sigma_k). \quad \text{(A.23)}$$

Substituting Eq. A.23 into Eq. A.22, it follows that:

$$\tau_k \frac{d}{dt}\langle\sigma_k\rangle = -\langle\sigma_k\rangle + \tanh\left(\frac{\langle E_k \rangle}{k_B T}\right), \quad \text{(A.24)}$$

where

$$\frac{1}{\tau_k} = \bar{W}(\sigma_k \to -\sigma_k) + \bar{W}(-\sigma_k \to \sigma_k)$$

$$= \frac{2\pi}{\hbar} \sum_{\alpha,\beta}^{\alpha\neq\beta} \rho_{\beta\beta}^{(B)} |\langle\alpha|\hat{F}_i|\beta\rangle|^2 [\delta(E_\alpha - E_\beta + 2\langle E_i\rangle) + \delta(E_\alpha - E_\beta - 2\langle E_i\rangle)] \tag{A.25}$$

Eq. A.25 is the quantum mechanical expression for the kinetic time scale of the present model, where the summation carries out the configurations of the heat bath according to the Fermi golden rule.

A.4. *Thermodynamic activation using Eyring equation*

The general form of the Eyring equation is given as:

$$k = \frac{k_B T}{h} e^{-\frac{\Delta G^\dagger}{k_B T}} = \frac{k_B T}{h} e^{\frac{\Delta S^\dagger}{k_B}} \cdot e^{-\frac{\Delta H^\dagger}{k_B T}} \tag{A.26}$$

where $\Delta G^\dagger$ refers to the Gibbs free energy of activation and that relation $\Delta G^\dagger = \Delta H^\dagger - T\Delta S^\dagger$ is used. The equation, derived from transition state theory (or activated-complex theory), describes the dependence of temperature on reaction rate via free energy barrier, similar to the empirical Arrhenius equation. Thus, we can assess the thermodynamic parameters of activation using the linear form of the

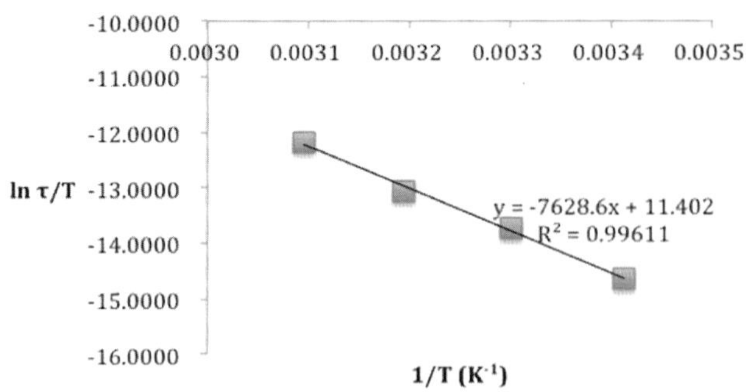

Figure A1: Eyring plot for the prion protein Sup35. The symbols denote the data points given by the fit result, and the solid line refers to a linear regression to the data. The inset presents the slope and the intercept, together with the goodness of the linear regression.

equation:

$$\ln\frac{k}{T} = \ln\frac{k_B}{h} + \frac{\Delta S^{\dagger}}{k_B} - \left(\frac{\Delta H^{\dagger}}{k_B}\right)\cdot\frac{1}{T}, \tag{A.27}$$

where the entropy ($\Delta S^{\dagger}$) and the enthalpy ($\Delta H^{\dagger}$) of activation can be obtained, respectively, from the intercept and the slope of the fit curve on a ($\ln\frac{k}{T}$) versus $\frac{1}{T}$ plot. In the case of the prion protein Sup35, we plot ($\ln\frac{\tau}{T}$) versus $\frac{1}{T}$, as shown in Fig. A1.

References

1. Anfinsen, C. B. Principles that govern folding of protein chains. *Science* **181**(4096), 223–230 (1973).
2. Chiti, F. and Dobson, C. M. Protein misfolding, functional amyloid, and human disease. *Annu. Rev. Biochem.* **75**, 333–366 (2006).
3. Knowles, T. P. J. and Buehler, M. J. Nanomechanics of functional and pathological amyloid materials. *Nature Nanotechnology* **6**(8), 469–479 (2011).
4. Tompa, P. and Friedrich, P. Prion proteins as memory molecules: an hypothesis. *Neuroscience.* **86**(4), 1037–1043 (1998).
5. Heinrich, S. U. and Lindquist, S. Protein-only mechanism induces self-perpetuating changes in the activity of neuronal Aplysia cytoplasmic polyadenylation element binding protein (CPEB). *Proc. Natl. Acad. Sci. USA.* **108**(7), 2999–3004 (2011).
6. Chen, M., Zheng, W., and Wolynes, P. G. Energy landscapes of a mechanical prion and their implications for the molecular mechanism of long-term memory. *Proc. Natl. Acad. Sci. USA.* **113**(18), 5006–5011 (2016).
7. Sunde, M. and Blake, C. C. F. From the globular to the fibrous state: protein structure and structural conversion in amyloid formation. *Q. Rev. Biophys.* **31**(1), 1–39 (1998).
8. Tycko, R. Progress towards a molecular-level structural understanding of amyloid fibrils. *Curr. Opin. Struct. Biol.* **14**(1), 96–103 (2004).
9. Luhrs, T., Ritter, C., Adrian, M., Riek-Loher, D., Bohrmann, B., Doeli, H., Schubert, D., and Riek, R. 3D structure of Alzheimer's amyloid-beta (1–42) fibrils. *Proc. Natl. Acad. Sci. U. S. A.* **102**(48), 17342–17347 (2005).
10. Anson, M. L. and Mirsky, A. E. The equilibrium between active native trypsin and inactive denatured trypsin, *J. Gen. Physiol.* **17**(3), 393–398 (1934).
11. Anson, M. L. and Mirsky, A. E. The equilibria between native and denatured hemoglobin in salicylate solutions and the theoretical consequences of the equilibrium between native and denatured protein, *J. Gen. Physiol.* **17**(3), 399–408 (1934).
12. Jackson, S. E. How do small single-domain proteins fold?, *Fold. Des.* **3**, R81–R91 (1998).
13. Privalov, P. L. and Khechinashvili, N. N. A thermodynamic approach to the problem of stabilization of globular protein structure: a calorimetric study. *J. Mol. Biol.* **86**(3), 665–684 (1974).
14. Bai, Y. W., Sosnick, T. R., Mayne L., and Englander, S. W. Protein-folding intermediates — native-state hydrogen-exchange. *Science* **269**(5221), 192–197 (1995).

15. Udgaonkar, J. B. and Baldwin, R. L. NMR evidence for an early framework intermediate on the folding pathway of ribonuclease A. *Nature* **335**(6192), 694–699 (1988).
16. Roder, H., Elove, G. A., and Englander, S. W. Structural characterization of folding intermediates in cytochrome c by H-exchange labeling and proton NMR. *Nature* **335**(6192), 700–704 (1988).
17. Plaxco, K. W., Simons, K. T., and Baker, D. Contact order, transition state placement and the refolding rates of single domain proteins. *J. Mol. Biol.* **277**(4), 985–994 (1998).
18. Zwanzig, R. Two-state models of protein folding kinetics. *Proc. Natl. Acad. Sci. USA.* **94**(1), 148–150 (1997).
19. Zwanzig, R. Simple model of protein folding kinetics. *Proc. Natl. Acad. Sci. USA.* **92**, 9801–9804 (1995).
20. Shiu, Y.-J., Jeng, U. S., Huang, Y.-S., Lai, Y.-H., Lu, H.-F., Liang, C.-T., Hso, I. J., Su, C.-H., Su, C., Chao, I., Su, A.-C., and Lin, S.-H. Global and local structural changes of cytochrome c and lysozyme characterized by a multigroup unfolding process. *Biophys. J.* **94**(12), 4828–4836 (2008).
21. Mayne, L. and Englander, S. W. Two-state vs. multistate protein unfolding studied by optical melting and hydrogen exchange. *Protein Sci.* **9**(10), 1873–1877 (2000).
22. Tsai, M.-Y. *Theoretical studies of protein folding: experimental and molecular dynamics insights into the design of phenomenological models*. PhD thesis, National Taiwan University, Taipei, Taiwan (2011).
23. Lewandowska, A., Oldziej, S., Liwo, A., and Scheraga, H. A. Beta-hairpin-forming peptides; models of early stages of protein folding. *Biophy. Chem.* **151**(1–2), 1–9 (2010).
24. Bryngelson, J. D. and Wolynes, P. G. Spin-glasses and the statistical-mechanics of protein folding. *Proc. Natl. Acad. Sci. USA.* **84**(21), 7524–7528 (1987).
25. Bryngelson, J. D. and Wolynes, P. G. Intermediates and barrier crossing in a random energy-model (with applications to protein folding). *J. Phys. Chem.* **93**(19), 6902–6915 (1989).
26. Bryngelson, J. D., Onuchic, J. N., Socci, N. D., and Wolynes, P. G. Funnels, pathways, and the energy landscape of protein-folding — a synthesis. *Proteins* **21**(3), 167–195 (1995).
27. Lindorff-Larsen, K., Piana, S., Dror, R. O., and Shaw, D. E. How fast-folding proteins fold. *Science.* **334**(6055), 517–520 (2011).
28. Zimm, B. H. and Bragg, J. K. Theory of the phase transition between helix and random coil in polypeptide chains. *J. Chem. Phys.* **31**, 526–535 (1959).
29. Lifson, S. Theory of helix-coil transition in polypeptides. *J. Chem. Phys.* **34**, 1963–1974 (1961).
30. Poland, D. and Scheraga, H. A. *Theory of Helix-coil Transitions in Biopolymers*. New York: Academic Press; 1970.
31. Doig, A. J., Chakrabartty, A., Klingler, T. M., and Baldwin, R. L. Determination of free energies of N-capping in alpha-helices by modification of the Lifson-Roig helix-coil theory to include N-capping and C-capping. *Biochemistry.* **33**, 3396–3403 (1994).
32. Thompson, P. A., Muñoz, V., Jas, G. S., Henry, E. R., Eaton W. A., and Hofrichter, J. The helix-coil kinetics of a heteropeptide. *J. Phys. Chem. B.* **104**, 378–389 (2000).
33. Wako, H. and Saito, N. Statistical mechanical theory of protein conformation. 1. general considerations and application to homopolymers. *J. Phys. Soc. Japan* **44**(6), 1931–1938 (1978).

34. Wako, H. and Saito, N. Statistical mechanical theory of protein conformation. 2. folding pathway for protein. *J. Phys. Soc. Japan* **44**(6), 1939–1945 (1978).
35. Pelizzola, A. Exactness of the cluster variation method and factorization of the equilibrium probability for the Wako-Saito-Munoz-Eaton model of protein folding. *J Stat. Mech. Theory E*, P11010 (2005).
36. Bruscolini, P. and Pelizzola, A. Exact solution of the Muñoz-Eaton model for protein folding. *Phys. Rev. Lett.* **88**(25), 258101 (2002).
37. Zamparo, M. and Pelizzola, A. Kinetics of the Wako-Saito-Munoz-Eaton model of protein folding. *Phys. Rev. Lett.* **97**, 068106 (2006).
38. Muñoz, V., Henry, E. R., Hofrichter, J., and Eaton, W. A. A statistical mechanical model for beta-hairpin kinetics. *Proc. Natl. Acad. Sci. USA.* **95**(11), 5872–5879 (1998).
39. Muñoz, V., Thompson, P. A., Hofrichter, J., and Eaton, W. A. Folding dynamics and mechanism of beta-hairpin formation. *Nature.* **390**(6656), 196–199 (1997).
40. Muñoz, V. and Eaton, W. A. A simple model for calculating the kinetics of protein folding from three-dimensional structures. *Proc. Natl. Acad. Sci. USA.* **96**(20), 11311–11316 (1999).
41. Krishna, M. M. G., Maity, H., Rumbley, J. N., Lin, Y., and Englander, S. W. Order of steps in the cytochrome c folding pathway: evidence for a sequential stabilization mechanism. *J. Mol. Biol.* **359**(5), 1410–1419 (2006).
42. Hu, W., Kan, Z.-Y., Mayne, L., and Englander, S. W. Cytochrome c folds through foldon-dependent native-like intermediates in an ordered pathway. *Proc. Natl. Acad. Sci. USA.* **113**(14), 3809–3814 (2016).
43. Panchenko, A. R., LutheySchulten, Z., and Wolynes, P. G. Foldons, protein structural modules, and exons. *Proc. Natl. Acad. Sci. USA.* **93**(5), 2008–2013 (1996).
44. Sugita, M., Matsuoka, M., and Kikuchi, T. Topological and sequence information predict that foldons organize a partially overlapped and hierarchical structure. *Proteins.* **83**(10), 1900–1913 (2015).
45. Tsai, M.-Y., Yuan, J.-M., Teranishi, Y., and Lin, S. H. Thermodynamics of protein folding using a modified Wako-Saito-Muñoz-Eaton model. *J. Biol. Phys.* **38**(4), 543–571 (2012).
46. Tsai, M.-Y., Yuan, J.-M., Yamaki, M., Lin, C.-K.,and Lin, S. H. Molecular dynamics insight into the diverse thermodynamic behavior of a beta-hairpin peptide. *J. Chin. Chem. Soc.* **60**(7), 915–928 (2013).
47. Liang, K. K., Hayashi, M., Shiu, Y. J., Mo, Y., Saho, J. S., Yan, Y. J., and Lin, S. H. Application of the generalized kinetic Ising model to the kinetics of protein folding. *J. Chin. Chem. Soc.* **50**(3A), 335–338 (2003).
48. Nobuhiro, G. Theoretical studies of protein folding. *Annu. Rev. Biophys. Bioeng.* **12**, 183–210 (1983).
49. Liang, K. K., Hayashi, M., Shiu, Y. J., Mo, Y., Shao, J. S., Yan, Y. J., and Lin, S. H. Thermodynamics and kinetics of protein folding: a mean field theory. *Phys. Chem. Chem. Phys.* **5**(23), 5300–5308 (2003).
50. Morozov, A. N., Shiu, Y. J., Liang, C. T., Tsai, M. Y., and Lin, S. H. Nonadditive interactions in protein folding: the Zipper model of cytochrome c. *J. Biol. Phys.* **33**(4), 255–270 (2007).
51. Glauber, R. J. Time-dependent statistics of the Ising model. *J. Math. Phys.* **4**(2), 294–307 (1963).

52. Xue, W.-F., Homans, S. W., and Radford, S. E. Systematic analysis of nucleation-dependent polymerization reveals new insights into the mechanism of amyloid self-assembly. *Proc. Natl. Acad. Sci. USA.* **105**(26), 8926–8931 (2008).
53. Serio, T. R., Cashikar, A. G., Kowal, A. S., Sawicki, G. J., Moslehi, J. J., Serpell, L., Arnsdorf, M. F., and Lindquist, S. L. Nucleated conformational conversion and the replication of conformational information by a prion determinant. *Science.* **289**(5483), 1317–1321 (2000).
54. Buell, A. K., Hung, P., Salvatella, X., Welland, M. E., Dobson, C. M., and Knowles, T. P. J. Electrostatic effects in filamentous protein aggregation. *Biophys. J.* **104**(5), 1116–1126 (2013).
55. Oosawa, F., Asakura, S., Hotta, K., Imai, N., and Ooi, T. G-F transformation of actin as a fibrous condensation. *J. Polym. Sci.* **37**(132), 323–336 (1959).
56. Hofrichter, J., Ross, P. D., and Eaton, W. A. Kinetics and mechanism of deoxyhemoglobin S gelation: a new approach to understanding sickle cell disease. *Proc. Natl. Acad. Sci. U. S. A.* **71**(12), 4864–4868 (1974).
57. Ferrone, F. A., Hofrichter, J., Sunshine, H. R., and Eaton, W. A. Kinetic studies on photolysis-induced gelation of sickle cell hemoglobin suggest a new mechanism. *Biophys. J.* **32**(1), 361–380 (1980).
58. Lee, J., Culyba, E. K., Powers, E. T., and Kelly, J. W. Amyloid-beta forms fibrils by nucleated conformational conversion of oligomers. *Nat. Chem. Biol.* **7**(9), 602–609 (2011).
59. Schreck, J. S., and Yuan, J. M. A kinetic study of amyloid formation: fibril growth and length distributions. *J. Phys. Chem. B* **117**(21), 6574–6583 (2013).
60. Bishop, M. F. and Ferrone, F. A. Kinetics of nucleation-controlled polymerization — a perturbation treatment for use with a secondary pathway. *Biophys. J.* **46**(5), 631–644 (1984).
61. Ferrone, F. A., Hofrichter, J., and Eaton, W. A. Kinetics of sickle hemoglobin polymerization. II. A double nucleation mechanism. *J. Mol. Biol.* **183**(4), 611–631 (1985).
62. Collins, S. R., Douglass, A., Vale, R. D., and Weissman, J. S. Mechanism of prion propagation: amyloid growth occurs by monomer addition. *Plos Biol.* **2**(10), 1582–1590 (2004).
63. Wegner, A., and Savko, P. Fregmentation of actin-filaments. *Biochemistry.* **21**(8), 1909–1913 (1982).
64. Andersen, C. B., Yagi, H., Manno, M., Martorana, V., Ban, T., Christiansen, G., Otzen, D. E., Goto, Y., and Rischel, C. Branching in amyloid fibril growth. *Biophys. J.* **96**(4), 1529–1536 (2009).
65. Amann, K. J. and Pollard, T. D. The Arp2/3 complex nucleates actin filament branches from the sides of pre-existing filaments. *Nat. Cell Biol.* **3**(3), 306–310 (2001).
66. Hong, L. and Yong, W.-A. Simple moment-closure model for the self-assembly of breakable amyloid filaments. *Biophys. J.* **104**(3), 533–540 (2013).
67. Michaels, T. C. T., Cohen, S. I. A., Vendruscolo, M., Dobson, C. M., and Knowles, T. P. J. Hamiltonian dynamics of protein filament formation. *Phys. Rev. Lett.* **116**(3), 038101 (2016).
68. Knowles, T. P. J., Waudby, C. A., Devlin, G. L., Cohen, S. I. A., Aguzzi, A., Vendruscolo, M., Terentjev, E. M., Welland, M. E., and Dobson, C. M. An analytical solution to the kinetics of breakable filament assembly. *Science* **326**(5959), 1533–1537 (2009).

69. Arosio, P., Beeg, M., Nicoud, L., and Morbidelli, M. Time evolution of amyloid fibril length distribution described by a population balance model. *Chem. Eng. Sci.* **78**(0), 21–32 (2012).
70. Vitalis, A. and Pappu, R. V. Assessing the contribution of heterogeneous distributions of oligomers to aggregation mechanisms of polyglutamine peptides. *Biophys. Chem.* **159**(1), 14–23 (2011).
71. Auer, S. and Kashchiev, D. Insight into the correlation between lag time and aggregation rate in the kinetics of protein aggregation. *Proteins.* **78**(11), 2412–2416 (2010).
72. Auer, S., Ricchiuto, P., and Kashchiev, D. Two-step nucleation of amyloid fibrils: omnipresent or not?, *J. Mol. Biol.* **422**(5), 723–730 (2012).
73. Ricchiuto, P., Brukhno, A. V., and Auer, S. Protein aggregation: kinetics versus thermodynamics. *J. Phys. Chem. B.* **116**(18), 5384–5390 (2012).
74. Wu, C. and Shea, J.-E. Coarse-grained models for protein aggregation. *Curr. Opin. Struct. Biol.* **21**(2), 209–220 (2011).
75. Carulla, N., Caddy, G. L., Hall, D. R., Zurdo, J., Gairi, M., Feliz, M., Giralt, E., Robinson, C. V., and Dobson, C. M. Molecular recycling within amyloid fibrils. *Nature.* **436**(7050), 554–558 (2005).
76. O'Nuallain, B., Shivaprasad, S., Kheterpal, I., and Wetzel, R. Thermodynamics of A beta(1–40) amyloid fibril elongation. *Biochemistry.* **44**(38), 12709–12718 (2005).
77. Kardos, J., Yamamoto, K., Hasegawa, K., Naiki, H., and Goto, Y. Direct measurement of the thermodynamic parameters of amyloid formation by isothermal titration calorimetry. *J. Biol. Chem.* **279**(53), 55308–55314 (2004).
78. Baldwin, A. J., Knowles, T. P. J., Tartaglia, G. G., Fitzpatrick, A. W., Devlin, G. L., Shammas, S. L., Waudby, C. A., Mossuto, M. F., Meehan, S., Gras, S. L., Christodoulou, J., Anthony-Cahill, S. J., Barker, P. D., Vendruscolo, M., and Dobson, C. M. Metastability of native proteins and the phenomenon of amyloid formation. *J. Am. Chem. Soc.* **133**(36), 14160–14163 (2011).
79. Thirumalai, D. and Reddy, G. Protein thermodynamics: are native proteins metastable?, *Nat. Chem.* 3(12), 910–911 (2011).
80. Schreck, J. S. and Yuan, J. M. Exactly solvable model for helix-coil-sheet transitions in protein systems. *Phys. Rev. E.* **81**(6), 061919 (2010).
81. van Gestel, J. and de Leeuw, S. W. A statistical-mechanical theory of fibril formation in dilute protein solutions. *Biophys. J.* **90**(9), 3134–3145 (2006).
82. Schreck, J. S. and Yuan, J.-M. A statistical mechanical approach to protein aggregation. *J. Chem. Phys.* **135**(23), 235102–235112 (2011).
83. Schreck, J. S. and Yuan, J.-M. Statistical mechanical treatments of protein amyloid formation. *Int. J. Mol. Sci.* **14**(9), 17420–17452 (2013).
84. Chiti, F., De Lorenzi, E., Grossi, S., Mangione, P., Giorgetti, S., Caccialanza, G., Dobson, C. M., Merlini, G., Ramponi, G., and Bellotti, V. A partially structured species of beta 2-microglobulin is significantly populated under physiological conditions and involved in fibrillogenesis. *J. Biol. Chem.* **276**(50), 46714–46721 (2001).
85. Chiti, F., Mangione, P., Andreola, A., Giorgetti, S., Stefani, M., Dobson, C. M., Bellottl, V., and Taddei, N. Detection of two partially structured species in the folding process of the amyloidogenic protein beta 2-microglobulin. *J. Mol. Biol.* **307**(1), 379–391 (2001).

86. Chatani, E., Ohnishi, R., Konuma, T., Sakurai, K., Naiki, H., and Goto, Y. Pre-steady-state kinetic analysis of the elongation of amyloid fibrils of beta(2)-microglobulin with tryptophan mutagenesis. *J. Mol. Biol.* **400**(5), 1057–1066 (2010).
87. Konuma, T., Chatani, E., Yagi, M., Sakurai, K., Ikegami, T., Naiki, H., and Goto, Y. Kinetic intermediates of beta(2)-microglobulin fibril elongation probed by pulse-labeling H/D exchange combined with NMR analysis. *J. Mol. Biol.* **405**(3), 851–862 (2011).
88. Gruebele, M. and Thirumalai, D. Perspective: reaches of chemical physics in biology. *J. Chem. Phys.* **139**(12), 121701 (2013).
89. Li, M. S., Klimov, D. K., Straub, J. E., and Thirumalai, D. Probing the mechanisms of fibril formation using lattice models. *J. Chem. Phys.* **129**(17), 175101–175110 (2008).
90. Tsai, M.-Y., Yuan, J.-M., and Lin, S.-H. Thermodynamic insight into protein aggregation using a kinetic ising model. *J. Chin. Chem. Soc.* **62**(1), 21–25 (2015).
91. Sabate, R., Villar-Pique, A., Espargaro, A., and Ventura, S. Temperature dependence of the aggregation kinetics of Sup35 and Ure2p yeast prions. *Biomacromolecules.* **13**(2), 474–483 (2012).
92. Williams, A. D., Shivaprasad, S., and Wetzel, R. Alanine scanning mutagenesis of A beta(1–40) amyloid fibril stability. *J. Mol. Biol.* **357**(4), 1283–1294 (2006).
93. Nobuhiro, G. Theoretical studies of protein folding. *Annu Rev Biophys Bioeng.* **12**, 183–210 (1983).
94. Narimoto, T., Sakurai, K., Okamoto, A., Chatani, E., Hoshino, M., Hasegawa, K., Naiki, H., and Goto, Y. Conformational stability of amyloid fibrils of beta-microglobulin probed by guanidine-hydrochloride-induced unfolding. *FEBS Lett.* **576**(3), 313–319 (2004).
95. Mishra, R., Sorgjerd, K., Nystrom, S., Nordigarden, A., Yu, Y.-C., and Hammarstrom, P. Lysozyme amyloidogenesis is accelerated by specific nicking and fragmentation but decelerated by intact protein binding and conversion. *J. Mol. Biol.* **366**(3), 1029–1044 (2007).
96. Lin, S. H., Alden, R. G., Islampour, R., Ma, H., and Villaeys, A. A. *Density Matrix Method and Femtosecond Processes*. Teaneck, NJ: World Scientific: 1991.

Folding and Misfolding of Amyloid-β40 and 42 in Alzheimer's Disease

7

Yu-Jen Chang and Yun-Ru Chen

Genomics Research Center, Academia Sinica, 128, Academia Road, Section 2, Nankang District, Taipei 115, Taiwan

The major component of senile plaques, the hallmark in the brain of patients with Alzheimer's disease (AD), is fibrillar amyloid-β(Aβ). Aβ is an intrinsically disordered peptide proteolyzed from amyloid precursor protein (APP). Several familial mutations that may cause the early onset of AD are within and around Aβ sequence. It demonstrates the importance of understanding folding and misfolding mechanisms of Aβ. Here, we summarize our previous works starting from examining folding stability of the initial state of Aβ42 and Aβ40, and comparing the relationship of folding stability among wild-type Aβ and familial AD (FAD) mutants to their misfolding properties. We also showed that zinc and aluminum ions destabilize Aβ and result in oligomer formation. Finally, we systematically examine the fibrillization kinetics and induced cytotoxicity among various mixtures of Aβ42/Aβ40.

1. Introduction

1.1. *Generation of Aβ*

Alzheimer's disease (AD) is the most prevalent neurodegenerative disease among those over the age of 65.[1] Two pathological hallmarks were found in brain biopsies of patients with AD. They are senile plaques comprising insoluble amyloid-β(Aβ) fibrils and neurofibrillary tangles composed of hyperphosphorylated tau. Evidences from genetics and biochemistry have supported that Aβ production and aggregation play critical roles in AD pathogenesis. However, the folding and misfolding properties of Aβ are still ambiguous. Amyloid precursor protein (APP) is a type I transmembrane protein that, when processed by α- and γ-secretases, generates soluble APP-α (APPα) and C83. This pathway is not amyloidogenic; whereas, when APP is processed by β- and γ-secretases, soluble APP-β (sAPPβ), Aβ, and C99 are generated. Several different Aβ isoforms ranging from 39 to 43 residues are generated after β- and γ-secretase cleavages. It is suggested that Aβ is produced from a longer sequence comprising 49 or 48 residues, then undergoes triple- or quadruple-residue trimming by γ-secretase into 40 or 42 residues. The trimming pattern is suggested to be Aβ49—Aβ46—Aβ43—Aβ40 and Aβ48—Aβ45—Aβ42—Aβ38.[2–4] Aβ is more frequently generated in early endosome where both secretases and APP are present.[5–7] Although Aβ deposits extracellularly, there are emerging evidences supporting the existence of intracellular Aβ and Aβ may be originated from intraneuronal space.[8]

Inheritable mutations in *APP* gene were found mostly around or within the Aβ region. The mutation causes early onset of AD and cerebral amyloid angiopathy (CAA).[9] There are two hot spots in Aβ mutations. One is located at N-terminal residues 6 or 7 and the other is within residues 21–23. The N-terminal mutants include English (H6R),[10] Tottori (D7N),[11] and Taiwan (D7H).[12] The mutations at the middle are Flemish (A21G),[13] Arctic (E22G),[14,15] Dutch (E22Q),[16,17] Italian (E22K),[18] and Iowa (D23N)[19] mutants. APP mutations near γ-secretase cleavage sites such as I716[20,21] and V717[21,22] could accelerate Aβ production to particularly the highly pathogenic Aβ42.[23–25]

1.2. *Aβ aggregation*

Aβ is an intrinsically disordered peptide that is natively unfolded, meaning it lacks intact secondary and tertiary structures. It can spontaneously aggregate into amyloid fibrils that are highly enriched in cross-β structures.[26] Kinetic studies showed *Aβ* fibrillization is a nucleation-dependent polymerization.[27] It first undergoes a rate-limiting step to form a nucleus, then the nucleus grows into fibrils through an elongation stage, until it reaches the steady state to form mature fibrils.[28,29] The process can be monitored biochemically by amyloid-binding dyes such as thioflavin T (ThT) and Congo red. ThT fluoresces upon binding to cross β-stands in fibrils and Congo red shows birefringence under polarized light upon binding.[30,31] Since the dyes specifically recognize fibrils, the oligomerization process is hindered under these biochemical assays. Recently, secondary nucleation events in the fibrillization pathway were reported.[32,33] Aβ aggregation is described as the combination of events: primary nucleation that describes the association between two monomers and elongation, and secondary nucleation which is the addition of monomers onto fibrils. Aβ40 has smaller rates for all events than Aβ42 especially in primary nucleation. Although Aβ40 has lower rate for secondary nucleation in comparison to Aβ42, aggregation of Aβ40 strongly depends on the secondary nucleation triggered by already formed fibril aggregates.[32,33]

The mature Aβ fibrils generally have a diameter of ~10 nm and the fibril length can be longer than 1 μm.[27] The fibril model of full-length Aβ40 contains two β-strands formed by residues 10–22 and 30–40. The two β-strands are linked by a bend/turn-like structure at residues 23–29.[34–37] A salt bridge between D23 and K28 is present.[34] In contrast, it has been suggested that Aβ42 fibrils consist of three parallel β-sheets. They are residues 12–18, 24–33, and 36–40.[38] The β-stands form in register parallel β-sheets that contribute to the cross-β spine. The β-strands are perpendicular to the fibril axis. Both straight and twisted Aβ fibrils can be found under electron microscopy.[39]

Emerging evidence indicates that prefibrillar oligomeric forms of Aβ are the primary pathological species rather than the mature fibrils.[40,41] However, it is difficult to study prefibrillar species due to their transient nature. Different preparation conditions were reported for Aβ oligomers and both antiparallel and parallel β-sheets were observed.[42–44] The reported oligomers include Aβ-derived diffusible ligands (ADDLs), Aβ56*, low-molecular-weight oligomers, spherical oligomers, amylospheroid, globulomers, and annular protofibrils.[27] Curvilinear protofibrils are also present prior to mature fibrils.

1.3. *Aβ* toxicity

Aβ aggregation plays a pivotal role in AD pathology, with possible mechanisms including impaired signal transductions, elevation of oxidative stress, disruption of calcium homeostasis in the cell, reduction of long-term potentiation, and inducement of inflammation.[45,46] Different isotypes and modifications of Aβ were found to have different degrees of toxicity. Pyro-glutamated[47] and nitrated[48] Aβ42 were found to have higher toxicity than unmodified Aβ species. The fact that abnormal behavior of transgenic AD mice was observed prior to formation of deposits raises a question about the source of Aβ toxicity.[49] Injection of specific Aβ assembly to normal young rats induced the memory deficit.[49] It has also been shown that prefibrillar oligomers elicit higher toxicity than fibrils in *in vitro* studies.[50,51] In addition, Aβ dimer is the dominant species purified from the cortices of AD patients and can induce hyperphosphorylation of tau and neurite degeneration.[52] Synthetic Aβ dimer can rapidly form toxic protofibrils.[53] Furthermore, the toxicity of Aβ oligomers can be assessed through the porelike structure that is feasible to penetrate membranes[54] and to induce calcium influx.[55] Overall, Aβ oligomers can induce synapse dysfunction followed by cascades of neurodegeneration,[56] whereas Aβ fibrils contribute to toxicity to a lesser extent, if any, and the toxicity depends on fibril morphology.[57]

1.4. *Native Aβ conformation*

Little is known about the native conformation of Aβ. Previously, a solution-state nuclear magnetic resonance (NMR) study showed that native monomeric Aβ possesses short residual β-strands in hydrophobic regions, i.e., a.a. 17–21 and a.a. 31–36.[58] Turn- or bend-like structures at 7–11 and 30–36 were also detected.[58] The residual β-strands are looser and shorter than the β-strands reported in mature fibrils. Limited proteolysis coupled with mass spectrometry revealed a protease-resistant segment from residues A21 to A30,[59] which may be a nucleation site for Aβ folding.[60] Bis-ANS (4,4'-Dianilino-1,1'-Binaphthyl-5,5'-Disulfonic Acid) binding sites are at the flanking sites of protease-resistant segment.[61] A summary of current structural information on native Aβ conformation is depicted in Fig. 1.

1.5. *Aβ40 and Aβ42*

Aβ40 and 42 are the main isoforms of Aβ resulting from γ-secretase cleavages. Aβ42 is more hydrophobic and insoluble due to the two extra hydrophobic amino acids, isoleucine and alanine, at the C terminus of Aβ42.[62] Aβ42 aggregates faster

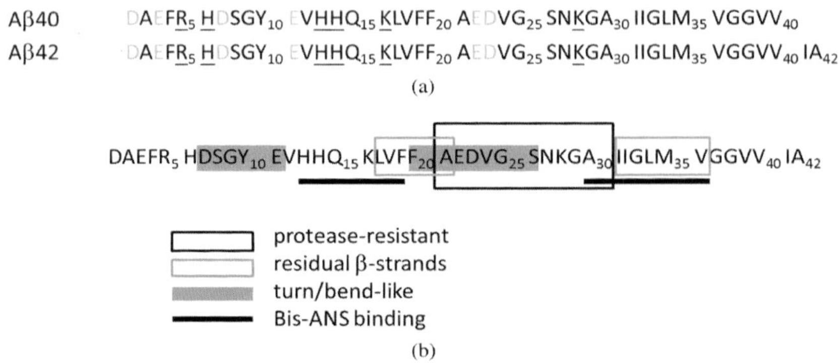

Figure 1: Primary sequence and residual structures of Aβ. (a) Primary sequence of Aβ40 and Aβ42. Acidic residues are in gray and basic residues are underlined. (b) Residual structures reported in Aβ prior to aggregation. The residue numbers are indicated.

than Aβ40 and is more detrimental.[62–64] Unequal distribution of Aβ40 and Aβ42 is present in diseased brains and fluids,[65,66] with Aβ40 about 10 times more abundant than Aβ42 in cerebrospinal fluids of AD patients.[67] Aβ40 predominately composes the CAA whereas Aβ42 is the main product in parenchymal plaques and has higher correlation with AD pathology.[68–71]

In solid-state NMR studies, both Aβ40 and Aβ42 fibrils generated in register parallel β-sheet structures,[72–74] but *Aβ*42 fibrils have been suggested to consist of three parallel β-sheets.[38] Cryo-EM study also suggested Aβ40 and Aβ42 fibrils share a similar protofilament structure.[75] The scanning transmission electron microscopy combined with mass-per-length analysis showed different numbers of peptide per cross-β-sheet.[75] In a photoinduced uncoupled cross-linking study, the distribution of Aβ oligomers was analyzed on sodium dodecyl sulfate–polyacrylamide gel electrophoresis (SDS-PAGE) after cross-linking and it was found that oligomeric species of Aβ40 are in similar amount of monomer, dimer, trimer, and tetramer as well as weaker intensity of pentamer and hexamer. However, Aβ42 contains small amount of monomer through trimer, more abundant tetramer through octamer, and some larger oligomers ranged from 30 to 60 kDa. The distinct patterns of oligomeric species in early stage of Aβ40 and Aβ42 depict that they have different mechanisms of fibrillization.[76] Data from ion mobility–mass spectrometry show that oligomer species of Aβ40 are mainly monomer, dimer, and tetramer, whereas Aβ42 adopts mainly dimer, "open" tetramer, hexamer, and dodecamer.[77] Other group indicated that Aβ40 oligomers can be separated into compact and extended conformation.[78]

1.6. *Mixture of Aβ40 and Aβ42*

Several studies have indicated that an elevated ratio of Aβ42/Aβ40, related to early onset FAD, causes higher toxicity and the combination has distinct aggregation properties from Aβ42 and Aβ40 alone.[64,79] Aβ40 and Aβ42 interact and monomeric Aβ40 has higher affinity than monomeric Aβ42 when binding to Aβ42 aggregates.[80] Addition of Aβ40 delays the fibril formation of Aβ42.[81] as well as inhibits the oligomerization of Aβ42, as detected by ion mobility–mass spectrometry.[82] Aβ42 to Aβ40 as low as 3/7 has been shown to have comparable neurotoxicity to Aβ42 alone.[79] In our study, we found formation of toxic and stable oligomers in equally mixed Aβ42 to Aβ40.[64] Structural analysis suggested no difference between the mixed and the pure species of Aβ40 and Aβ42 fibrils but coassembly of Aβ40 and Aβ42 alters kinetics by changing the pattern of oligomer formation.[83] In AD transgenic mice model, formation of Aβ plaques was decelerated by the coexpression of Aβ40.[84]

1.7. *Summary*

Aβ aggregation is highly implicated in AD and is the process most targeted for therapeutic remedies. Understanding folding and misfolding properties of Aβ is essential toward combating AD. Here, we summarized our previous works including examining folding stability of the initial state of Aβ42 and Aβ40 as well as FAD mutants and metal ion effect. We also correlated them to their oligomer- and fibril-forming properties. Finally, we characterized mixtures of Aβ42/Aβ40 and found novel toxic oligomers in the equally mixed Aβ species.

2. Materials and Methods

2.1. *Aβ peptide synthesis*

All Aβ peptides and short peptides were synthesized by fluorenylmethoxycarbonyl (Fmoc) solid-phase synthesis method and purified by reversed-phase high-performance liquid chromatography (RP-HPLC) followed by matrix-assisted laser desorption/ionization (MALDI) mass spectrometry to ensure the purity. In Sec. 3.1, Aβ was synthesized in University of California, Irvine. In Secs. 3.2–3.4, Aβ peptides were synthesized by the peptide synthesis core, Genomics Research Center, Academia Sinica, Taiwan. Purified peptides were dissolved in 50% acetonitrile/water, aliquoted, and lyophilized. In Sec. 3.2, reversed Aβ, denoted as Aβ(40-1), was purchased from rPeptide (Bogart, Georgia).

2.2. *Sample preparation*

In Sec. 3.1, Aβ at 10 mg/mL was denatured in 10 M urea containing 10 mM sodium phosphate buffer, pH 7.4, and followed by centrifugation at 15,000× g for 10 min at room temperature. Preparation of urea in phosphate buffer was quantified by a refractometer. Denatured Aβ stock was rapidly refolded in 10 mM sodium phosphate, pH 7.4, and instantly vortexed followed by centrifugation and filtration through a filter with 0.22 micrometer pore size. The final concentration of Aβ in supernatant was determined by ultraviolet (UV) absorption at 280 nm ($\varepsilon = 1280\,\mathrm{cm}^{-1}\,\mathrm{M}^{-1}$) based on the Edhoch equation.[85] For the result discussed in Sec. 3.2, Aβ was freshly obtained by unfolding Aβ in 8 M GdnHCl at 20 mg/mL containing 10 mM sodium phosphate buffer, pH 7.4, followed by rapid refolding in 10 mM sodium phosphate buffer, pH 7.4, centrifuging at 17,000× g for 20 min at 4°C, and collecting supernatant. The final percentage of GdnHCl was less than 3.3%. Aβ concentration was determined also by absorbance at 280 nm. Aβ was adjusted to the experimental concentrations with either the same phosphate buffer or the buffer containing 3 M GdnHCl. In Sec. 3.3, Aβ was prepared in 10 mM Tris-HCl, pH 7.4, instead of phosphate buffer due to the insolubility of metal ions in phosphate buffer. Aβ was denatured in 8 M GdnHCl and rapidly refolded into in Tris buffer at ~1 mg/mL followed by centrifugation at 17,000 × *g* at 4°C for 15 min. The supernatant was collected and quantified by absorbance at 280 nm. Final Aβ concentration in Sec. 3.3 was 25 μM. In Sec. 3.4, Aβ peptides at 2.5 mg/mL in hexafluoroisopropanol (HFIP) were prepared for dissociating pre-aggregates followed by evaporation of HFIP in vacuum. Aβ was then dissolved in anhydrous dimethyl sulfoxide (DMSO) at 10 mg/mL as DMSO stock for Aβ. The mixtures of Aβ42 and Aβ40 were prepared in this step by mixing DMSO stocks for Aβ40 and Aβ42. Final Aβ concentration was 50 μM in all conditions after Aβ was diluted to 10 mM sodium phosphate buffer, pH 7.4. In far-UV circular dichrosim (CD) assay, we used NaOH instead of DMSO to prepare Aβ due to the intolerance of DMSO in far-UV CD with high UV absorption. For the CD samples, pre-aggregates were dissociated by addition 100 μL of 3 mM NaOH to Aβ, 0.1 mg, sonication for 1 min, and lyophilization. Two hundred microliters of 50 mM sodium phosphate buffer, pH 7.2, was used to dissolve lyophilized Aβ followed by centrifugation at 17,000× g for 10 min at 4°C and quantification by absorbance at 280 nm. The final solution was neutral, around pH 7.4.

2.3. *Native and SDS-PAGE*

Aβ conformers were analyzed by native PAGE performing on a commercial Tris-Tricine precast gel (BioRad, Hercules, California, United Sates) operating at 4°C. Samples for native PAGE were kept under native form without heating and denaturing by SDS and β-mercaptoethanol. Aβ conformers were transferred to a nitrocellulose (NC) membrane and stained by *Direct Blue 71* (Sigma, St. Louis, Missouri, United States). In Sec. 3.2, we used 15% Bis-Tris gels without SDS to perform native PAGE on which freshly prepared Aβ samples were electrophoresed for ~5 h at 4°C. Two methods were adopted to detect Aβ species on the gel: one was fluorescently stained by *SYPRO Ruby* (Invitrogen Inc., Carlsbad, CA) followed by scanning gels in a *Typhoon 9400* fluorescence imager (GE Healthcare Inc., Piscataway, NJ). The other was to transfer Aβ species to polyvinylidene fluoride (PVDF) membrane followed by performing western blotting immunoblotted by antibody *6E10* (Chemicon Inc., Billerica, MA) recognizing Aβ sequence residues from 1 to 17.

2.4. *Extrinsic and intrinsic fluorescence of Aβ*

In Sec. 3.1, extrinsic fluorescence of full-length Aβ and Aβ fragments mixed with different concentrations of urea containing 5 μM Bis-ANS (4,4′-dianilino-1,

1′-binaphthyl-5, 5′-disulfonic acid dipotassium salt; Sigma) was measured by using a *SPEX Fluorolog* spectrofluorometer (Jobin Yvon) with excitation at 400 nm and emission from 450 to 600 nm. In Secs. 3.2 and 3.3, experiments were performed in a *FluoroMax-3* spectrofluorometer (Horiba Jobin Yvon, Kyoto, Japan). In Secs. 3.2 and 3.3, spectra of Bis-ANS at 5 μM mixed with Aβ variants were excited at 400 nm and monitored from 450 to 550 nm at room temperature. For intrinsic fluorescence, emission of the single tyrosine on Aβ was recorded from 290 to 360 nm with excitation at 270 nm.

2.5. *Far-UV circular dichroism spectroscopy*

In Sec. 3.1, far-UV CD spectroscopy was performed on a *Jasco Model J-800* (Jasco Inc., Tokyo, Japan) spectropolarimeter and a cylindrical cell (Hellma) with the path length of 1 cm. CD spectra of Aβ samples in different concentrations of urea were measured from 190 to 250 nm. In Secs. 3.2, 3.3, and 3.4, CD spectroscopy was performed on a *Jasco J-815* spectropolarimeter (Jasco Inc.) using a cuvette with 1 mm path length. In Secs. 3.2 and 3.4, the spectra were recorded from 200 to 250 nm and in Sec. 3.3 the spectra were recorded from 202 to 250 nm. All spectra were processed through the subtraction of background.

2.6. *Folding stability measurement by urea/GdnHCl denaturation and data fitting*

Aβ concentrations were prepared at 50, 25, and 12.5 μM. Titration was made by adding the equal concentration of unfolded Aβ in 8 M urea to the folded Aβ in less than 0.2 M urea and ended in 8 M urea. Reversed titration was performed by the addition of folded Aβ to unfolded Aβ and ended in 0.08 M urea. Each of the experiments was completed within 1 h to ensure no aggregation occurred. Recording of Bis-ANS fluorescence was performed with the emission wavelength from 450 to 600 nm and the intensity at 508 nm for Aβ40 and 500 nm for Aβ42 were normalized and plotted against the urea concentration. When urea denaturation was monitored in far-UV CD, we chose the signals at 220 nm to be averaged, normalized to the signal of native Aβ, and plotted against urea concentration. All denaturation experiments of Secs. 3.1–3.3 were performed at 25°C.

In Sec. 3.1, *Kaleidograph 3.0* (Synergy Software Inc., Schaumburg, IL) or *Igor Pro 5* (Wave Matrix Inc., Lake Oswego, Oregon, United States) was adopted for analyzing the denaturation data. On the basis of Santoro and Bolen study,[86] a two-state model (N $\leftrightarrows$ U) was used to describe the mechanism of the denaturation of Aβ40. To describe the mechanism, total signal is separated into two parts: the fraction of folded species N,f_N, and the fraction of unfolded species U,f_U, with P representing the signal. Slopes are denoted as m_N and m_U representing for slopes of pre- and posttransition, respectively, into Eq. 1.

$$\begin{aligned} P_{\text{total}} &= f_N P_N + f_U P_U \\ &= f_N(P_{N0} + m_N[urea]) + f_U(P_{U0} + m_U[urea]) \end{aligned} \tag{1}$$

In the two-state model, sum of f_N and f_U should be 1, i.e., $f_N + f_U = 1$.

Since there are only two fractions, folded and unfolded monomer, in a two-state mechanism (N $\leftrightarrows$ U), the equilibrium constant K_{UN}, is

$$K_{UN} = \frac{[U]}{[N]} = f_U/(1 - f_U) \tag{2}$$

and

$$K_{UN} = \exp(-\Delta G_{UN}/RT) \tag{3}$$

In Eq. (3), several parameters are listed: the gas constant, R; absolute temperature, T; and the free energy between folded and unfolded states, ΔG_{UN}. Since ΔG_{UN} is linearly correlated with the denaturant concentration, i.e., urea in our experiments, we used Eq. (4) to describe the linearity of the free energy in the presence of denaturant.

$$\Delta G_{UN} = \Delta G_{UN}^{H_2O} - m[urea] \tag{4}$$

To obtain $\Delta G_{UN}^{H_2O}$, the Gibbs free energy of unfolding with denaturant, the fitting is made by incorporating from Eqs. (1) to (4).

From Eqs. (5) and (6), we can access the solvent-exposed surface area (ΔASA) and the estimated number of residues:

$$m_{cal/M/mol} = 374 + 0.11(\Delta \text{ASA}) \tag{5}$$

$$\Delta \text{ASA} = -907 + 93\,(\#\text{residue}) \tag{6}$$

In a two-state equilibrium between a folded trimer and three unfolded monomers, $N_3 \leftrightarrows 3U$, or a folded tetramer and four unfolded monomers, $N_4 \leftrightarrows 4U$, the equilibrium constants are obtained from Eq. 7 and Eq. 8, respectively:

$$K_{UN_3} = \frac{[U]^3}{[N_3]} = 3P_t^2 f_U^3/(1 - f_U) \tag{7}$$

$$K_{UN_4} = \frac{[U]^4}{[N_4]} = 4P_t^3 f_U^4/(1 - f_U) \tag{8}$$

The total subunit concentration is denoted as P_t in Eqs. (7) and (8). In the model of $N_3 \leftrightarrows 3U$, we use Eq. (10) to describe f_U by solving the third-order equation.

$$\begin{aligned} f_U = {} & [0.5c + (13^{1/2}/18)c(4c + 27)^{1/2}]^{1/3} \\ & -c/\{3[0.5c + (13^{1/2}/18)c(4c + 27)^{1/2}]^{1/3}\} \end{aligned} \tag{9}$$

$$c = K_{UN_3}/(3P_t) \tag{10}$$

In the model of $N_4 \leftrightarrows 4U$, we use Eq. (11) to describe f_U by solving the fourth-order equation.

$$f_U = \{-(s^{1/2}) + [-s + 2/(as^{1/2})]^{1/2}\}/2 \tag{11}$$

$$\begin{aligned} s = {} & -4\{2/[27a + (729a^2 + 6912a^3)^{1/2}]\}^{1/3} \\ & +\{1/[3a\{2/[27a + (729a^2 + 6912a^3)^{1/2}]\}^{1/3}]\} \end{aligned} \tag{12}$$

$$a = 4P_t^3/\exp(K_{UN_4}) \tag{13}$$

We fitted the denaturation data of Aβ42 by combining Eqs. 1, 4, 7, 9, and 10 for the $N_3 \leftrightarrows 3U$ model and Eqs. 1, 4, 7, 9, and 10 for the $N_4 \leftrightarrows 4U$ model. We also calculated the parameters including $\Delta G_{UN_3}^{H_2O}$, $\Delta G_{UN_4}^{H_2O}$, *m*-value, f_U, f_{trimer}, f_{tetramer}, and m_N and m_U, the slopes of pre- and posttransitions, respectively. To obtain more accurate values of $\Delta G_{UN}^{H_2O}$ and *m*-value, in *Igor Pro 5*, we set them as global parameters fitting to various data sets and had single result while the other parameters varied.

When we described the Aβ42 folding using three-state model, $N_3 \leftrightarrows 3M \leftrightarrows 3U$ or $N_4 \leftrightarrows 4M \leftrightarrows 4U$, we linearly extrapolated the plot of free energies against urea concentration and thus obtained $\Delta G_{UT}^{H_2O}$, the free energy in the absence of denaturant. In this three-state model, the first transition represents the dissociation of multimers, i.e., $N_3 \leftrightarrows 3M$ or $N_4 \leftrightarrows 4M$. Using Eq. (14), the monomer fraction denoted as f_M can be calculated from the experimental data.

$$f_M = (\lambda_{obs} - \lambda_T)/(\lambda_M - \lambda_T) \tag{14}$$

In this equation, the observed signal is described as λ_{obs}, and λ_T and λ_M are trimeric/tetrameric and monomer signals, respectively. From the observance of denaturation curve, we use the signal at 2 M urea as the monomeric signal. Combining Eqs. 14, 7, and 8, we can calculate various $\Delta G_{\text{dissociation}}$, free energies of multimer dissociation, against different concentrations of urea.

In the denaturation curve, the second transition exhibits the monomer unfolding, i.e., M $\leftrightarrows$ U. Parallel to the previous concept, unfolded fraction, f_U, is described in Eq. (15).

$$f_U = (\lambda_{obs} - \lambda_M)/(\lambda_U - \lambda_M) \tag{15}$$

We plotted the free energies of unfolding, ΔG_{UM}, against the concentration of urea with the combination of Eqs. (15) and (2). We can obtain the free energy without denaturant, $\Delta G_{UN}^{H_2O}$, and the m-value with the linear fitting of Eq. (4).

For denaturation data in Sec. 3.2, *Kaleidograph 3.5* (Synergy Software) was adopted to perform the fitting with the model of a two-state equilibrium folding mechanism (N $\leftrightarrows$ U). The fitting equation is shown as Eq. (16), which is the combination of Eqs. 1–4.

$$\begin{aligned} P_{total} = {} & \{(P_N + m_N[\text{GdnHCl}]) + (P_U + m_U[\text{GdnHCl}]) \\ & \times \exp[-(\Delta G_{UN}^{H_2O} - m[\text{GdnHCl}])/RT]\}/ \\ & \{1 + \exp[-(\Delta G_{UN}^{H_2O} - m[\text{GdnHCl}])/RT]\} \end{aligned} \tag{16}$$

In Eq. (16), P_{total} is the observed fluorescence intensity contributed by the folded, N, and the unfolded protein. Other parameters represent same meaning as previously described. In the denaturation with 5% trifluoroethanol (TFE), $\Delta G_{UN}^{H_2O}$ is replaced by $\Delta G_{UN}^{(5\%TFE)}$, which represents the Gibbs free energy of folding without denaturant.

In Sec. 3.3, all solutions were prepared in 10 mM Tris-HCl, pH 7.4. Denaturation was performed by titrating unfolded Aβ with ~6.2 M GdnHCl to folded Aβ with less than 0.1 M GdnHCl. Both unfolded and fold Aβ solutions contained equal amount of metal ions at indicated concentrations, Aβ at 25 μM and Bis-ANS at 5 μM. Emission at 490 nm with excitation at 400 nm was collected, averaged, normalized, and plotted against GdnHCl concentration.

2.7. *Aggregation assay and dot blotting*

In Sec. 3.1, to monitor the formation of amyloid fibrils, ThT assay was performed in enzyme-linked immunosorbent assay (ELISA) plate monitored by a fluorescence plate reader (*XS*; Molecular Devices Inc., Toronto, Canada). In Sec. 3.1, Aβ peptides were refolded in 10 mM phosphate buffer, pH 7.4, with 0.02% sodium azide, and all samples had equal Aβ concentration at 50 μM but with indicated urea concentrations. The samples were agitated by a Teflon-coated micro stir bar at 300 rpm incubated at room temperature. At indicated time 10 μL of each Aβ sample was combined with 90 μL of ThT solution, 30 μM, for monitoring ThT signal. Fluorescence with excitation at 442 nm and emission at 485 nm was recorded. The emission in various urea concentrations was subtracted from background and used to describe the fibrillary content. Formation and transition of oligomeric species in fibrillization were monitored by using dot blotting in which 2 μL of the samples from the indicated time was blotted on the NC membrane. The membrane was either stained by *Direct Blue 71* or immunoblotted with A11 antibody that highly recognized prefibrillar oligomer.[87]

In Secs. 3.2 and 3.3, ThT assay was performed in a *SpectraMax M5* fluorescence plate reader (Molecule Devices Inc.) at 25°C. Preparation of 25 μM Aβ was described previously with addition of 50 μM ThT in native form or in the presence of 3 M GdnHCl. Samples were quiescently incubated and agitated for 5 s followed by monitoring the emitted fluorescence at 485 nm with excitation at 442 nm every 20 min for 90 h. All measurements were processed through background subtraction, plotted against time, and fitted by *OriginPro 8.0* (OriginLab Inc., Northampton, MA) with the model of amyloid nucleation-elongation, which is described in Eq. (17).[88]

$$Y = (y_i + m_i x) + (y_f + m_f x)/\{1 + \exp[-(x - x_0)/\tau]\} \tag{17}$$

In Eq. (17), Y stands for the observed intensity. The parameters m_i and m_f represent the slopes of lag and stationary phase while y_i and y_f for the signal contributions of lag and stationary phase, respectively. Parameter x_0 represents needed time reaching half of the maximum fluorescence. To describe the fibril formation, we use $1/\tau$ to correlate with the apparent rate constant, k_{app} and calculated lag time is $x_0 - 2\tau$.

In Sec. 3.3, Aβ40 at 25 μM was prepared in Tris buffer with 25 μM ThT and indicated concentrations of different metal ions. The samples were incubated in a microplate, monitored at 25°C, and quiescently incubated. After the samples were agitated for 10 s, fluorescence was monitored for 100 h with excitation at 442 nm, emission at 485 nm, and interval of 1 h. In Sec. 3.3, dot blotting was also performed by blotting 2 μL the samples to NC membrane followed by immunoblotting of A11 antibody. In Sec. 3.4, ThT assay was performed in the presence of 5 μM ThT in Aβ solution and incubated agitatedly in Eppendorf tubes at 25°C. ThT fluorescence was monitored using an ELISA microplate reader *SpectraMax M5* with emission at 485 nm and excitation at 442 nm. Each curve was averaged from more than three independent replicates and plotted with standard deviation. In Sec. 3.4, replicates of NC membrane blotted with 2 μL Aβ samples were collected at indicated time and were immunoblotted by A11, 6E10, and OC[89] antibodies, or stained by *Direct Blue 71*.

2.8. *Analytical ultracentrifugation*

Sedimentation velocity (SV) was assayed by using a *ProteomeLab*™ *XL-I* centrifuge (Beckman Coulter Inc., Brea, CA) with an *An-60Ti* rotor. Aβ variants and Aβ (40-1) were all freshly prepared as previous description and loaded into a 12-mm aluminum double-sector centerpiece. Centrifugation was carried out at 60,000 rpm for at least 20 h at 4°C. UV absorption was adopted to continuously observe the moving species due to ultracentrifugation with the frequency of scanning once every 15 min and a fixed interval of 0.003 cm. The software *SEDFIT* (National Institute of Health (NIH), Bethesda, Maryland, United States) was used to analyze SV data while *SEDNTERP* was used to calculate the parameters of partial specific volume, solvent density, and solvent viscosity.

2.8.1. *GdnHCl denaturation with/without TFE*

The denaturation of Aβ variants was monitored at room temperature when unfolded Aβ variants (with ~5.8 M GdnHCl) were titrated manually into folded Aβ species. Both folded and unfolded samples had 5 μM Bis-ANS, the same Aβ concentrations, 50, 25, and 12.5 μM, and 5% TFE in the parallel part of denaturation. Each titration of unfolded species had an ~30 s interval to achieve equilibrium. To avoid aggregation, each experiment of denaturation was finished within 1 h. Averaged Bis-ANS fluorescence at 500 nm was normalized to the fluorescence from unfolded species.

2.8.2. *Transmission electron microscopy*

In Sec. 3.2, each sample was deposited on a glow-discharged 400-mesh Formvar carbon-coated copper grid (EMS Inc., Hatfield, PA) and incubated for 3 min. The grid was water-rinsed, negatively stained using 2% uranyl acetate solution, and observed in a *Hitachi H-7000* TEM (Hitachi Inc., Chiyoda-ku, Tokyo, Japan) with an accelerating voltage of 75 kV. In Sec. 3.4, after centrifugation at $17,000 \times g$ for 10 min, the samples were fractioned into supernatants and pellets, which were resuspended in phosphate buffer later. Both fractions were blotted onto 400-mesh Formvar carbon-coated copper grids (EMS Inc.), incubated for 5 min, stained by

2% uranyl acetate solution, and rinsed by milliQ water. Morphology was observed using a *Hitachi H-7000* TEM with an accelerating voltage of 75 kV.

2.8.3. *Titration of metal ion to Aβ for equilibrium binding*

Metal ion solution was prepared at either 2 mM for $ZnCl_2$, $CuCl_2$, and $FeCl_3$ or 20 mM for $AlCl_3$. Titration of various metal ions was performed with Aβ at 25 μM and limit of dilution factor was 0.9. Three replicates were assayed for each metal and corrected by the dilution factor. The signals from Bis-ANS fluorescence at 490 nm, tyrosine fluorescence at 305 nm, and CD at 216 nm were processed, normalized, and plotted against the metal ion concentration. Atomic absorption (AA) spectroscopy was performed to ensure the quantity of aluminum atom in the solution.

2.8.4. *Size exclusion chromatography (SEC)*

The SEC was performed after samples were collected, centrifuged at 17,000× g for 10 min, and injected with 50 μL of the supernatants into *Superdex 200 5/150* (GE Healthcare Biosciences, Boston, Massachusetts, United States). The flow rate of fast protein liquid chromatography (FPLC) was 0.2 mL/min in 10 mM sodium phosphate buffer, pH 7.4. Conformer detection was conducted by the absorbance at 280 nm.

2.8.5. *PICUP and western blotting*

In Sec. 3.4, during aggregation Aβ from the five mixing ratios were taken out at indicated time. PICUP was performed by mixing 9 μL sample with 0.5 μL of 1 mM Ru(Bpy) and 0.5 μL of 20 mM ammonium persulfate (APS). Mixtures were cross-linked by exposing to blue light LED for 30 s and stopped by mixing with SDS-PAGE sample buffer prior to loading to a 13.3% Tris-Tricine separating gel with 10% and 4% stacking portions. Aβ conformers on the gel were transferred and subjected to western blotting using 6E10 antibody.

2.8.6. *Neurite degeneration and cytotoxicity*

All cells involved in the neurite degeneration and cytotoxicity studies were cultured at 37°C with humidified 5% CO_2 atmosphere. The growth medium for mouse neuroblastoma Neuro2a (N2a) (ATCC number CCL-131) cells was minimum essential medium (MEM) containing 10% fetal bovine serum (FBS), 1% penicillin/streptomycin, 2 mM L-glutamine, and 1 mM sodium pyruvate. A total of 150,000 N2a cells per well were incubated in a six-well plate for 16 h, transfected with an enhanced green fluorescent protein (EGFP) for 6 h and incubated with differentiation media for 3 days by discarding the growth media. The differentiation media was MEM containing 10 μM trans-retinoic acid and 1% FBS. Neurite length assay was performed by incubating Aβ mixtures of various ratios at 5 μM for 1 day followed by observation of the neurite lengths using a fluorescence microscope (*IX71*; Olympus, Shinjuku, Tokyo, Japan). Analysis was conducted by calculating neurite length in the fluorescent images using the NeuronJ function in ImageJ software (NIH). A differentiated neuron was defined as a cell with neurite twice longer than the cell body. To enhance the accuracy, more than 200 cells were collected in each condition.

The human neuroblastoma BE(2)-C cells (ATCC number CRL-2268) were conducted to MTT (3-(4,5-dimethylthiazol-2-yl)-2,5-diphenyltetrazolium bromide) assay differentiating the cytotoxicity from the five ratios of Aβ mixtures. Cells were cultured in RPMI medium with 10% FBS. A total of 60,000 cells were seeded in a 96-well plate for 24 h, followed by discarding growth media, washing cells by RPMI medium without FBS, adding 40 μL RPMI medium, mixing with 10 μL Aβ samples, and incubating for another 24 h. MTT at final 0.5 mg/mL was incubated with cells for 3 h. Absorbance was monitored at 570 and 690 nm by a *SpectraMax M5* after discarding MTT-contained medium and lysing formazan crystals by DMSO. Five replicates of absorbance at 570 nm subtracted from 690 nm were averaged and normalized to the buffer control.

3. Results

3.1. *Distinct folding stability of Aβ40 and Aβ42*

Since Anfinsen's discovery on reversible folding of RNase, it is believed that proper protein folding leads to a correct native conformation that gives the protein its cellular function. However, proteins with long intrinsic disordered segments are present from 25% to 41% in the proteome of eukaryotes.[90] This disorder hinders the understanding of residual structure existing in natively unfolded proteins. Hence, the protein stability attributed from entropy and enthalpy is difficult to examine. In 2006, we developed a new method to examine Aβ stability. First, we examined the assembly of native Aβ40 and Aβ42 by native PAGE. Aβ samples were prepared from rapidly refolding from Aβ stock dissolved in urea solution, as described in Sec. 2. High concentration of denaturant is known to denature preformed aggregates in the sample solution. We found native Aβ40 migrates as a single species indicating it exits as a monomer, but Aβ42 migrates as two bands, a monomer at

~4 kDa and a trimer/tetramer at ~14.4 kDa. We also used an extrinsic fluorescence probe, Bis-ANS, commonly used in protein folding studies to detect folding intermediates to probe native and unfolded states of Aβ40 and Aβ42. Interestingly, we found Aβ40 and Aβ42 in native buffer have significantly higher Bis-ANS fluorescence emission than Aβ in high concentration of denaturants such as urea and GdnHCl. The maximum wavelength of Bis-ANS emission for native Aβ40 is ~513 nm, and for Aβ42 is ~495 nm. At 513 nm, the maximal emission signal for native Aβ40 is $\sim 1 \times 10^5$, whereas that for the unfolded Aβ40 is $\sim 0.3 \times 10^5$. The difference between folded and unfolded Aβ40 signal is more than three times. In the case of Aβ42, the native Aβ42 is $\sim 9.2 \times 10^5$ and unfolded Aβ42 is $\sim 0.72 \times 10^5$. The difference between folded and unfolded signals is more than 12 times. The property allows us to differentiate native Aβ and unfolded Aβ and further examine their stability in the pseudoequilibrium state prior to aggregation. As Aβ aggregates on a time scale of days, the pseudoequilibrium can be assumed in the first hour due to fast equilibrium between native and unfolded Aβ. The differences between native and unfolded Aβ were also revealed by far-UV CD. The spectrum of native Aβ40 contains more degree in ellipticity than that of the unfolded one even though native Aβ40 is enriched with random coils. In contrast, native Aβ42 comprises β-stands with a clear spectral minimum at 216 nm. The spectrum becomes random coils in high concentration of urea. Aβ denaturation monitored by CD showed a similar pattern to that by Bis-ANS fluorescence. We also found Bis-ANS only binds to Aβ peptides possessing residues 13–24 and 25–36, but not 1–12, 7–18, or 19–30. The Bis-ANS binding sites should be situated at the flanking hydrophobic regions of the protease-resistant core, which is reported to be residues 21–30.[59,91]

On further examination of the signal changes upon urea denaturation, we found Aβ40 has a small pretransition state, i.e., <0.75 M urea, and a single transition between 0.75 and 4 M urea. The pretransition and posttransition states, i.e., >4 M, are affected by urea in a linear manner. The transition showed a cooperative change that is due to structural changes of Aβ. Aβ40 at 50, 25, or 12.5 μM all adopt the same transition, indicating no dissociation occurred from a multimer. The result is consistent with the single band observed on native PAGE. The data in urea denaturation can be accurately fit by a two-state folding equilibrium model in which a native monomer and an unfolded monomer are present at equilibrium.

$$\mathrm{N} \leftrightarrows \mathrm{U}$$

The calculated Gibb's free energy $\Delta G^{\mathrm{UN}}_{(\mathrm{H2O})}$ is 1.1 kcal/mol and the *m*-value is 0.7 kcal/mol/M. On examination of equilibrium denaturation of Aβ42 at 50, 25, or 12.5 μM, we found that the transition changed. Aβ42 showed concentration dependence in the denaturation curves. The midpoint of transition shifted from ~1.4 M urea for 50 μM, ~1.1 M urea for 25 μM, to ~0.6 M urea for 12.5 μM. The data also show one apparent cooperative transition suggesting Aβ42 also adopts a two-state mechanism. We globally fit the data to a model of trimer and unfolded monomer

$$\mathrm{N3} \leftrightarrows 3\mathrm{U}$$

or tetramer and unfolded monomer model

$$\mathrm{N4} \leftrightarrows 4\mathrm{U}$$

and obtained a $\Delta G^{\mathrm{UN}}_{(\mathrm{H_2O})}$ of ~14.9 kcal/mol and an m-value of ~2.8 kcal/M/mol for timer model and $\Delta G^{\mathrm{UN}}_{(\mathrm{H_2O})}$ of ~21.6 kcal/mol and an m-value of ~3.4 kcal/M/mol for tetramer model. Alternatively, we can fit the data to a three-state trimer or tetramer model as below with comparable residuals.

$$\mathrm{N3} \leftrightarrows 3\mathrm{M} \leftrightarrows 3\mathrm{U}$$

or

$$\mathrm{N4} \leftrightarrows 4\mathrm{M} \leftrightarrows 4\mathrm{U}$$

It is possible that a second transition is present at higher urea concentration due to abnormally high percentage of unfolded species in native condition if we extrapolate the posttransition back to water in the two-state models. In the three-state model, a transition of multimer dissociation occurs from 0 to ~2 M urea and an unfolding of a monomeric intermediate occurs from ~2 to 4 M urea. In this model, we obtained $\Delta G^{\mathrm{dissociation}}_{(\mathrm{H_2O})}$ for trimer as ~16 kcal/mol and for tetramer as ~23.2 kcal/mol. The monomer unfolding $\Delta G^{\mathrm{UN}}_{(\mathrm{H_2O})}$ was ~5.8 kcal/mol for the second transition. The study demonstrated the distinct equilibrium folding properties of Aβ40 and Aβ42.

By further examining the oligomerization and fibrillization of Aβ40 and Aβ42 in various urea concentrations that affected the initial conformation of Aβ, we found they oligomerize and fibrillize progressively at native condition; however, the fibrillization capability decreased upon increasing urea concentration. This indicates the importance of the native Aβ conformation. Interestingly, by dot blot probed by a conformational-specific antibody, A11, that recognizes amyloid oligomers, we found amyloid oligomers only appear in the native condition. They do not appear in the aggregation pathway of Aβ42 in 2 M urea, even Aβ42 is still capable of forming fibrils. The result showed A11-positive oligomer may not be obligatory for the fibrillization pathway. Together, an illustration of the results is plotted in Fig. 2.

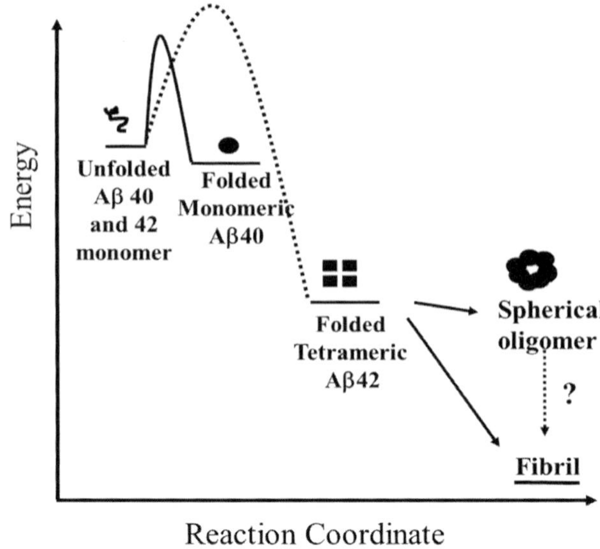

Figure 2: Energy landscape of Aβ40 and Aβ42. The figure is modified from our previous work.[92]

3.2. *Folding stability of Aβ is an important determinant of the nucleation kinetics in fibrillization*

On the basis of the successful stability examination of Aβ40 and Aβ42, we hypothesized that hidden folding mechanisms are embedded in the early stage of Aβ aggregation. Genetic evidences from nature provide us a good way to understand the underlying mechanism, i.e., to compare the wild-type Aβ and the single mutants found in early onset of AD (FAD). In 2011, we used the aforementioned equilibrium folding techniques to examine the folding stability of Aβ40 variants. They were the clinically related familial mutants, i.e., Aβ harboring A21G (Flemish), E22G (Arctic), E22K (Italian), E22Q (Dutch), and D23N (Iowa) mutants. To simplify the result and reduce complication, we compare the wild type and variants using Aβ40.

3.2.1. *FAD Aβ40 variants are monomeric*

To ensure we still obtained monomeric Aβ40 in the variants, we used both native PAGE and SV of analytical ultracentrifugation (AUC) to study their assemblies. Aβ preparation was carried out in rapid refolding from GdnHCl containing buffer. We changed the denaturant to GdnHCl as it provides a better pretransition in denaturation study than urea. Both native PAGE and AUC data showed that all Aβ40 variants are monomeric. SV experiments based on particle diffusion provided good resolution to analyze species distribution without a matrix effect. It was found that only one symmetric peak located between 0.31 and 0.38 S corresponding to molecular masses between 3121 and 4213 Da was observed. No oligomer was observed up to 5.5 S (approximately 220 kDa, based on molecular weight estimation). The frictional ratio was from 1.53 to 1.63, indicating Aβ variants had extended, rodlike shapes. The sedimentation result from Aβ variants is similar to that of a reverse sequence of Aβ peptide Aβ(40-1), which has been shown to lose the aggregation property.[81,93,94]

3.2.2. *Conformation and stability of FAD Aβ40 variants*

In far-UV CD spectra, we measured the native and unfolded Aβ, i.e., Aβ in buffer containing 3 M GdnHCl. They had random coil-like conformation. However, after the native spectrum was subtracted from the unfolded spectra for each variant, the differential spectra showed β-strand content resided in the residual structure of Aβ variants. The degree of β-strands was indifferentiable among the variants. However, we were able to differentiate the surface hydrophobic clusters on each Aβ variant by Bis-ANS fluorescence. The Bis-ANS fluorescence intensity was in order of E22K > E22Q > E22G > D23N > WT > A21G, indicating E22 and D23 mutants had larger hydrophobic exposure on the protein surface. Aβ variants were further examined with and without 5% TFE by GdnHCl denaturation probed using Bis-ANS. TFE has been shown to help in stabilizing protein conformation.[95–97] All Aβ variants adopt two-state equilibrium folding model (N $\leftrightarrows$ U), as reported previously for wild-type Aβ40. In the absence of TFE, the transitions of all variants were with similar slopes with posttransition starting around 1.25–1.5 M GdnHCl. The mid-point, $[\text{GdnHCl}]_{1/2}$, of the transitions was in the order of A21G > WT > D23N > E22 mutants. The denaturation curves of E22 mutants overlapped. Owing to short pretransitions, we were unable to fit the data. In the presence of 5% TFE, the denaturation curves of WT, A21G, and D23N were more classic. We were able to fit them well to a two-state model with ΔG_{UN} in the presence of 5% TFE to be 2.04, 4.02, and 1.64 kcal/mol and the *m*-values for cooperativity to be 2.84, 3.99, and 1.85 kcal/mol/M for WT, A21G, and D23N, respectively. The E22 mutants cannot be fitted due to lack of pretransitions. The calculated and experimental $[\text{GdnHCl}]_{1/2}$ matched well, indicating the fitting is precise. The result showed the order of stability among the variants is A21G > WT > D23N > E22 mutants.

To understand the effect of Aβ monomer stability to its aggregation, we first performed ThT assay to monitor fibrillization. All species fibrillized through a nucleation-dependent pathway with a lag time. The order of the lag time is A21G > WT > E22 mutants > D23N where the Flemish mutant A21G showed a much slower aggregation. The order is the same with the order of stability except for D23N. The results showed that the more stable Aβ variants the slower the aggregation. D23N does not follow the rule. This may be due to breaking the important salt bridge D23-K28 in Aβ. Interestingly, E22 mutants adopting the same denaturation curve have also nearly the same lag time in fibrillization. Therefore, based on our study an energy landscape is illustrated in Fig. 3b. The less stable variant is more likely to cross the energy barrier to a downhill landscape of protein misfolding. This provides a fundamental mechanism to explain different aggregation kinetics resided

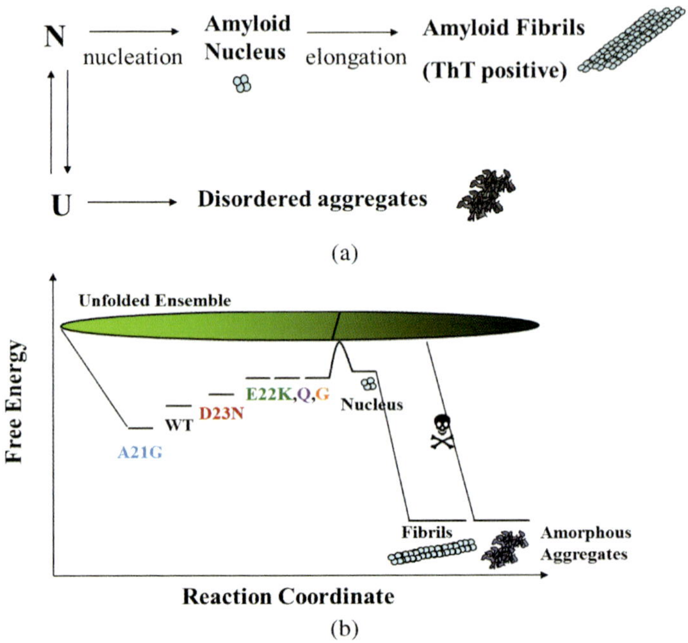

Figure 3: Proposed model for Aβ folding and aggregation mechanisms. (a) Aβ40 exists as two species at equilibrium (N and U) prior to the aggregation. Native Aβ undergoes fibrillization through a nucleation step and forms ordered amyloid fibrils. Losing the native conformation, the unfolded Aβ is prone to form disordered, amorphous aggregates. (b) Schematic free-energy landscape of Aβ40 variants. The folding stability of Aβ variants predominantly determines the kinetics of nucleation in the fibrillization process. The less stable the species is easier to cross the activation barrier toward the pathological states.

in FAD Aβ mutants. Overall, the result demonstrated that the folding stability of Aβ is an important determinant of the nucleation kinetics.

3.2.3. *Influence of fibril formation in denaturing condition*

We further studied the fibril formation capability in the presence of various denaturant conditions by ThT assay and TEM imaging. We found in the more native condition, that is, Aβ in its pretransition state, the more fibril can form. Fewer fibrils formed with increasing GdnHCl. In the posttransition state, e.g., Aβ in 3 M GdnHCl, little fibrils can be found but there is increasing amounts of nonfibrillar, amorphous-like aggregates. This is showing the unfolded Aβ is more prone to form amorphous aggregates (lower pathway in Fig. 3a). The fast aggregating Aβ species especially D23N can still form fibrils in 3 M GdnHCl. The phenomenon may be due to its fast aggregation property driving few native Aβ population in 3 M GdnHCl to form fibrils (upper pathway in Fig. 3a). The event further results in driving the unfolded state to native state through the two-state pseudoequilibrium.

3.3. *Metal ion effects on Aβ stability and aggregation*

Aβ is able to bind to specific metal ions including zinc (Zn), copper (Cu), and iron (Fe) that are present in the disease lesions.[98,99] In the senile plaques, it was reported to contain high level of Cu (25 μg/g, ~393 μM), Zn (69 μg/g, ~1055 μM), and Fe (52 μg/g, ~940 μM).[100,101] In neuropil of AD patients, it was reported to have Zn, elevated from 346 to 786 μM, Cu from 69 to 304 μM, and Fe from 338 to 695 μM.[100,101] Aluminum (Al) was also detected in the senile plaques.[102] Imbalance of cellular Zn/Cu homeostasis may modulate AD pathology.[98,99] To understand the metal ion effects on Aβ stability and aggregation, we examined systematically the effects of Zn^{2+}, Cu^{2+}, Fe^{3+}, and Al^{3+}, on early and aggregated stages of full-length Aβ40.

3.3.1. *Zn^{2+} and Al^{3+} increase hydrophobic exposure of Aβ40 monomer and destabilize Aβ stability*

We first examined the metal ion effect on native Aβ conformation by Bis-ANS fluorescence. Bis-ANS spectra of Aβ40 at 25 μM in the absence and presence of Zn^{2+}, Cu^{2+}, Fe^{3+}, and Al^{3+} (200 μM of Zn^{2+}, Cu^{2+}, Fe^{3+}, and 500 μM of Al^{3+}) were examined. In the presence of Zn^{2+} and Al^{3+}, the Bis-ANS intensity showed ~15 and ~30 times of increase, respectively. In contrast, the Bis-ANS intensity decreased for ~2.5-fold in the presence of Cu^{2+}. Fe^{3+} did not change much on

Bis-ANS fluorescence. Since Bis-ANS reports the hydrophobic clusters exposed on protein surfaces and probes Aβ conformation at the early stage, the data suggest Zn^{2+} and Al^{3+} specifically increase hydrophobic exposure of Aβ40 monomer, whereas Cu^{2+} decreases it.

We then investigate the effects of the metal ions on Aβ stability. GdnHCl denaturation was conducted to examine Aβ stability in the presence of Zn^{2+}, Cu^{2+}, Fe^{3+}, and Al^{3+}. The Bis-ANS intensity of Aβ with and without metal ions was collected in various GdnHCl concentrations. The samples were prepared in Tris buffer due to the precipitation problem observed in phosphate buffer. GdnHCl denaturation of Aβ in the absence of metals, an apparent two-state equilibrium mechanism ($N \rightleftarrows U$), was observed as described in our previous result.[61] There is a pre-transition, <0.5 M GdnHCl, followed by a single transition between 0.5 and 1.5 M GdnHCl, and a posttransition > 1.5 M GdnHCl. In the presence of 25 μM Cu^{2+} or 50 μM Fe^{3+}, the denaturation curves do not change. The mid-point of transition $[GdnHCl]_{1/2}$ is still at ~1 M. In contrast, with 25 μM Zn^{2+} or 100 μM Al^{3+}, the $[GdnHCl]_{1/2}$ shifted to ~0.4 and ~0.6 M, respectively. Both of the pre-transitions were lost with Zn^{2+} or Al^{3+}. The results demonstrated that Zn^{2+} and Al^{3+} significantly destabilized Aβ. Other ions that do not bind to Aβ, including Fe^{2+}, Ca^{2+}, Mg^{2+}, and Na^{+}, had no effect on Aβ stability.

3.3.2. *Zn^{2+}* and *Al^{3+}* induce *Aβ* oligomerization

To further understand the metal ion effect on Aβ oligomerization and fibrillization, ThT assay and dot blot were used. In the absence of metal ions, Aβ aggregation follows a nucleation-dependent mechanism. Aβ nucleates at a lag time of 20 h, elongated to reach a steady state at ~60 h. In the presence of Zn^{2+}, ThT intensity increased readily without lag time and a plateau was researched after 20 h. The final ThT intensity was much lower in comparison with that of Aβ alone. The other three metal ions showed inhibitive effects in a dose-dependent manner. At 5 μM, Cu^{2+} prolonged the lag phase to over 60 h and the ThT fluorescence was decreased. In higher dose, Cu^{2+} completely suppressed Aβ fibrillization. Fe^{3+} and Al^{3+} at 50 μM both prolonged the lag phase to ~40 h. Al^{3+} has a strong inhibition effect above 250 μM that may be primarily due to the induced acidic pH.

We further used dot blotting to track oligomer appearance. The oligomers can be probed by anti-amyloid oligomer antibody, A11, which specifically recognizes oligomer conformation in many amyloid proteins.[87] In the presence of Zn^{2+}, A11-positive Aβ oligomers were more abundant and appeared immediately, whereas, they were absence in Cu^{2+} and Fe^{3+}. Al^{3+} induced the oligomers in

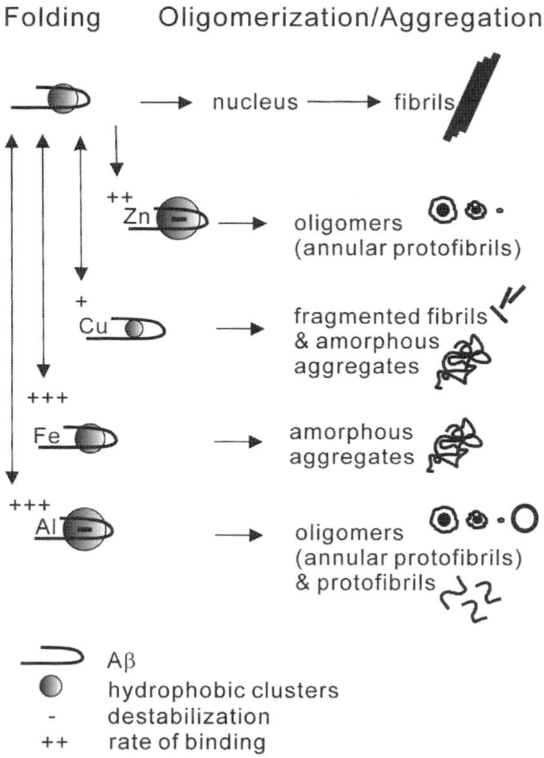

Figure 4: Proposed folding and aggregation mechanisms of Aβ in the presence of Zn^{2+}, Cu^{2+}, Fe^{3+}, and Al^{3+}. The hydrophobic exposed clusters of Aβ conformation, destabilization, and rate of the metal ion binding are illustrated. Aβ without ions underwent a nucleation-dependent pathway. In the presence of Zn^{2+}, annular protofibrils were preferentially formed. In the presence of Cu^{2+}, Fe^{3+}, and Al^{3+}, Aβ partitioned between fibrils and other aggregates such as amorphous aggregates. Destabilization of Aβ induced annular oligomer formation.

a time-dependent manner. TEM imaging showed that without metal ions, Aβ formed abundant long straight fibrils, but in the presence of Zn^{2+} massive amounts of heterogeneous oligomers were observed. In the presence of Cu^{2+} and Fe^{3+}, the amount of fibrils decreased and amorphous aggregates increased. The result showed destabilization of Aβ induced oligomer formation. Overall, our result showed the four metal ions induce distinct aggregation pathways. The mechanism is illustrated in Fig. 4.

3.4. *Coexistence of an equal amount of Aβ40 and Aβ42 forms structurally stable and toxic oligomers through a distinct pathway*

Due to the interest of Aβ42 to Aβ40 ratio implicated in AD pathogenesis, we systematically examined and characterized the aggregation pathways of different Aβ42 to Aβ40 mixing ratios. We designed five ratios of Aβ42 to Aβ40 indicated as 10/0 (Aβ42 alone), 9/1, 5/5, 1/9, and 0/10 (Aβ40 alone) and kept the total Aβ concentration at 50 μM.

3.4.1. *Coincubation of equal amount of Aβ42 and Aβ40 showed lowest level of ThT signal that contains amyloid oligomers*

We first adopted a ThT assay to elucidate the differences of fibril formation in the various mixing ratios of Aβ42 to Aβ40. The results showed that Aβ42 alone aggregated very rapidly and that no obvious lag phase can be observed and ThT signal saturated after 10 h of incubation. By contrast, the lag time of Aβ40 alone was about 17 h and ThT signal saturated after 30 h. The ratio of 9/1 of Aβ42/Aβ40 was unable to be differentiated from Aβ42 alone in the lag time and elongation phase. On the contrary, elongated lag time was observed from 17 h (Aβ40 alone) to 80 h after addition of 10% Aβ42 to Aβ40 (in the ratio 1/9 of Aβ42/Aβ40). Interestingly, ThT signal from equally mixed Aβ42 and Aβ40 only slightly increased and saturated after ~6 h until the end of incubation. The ThT level was only ~14.3% of that of Aβ42 alone. The result indicates that least fibrils could be found in the equally mixed Aβ among the ratios we examined.

The soluble and insoluble parts of the samples at 6 h incubation and end-point were imaged under TEM. In the samples of 6 h incubation, short protofibrils of ~20 to ~90 nm in length were found in the supernatants of Aβ42-dominant species (i.e., 10/0 and 9/1) but not in the Aβ40-dominant species. Round-shaped oligomers with ~6 to 20 nm in diameter were the major component in the ratio of 5/5 samples while some short protofibrils were also found. Longer fibrils were predominantly found in the insoluble portions of Aβ42-dominant species. For end-point products, both supernatant and pellet Aβ42 alone contain dominantly short filaments that were more bundled in the pellet. In the 9/1 ratio, curvy filaments were primarily found in the supernatants and the pellets contained wider and shorter filaments. Corresponding to ThT assay, the supernatants from the equally mixed sample had similar oligomers that were abundant and spherical at 6 h while short filaments were still observed. The diameter of these spherical oligomers was from 6 to ~20 nm; they were dominated by the species with a diameter from 12 to 15 nm. The fact that spherical oligomers can be found in equally mixed Aβ sample as early as 6 h indicates the relatively fast formation of the species. The result is consistent with our ThT data.

In dot blot analysis, we used different conformational antibodies to probe the epitope changes during aggregation. The different Aβ antibodies used were A11, used commonly for amyloid prefibrillar oligomers[87], OC, aimed for fibrils and fibrillar oligomers;[89] and 6E10, used against Aβ1-17. *Direct Blue* staining was used to detect the equal peptide dotting. A11 signal of Aβ42 alone progressively increased after 6 h incubation, which correlated with the short lag phase presented in ThT assay. A11 signal was much enhanced in the presence of Aβ40. In the ratio of 0/10, A11 signal gradually strengthened after 6 h and saturated after 29 h, which correlated the ~17 h of lag phase found in ThT result. Interestingly, the correlation was nearly absence in the ratios of 9/1, 5/5, and 1/9 between A11 and ThT results. Moreover, the ratio of 5/5 had weak ThT intensity whereas A11-positive oligomer was rapidly generated and saturated from the starting point. From the OC blot, we did not observe significant changes in the Aβ42-dominant samples, i.e., 10/0 and 9/1. Contrasts to A11 antibody, the OC signals were dominantly stronger in the presence of Aβ42. However, the OC signal was found elevated in the late stage of incubation for the 1/9 and 0/10 samples. In the equally mixed Aβ sample, low OC signal was maintained to the end of the experimental time. It also indicates the ratio of 5/5 had the least fibrillar structure among the ratios we tested.

3.4.2. *Aβ42 oligomers were stabilized by addition of Aβ40 during aggregation*

The aggregated samples at various time points were collected and subjected to SEC chromatography. By injecting supernatants of the aggregated Aβ samples at various incubation times after centrifugation into SEC, we observed two major peaks. One is in the void volume standing for particles larger than 250 kDa, namely the oligomer peak, and the other is Aβ monomer, namely the monomer peak. The oligomer peak was immediately observed at time 0 without incubation in the ratio of Aβ42-dominant samples (10/0 and 9/1) and the equally mixed one (5/5). However, in the samples of 1/9 and 0/10 the oligomer peaks are very little and had negligible difference, but the monomer peaks reduced gradually during incubation. The monomer peak is the major population in the Aβ40-dominant samples. The differences demonstrate these species undergo different aggregation pathways, especially that the Aβ42- and Aβ40-dominant ones adopt distinct routes. In Aβ42 alone, the oligomer peak diminished pronouncedly after 104 h incubation, indicating formation of insoluble fibrils being precipitated before entering SEC column. In contrast, the intensity of oligomer peaks from 9/1 and 5/5 samples did not decrease through incubation time. The results demonstrated that oligomers are stabilized through addition of Aβ40 to Aβ42.

In far-UV CD study, we measured the secondary structure changes of Aβ samples in 6 consecutive days and compared the differences between day 0 and day 6. We observed β-sheet-rich spectra with a minima at 216 nm gradually shifted to 218 nm in the ratios of 10/0 and 9/1. In Aβ40-dominant species, the conformation changes from random coil-like to β-sheet-rich structure. Meanwhile, the spectra of 5/5 had nearly no change from day 0 to day 6 with minima remained at 217 nm. The results correlated with the SEC data that large aggregates likely enriched in β-strands were shown in Aβ42-dominant species at the beginning but not in Aβ40-dominant species. Also, the conformation of 5/5 species does not change much during aggregation.

3.4.3. *The oligomer population changes in the intermediate stages of fibrillization*

We further captured the oligomer assemblies by PICUP assay and examined by Western blot. The samples were collected at 0, 6, 80, and 104 h incubation. In the freshly prepared or 6 h, noncross-linked samples, Aβ42-dominant species and the equally mixed sample had more tetramers or paranuclei migrated at ~17 to ~25 kDa. This result is consistent with previous reports.[76,77] After cross-linking, they smeared from 11 to 95 kDa, but the Aβ40-dominant species showed ladders of monomer, dimer, trimer, and multimers. After 80 and 104 h incubation, Aβ42 alone contained larger aggregates stocked in the well. The 9/1 and 5/5 samples showed different assemblies in comparison to Aβ42 alone (10/0) and Aβ40-dominant species. The Aβ40-dominant samples showed mostly aggregates stocked in the wells without oligomers. This correlates with the SEC analysis showing that Aβ40-dominant species undergo a direct transformation from monomer to insoluble fibrils.

3.4.4. *Coincubation of Aβ40 and Aβ42 reveal novel Aβ oligomers with highest toxicity*

Next, we examined the toxicity of different Aβ aggregates by treating 5 μM of Aβ species to differentiated mouse neuroblastoma N2a cells and examined the level of shortened neurites. In order to be observable under fluorescence microscope, EGFP plasmid was transfected to N2a cells prior to Aβ treatment. The results showed that Aβ42-dominant species have more profound effect than the Aβ40-dominant species. Among them, the 5/5 species caused significantly higher toxicity than others. Furthermore, we examined the cell viability treated with Aβ aggregates by MTT assay using human neuroblastoma BE(2)-C cells and observed that the 5/5 species is indeed the most toxic species among the five ratios we examined. Both results from neurite length assay and MTT assay confirmed that the oligomers formed in equally mixed of Aβ42 and Aβ40 were the most toxic species among the five ratios examined.

Overall, we studied the aggregation kinetics of co-existence of Aβ42 and Aβ40. The mechanism are summarized and illustrated in Fig. 5. Different intermediates and fibrils are represented in type I, II, and III.

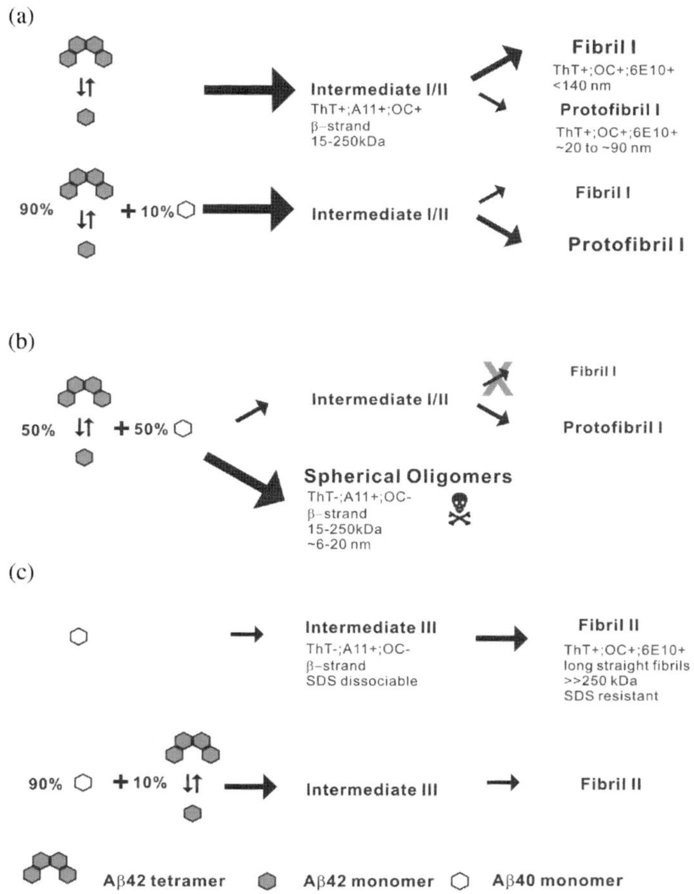

Figure 5: Aggregation pathways of Aβ42 and Aβ40 mixtures. Aggregation pathways of (a) Aβ42 alone and Aβ42/Aβ40 as 9/1, (b) equimolar Aβ42/Aβ40, and (c) Aβ40 alone and Aβ42/Aβ40 as 1/9. Different species are illustrated. The arrows indicate aggregation rate. The figure is modified from previous reference.[64]

4. Discussion

Understanding fundamental properties of Aβ is critical toward combating AD, especially the properties that influence the Aβ oligomerization and fibrillization process. In our studies, we found distinct conformation of native Aβ40 and Aβ42 and they behave differently. Aβ40 exists as an unstable monomer population with a folding free energy of 1.1 kcal/mol described from a two-state model (N $\leftrightarrows$ U). It contains a large fraction of random coil but is not completely unfolded. Aβ42 adopts a trimer/tetramer to the monomer unfolding mechanism in either two-state (N3 $\leftrightarrows$ 3U or N4 $\leftrightarrows$ 4U) or three-state models (N3 $\leftrightarrows$ 3N $\leftrightarrows$ 3U or N4 $\leftrightarrows$ 4N $\leftrightarrows$ 4U). Comparing FAD Aβ40 mutant monomers, we found the more stable the variant the slower the fibrillization, especially the nucleation step. The result demonstrated that folding stability of Aβ is an important determinant of the nucleation kinetics in fibrillization. The only exception is D23N mutant that may be due to breaking of the important salt bridge between D23 and K28. Unexpectedly, the A21G mutant is more stable than wild-type one and aggregates slowly. This result indicates that the early onset FAD harboring A21G mutant is due to other factors such as twofold γ-secretase cleavage[103] and less degradation.[104] Furthermore, the destabilization caused by Zn^{2+} or Al^{3+} binding results in large exposure of hydrophobic clusters on Aβ surface and leads to formation of annular protofibrils. In addition, manipulating the ratio of Aβ42 to Aβ40 results in different aggregation pathways. Novel toxic oligomers can form in equally mixed Aβ species. Similar ratios have been found in leptomeningeal and parenchymal vessels in patients with AD.[105]

For structured proteins, the level of native structure governs the protein function. Our study provides for the first time that the level of foldness for intrinsically disordered peptide-like Aβ is important for its aggregation properties. The level of aggregation decreased when the native conformation decreased. Both denatured Aβ40 and Aβ42 are unable to fibrilize; however, some fast aggregating FAD mutants are still able to form fibrils in denaturing condition, which may be due to rapid equilibrium of unfolded species and native species that are drawn to fibril formation. It is shown that oligomerization of Aβ is more significantly affected by denaturant. Only the native folded Aβ is capable of forming oligomers. This indicates that the native structure of Aβ is essential for AD pathogenesis. Overall, our studies systematically examined folding and misfolding mechanisms of Aβ under various conditions that could occur pathologically and provide insights for future therapeutic design for AD.

References

1. Querfurth, H. W. and LaFerla, F. M. Alzheimer's disease. *N. Engl. J. Med.* **362**, 329–344 (2010).
2. Qi-Takahara, Y. *et al.* Longer forms of amyloid β protein: Implications for the mechanism of intramembrane cleavage by γ-secretase. *J. Neurosci.* **25**, 436–445 (2005).
3. Takami, M. *et al.* γ-Secretase: Successive tripeptide and tetrapeptide release from the transmembrane domain of β-carboxyl terminal fragment. *J. Neurosci.* **29**, 13042–13052 (2009).
4. Olsson, F. *et al.* Characterization of intermediate steps in amyloid beta (Aβ) production under near-native conditions. *J. Biol. Chem.* **289**, 1540–1550 (2014).
5. Cataldo, A. M. *et al.* Endocytic pathway abnormalities precede amyloid β deposition in sporadic Alzheimer's disease and down syndrome: Differential effects of APOE genotype and presenilin mutations. *Am. J. Pathol.* **157**, 277–286 (2000).
6. Cataldo, A. M. *et al.* Abeta localization in abnormal endosomes: Association with earliest Abeta elevations in AD and Down syndrome. *Neurobiologia.* **25**, 1263–1272 (2004).
7. Kinoshita, A. *et al.* Demonstration by FRET of BACE interaction with the amyloid precursor protein at the cell surface and in early endosomes. *J. Cell Sci.* **116**, 3339–3346 (2003).
8. LaFerla, F. M., Green, K. N., and Oddo, S. Intracellular amyloid-[beta] in Alzheimer's disease. *Nat. Rev. Neurosci.* **8**, 499–509 (2007).
9. Selkoe, D. J. and Podlisny, M. B. Deciphering the genetic basis of Alzheimer's disease. *Annu. Rev. Genomics Hum. Genet.* **3**, 67–99 (2002).
10. Hori, Y. *et al.* The Tottori (D7N) and English (H6R) familial Alzheimer disease mutations accelerate Abeta fibril formation without increasing protofibril formation. *J. Biol. Chem.* **282**, 4916–4923 (2007).
11. Wakutani, Y. *et al.* Novel amyloid precursor protein gene missense mutation (D678N) in probable familial Alzheimer's disease. *J. Neurol. Neurosurg. Psychiatry.* **75**, 1039–1042 (2004).
12. Chen, W. T. *et al.* Amyloid-beta (Abeta) D7H mutation increases oligomeric Abeta42 and alters properties of Abeta-zinc/copper assemblies. *PLoS One.* **7**, e35807 (2012).
13. Hendriks, L. *et al.* Presenile dementia and cerebral haemorrhage linked to a mutation at codon 692 of the beta-amyloid precursor protein gene. *Nat. Genet.* **1**, 218–221 (1992).
14. Kamino, K. *et al.* Linkage and mutational analysis of familial Alzheimer disease kindreds for the APP gene region. *Am. J. Hum. Genet.* **51**, 998–1014 (1992).
15. Nilsberth, C. *et al.* The 'Arctic' APP mutation (E693G) causes Alzheimer's disease by enhanced Abeta protofibril formation. *Nat. Neurosci.* **4**, 887–893 (2001).
16. Levy, E. *et al.* Mutation of the Alzheimer's disease amyloid gene in hereditary cerebral hemorrhage, Dutch type. *Science.* **248**, 1124–1126 (1990).
17. Van Broeckhoven, C. *et al.* Amyloid beta protein precursor gene and hereditary cerebral hemorrhage with amyloidosis (Dutch). *Science.* **248**, 1120–1122 (1990).
18. Rossi, G. *et al.* Fatal familial insomnia: Genetic, neuropathologic, and biochemical study of a patient from a new Italian kindred. *Neurology.* **50**, 688–692 (1998).

19. Grabowski, T. J., Cho, H. S., Vonsattel, J. P., Rebeck, G. W., and Greenberg, S. M. Novel amyloid precursor protein mutation in an Iowa family with dementia and severe cerebral amyloid angiopathy. *Ann. Neurol.* **49**, 697–705 (2001).
20. Eckman, C. B. *et al.* A new pathogenic mutation in the APP gene (I716V) increases the relative proportion of Aβ42(43). *Hum. Mol. Genet.* **6**, 2087–2089 (1997).
21. De Jonghe, C. *et al.* Pathogenic APP mutations near the γ-secretase cleavage site differentially affect Aβ secretion and APP C-terminal fragment stability. *Hum. Mol. Genet.* **10**, 1665–1671 (2001).
22. Murrell, J. R., Hake, A. M., Quaid, K. A., Farlow, M. R., and Ghetti, B. Early-onset Alzheimer disease caused by a new mutation (v717l) in the amyloid precursor protein gene. *Arch. Neurol.* **57**, 885–887 (2000).
23. Felsenstein, K. M., Hunihan, L. W., and Roberts, S. B. Altered cleavage and secretion of a recombinant beta–APP bearing the Swedish familial Alzheimer's disease mutation. *Nat. Genet.* **6**, 251–256 (1994).
24. Maruyama, K. *et al.* Familial Alzheimer's disease-linked mutations at Val717 of amyloid precursor protein are specific for the increased secretion of A beta 42(43). *Biochem. Biophys. Res. Commun.* **227**, 730–735 (1996).
25. Suzuki, N. *et al.* An increased percentage of long amyloid beta protein secreted by familial amyloid beta protein precursor (beta APP717) mutants. *Science.* **264**, 1336–1340 (1994).
26. Rambaran, R. N. and Serpell, L. C. Amyloid fibrils: Abnormal protein assembly. *Prion.* **2**, 112–117 (2008).
27. Roychaudhuri, R., Yang, M., Hoshi, M. M., and Teplow, D. B. Amyloid beta-protein assembly and Alzheimer disease. *J. Biol. Chem.* **284**, 4749–4753 (2009).
28. Lomakin, A., Chung, D. S., Benedek, G. B., Kirschner, D. A., and Teplow, D. B. On the nucleation and growth of amyloid beta-protein fibrils: Detection of nuclei and quantitation of rate constants. *Proc. Natl. Acad. Sci. USA.* **93**, 1125–1129 (1996).
29. Frieden, C. Protein aggregation processes: In search of the mechanism. *Protein Sci.* **16**, 2334–2344 (2007).
30. Puchtler, H., Sweat, F., and Levine, M. On the binding of Congo red by amyloid. *J. Histochem. Cytochem.* **10**, 355–364 (1962).
31. Wolman, M. and Bubis, J. J. The cause of the green polarization color of amyloid stained with Congo red. *Histochemie.* **4**, 351–356 (1965).
32. Cohen, S. I. A. *et al.* Proliferation of amyloid-β42 aggregates occurs through a secondary nucleation mechanism. *Proc. Natl. Acad. Sci.* **110**, 9758–9763 (2013).
33. Meisl, G. *et al.* Differences in nucleation behavior underlie the contrasting aggregation kinetics of the Aβ40 and Aβ42 peptides. *Proc. Natl. Acad. Sci.* **111**, 9384–9389 (2014).
34. Tycko, R. Molecular structure of amyloid fibrils: Insights from solid-state NMR. *Q. Rev. Biophys.* **39**, 1–55 (2006).
35. Torok, M. *et al.* Structural and dynamic features of Alzheimer's Abeta peptide in amyloid fibrils studied by site-directed spin labeling. *J. Biol. Chem.* **277**, 40810–40815 (2002).
36. Williams, A. D. *et al.* Mapping abeta amyloid fibril secondary structure using scanning proline mutagenesis. *J. Mol. Biol.* **335**, 833–842 (2004).
37. Guo, J. T., Wetzel, R., and Xu, Y. Molecular modeling of the core of Abeta amyloid fibrils. *Proteins.* **57**, 357–364 (2004).

38. Xiao, Y. L. *et al.* A beta(1-42) fibril structure illuminates self-recognition and replication of amyloid in Alzheimer's disease. *Nat. Struct. Mol. Biol.* **22**, 499–U497 (2015).
39. Kodali, R. and Wetzel, R. Polymorphism in the intermediates and products of amyloid assembly. *Curr. Opin. Struct. Biol.* **17**, 48–57 (2007).
40. Hardy, J. and Selkoe, D. J. The amyloid hypothesis of Alzheimer's disease: Progress and problems on the road to therapeutics. *Science.* **297**, 353–356 (2002).
41. Glabe, C. C. Amyloid accumulation and pathogensis of Alzheimer's disease: Significance of monomeric, oligomeric and fibrillar Abeta. *Subcell Biochem.* **38**, 167–177 (2005).
42. Yu, L. *et al.* Structural characterization of a soluble amyloid β-peptide oligomer. *Biochemistry.* **48**, 1870–1877 (2009).
43. Cerf, E. *et al.* Antiparallel β-sheet: A signature structure of the oligomeric amyloid β-peptide. *Biochem. J.* **421**, 415–423 (2009).
44. Liu, P. *et al.* Quaternary structure defines a large class of amyloid-β oligomers neutralized by sequestration. *Cell Reports.* **11**, 1760–1771 (2015).
45. Mattson, M. P. Pathways towards and away from Alzheimer's disease. *Nature.* **430**, 631–639 (2004).
46. Weiner, H. L. and Frenkel, D. Immunology and immunotherapy of Alzheimer's disease. *Nat. Rev. Immunol.* **6**, 404–416 (2006).
47. Moore, B. D., Rangachari, V., Tay, W. M., Milkovic, N. M., and Rosenberry, T. L. Biophysical analyses of synthetic amyloid-β(1-42) aggregates before and after covalent cross-linking. Implications for deducing the structure of endogenous amyloid-β oligomers. *Biochemistry.* **48**, 11796–11806 (2009).
48. Kummer, Markus P. *et al.* Nitration of tyrosine 10 critically enhances amyloid beta aggregation and plaque formation. *Neuron.* **71**, 833–844 (2011).
49. Lesne, S. *et al.* A specific amyloid-beta protein assembly in the brain impairs memory. *Nature.* **440**, 352–357 (2006).
50. Dahlgren, K. N. *et al.* Oligomeric and fibrillar species of amyloid-beta peptides differentially affect neuronal viability. *J. Biol. Chem.* **277**, 32046–32053 (2002).
51. Ahmed, M. *et al.* Structural conversion of neurotoxic amyloid-beta(1–42) oligomers to fibrils. *Nat. Struct. Mol. Biol.* **17**, 561–567 (2010).
52. Jin, M. *et al.* Soluble amyloid beta-protein dimers isolated from Alzheimer cortex directly induce Tau hyperphosphorylation and neuritic degeneration. *Proc. Natl. Acad. Sci. USA.* **108**, 5819–5824 (2011).
53. O'Nuallain, B. *et al.* Amyloid beta-protein dimers rapidly form stable synaptotoxic protofibrils. *J. Neurosci.* **30**, 14411–14419 (2010).
54. Kayed, R. *et al.* Permeabilization of lipid bilayers is a common conformation-dependent activity of soluble amyloid oligomers in protein misfolding diseases. *J. Biol. Chem.* **279**, 46363–46366 (2004).
55. Demuro, A. *et al.* Calcium dysregulation and membrane disruption as a ubiquitous neurotoxic mechanism of soluble amyloid oligomers. *J. Biol. Chem.* **280**, 17294–17300 (2005).
56. Selkoe, D. J. Soluble oligomers of the amyloid [beta]-protein impair synaptic plasticity and behavior. *Behav. Brain Res.* **192**, 106–113 (2008).
57. Petkova, A. T. *et al.* Self-propagating, molecular-level polymorphism in Alzheimer's beta-amyloid fibrils. *Science.* **307**, 262–265 (2005).

58. Hou, L. *et al.* Solution NMR studies of the A beta(1-40) and A beta(1-42) peptides establish that the Met35 oxidation state affects the mechanism of amyloid formation. *J. Am. Chem. Soc.* **126**, 1992–2005 (2004).
59. Lazo, N. D., Grant, M. A., Condron, M. C., Rigby, A. C., and Teplow, D. B. On the nucleation of amyloid beta-protein monomer folding. *Protein Sci.* **14**, 1581–1596 (2005).
60. Grant, M. A. *et al.* Familial Alzheimer's disease mutations alter the stability of the amyloid beta-protein monomer folding nucleus. *Proc. Natl. Acad. Sci. USA.* **104**, 16522–16527 (2007).
61. Chen, Y. R. and Glabe, C. G. Distinct early folding and aggregation properties of Alzheimer amyloid-beta peptides Abeta40 and Abeta42: Stable trimer or tetramer formation by Abeta42. *J. Biol. Chem.* **281**, 24414–24422 (2006).
62. Snyder, S. W. *et al.* Amyloid-beta aggregation: Selective inhibition of aggregation in mixtures of amyloid with different chain lengths. *Biophys. J.* **67**, 1216–1228 (1994).
63. McGowan, E. *et al.* Aβ42 Is essential for parenchymal and vascular amyloid deposition in mice. *Neuron.* **47**, 191–199 (2005).
64. Chang, Y. J. and Chen, Y. R. The coexistence of an equal amount of Alzheimer's amyloid-beta 40 and 42 forms structurally stable and toxic oligomers through a distinct pathway. *FEBS J.* **281**, 2674–2687 (2014).
65. Suzuki, N. *et al.* High tissue content of soluble beta 1-40 is linked to cerebral amyloid angiopathy. *Am. J. Pathol.* **145**, 452–460 (1994).
66. Prelli, F., Castano, E., Glenner, G. G., and Frangione, B. Differences between vascular and plaque core amyloid in Alzheimer's disease. *J. Neurochem.* **51**, 648–651 (1988).
67. Seubert, P. *et al.* Isolation and quantification of soluble Alzheimer's beta-peptide from biological fluids. *Nature.* **359**, 325–327 (1992).
68. Iwatsubo, T. *et al.* Visualization of A beta 42(43) and A beta 40 in senile plaques with end-specific A beta monoclonals: Evidence that an initially deposited species is A beta 42(43). *Neuron.* **13**, 45–53 (1994).
69. Miller, D. L. *et al.* Peptide compositions of the cerebrovascular and senile plaque core amyloid deposits of Alzheimer's disease. *Arch. Biochem. Biophys.* **301**, 41–52 (1993).
70. Akiyama, H. *et al.* Variable deposition of amyloid beta-protein (Abeta) with the carboxy-terminus that ends at residue valine40 (Abeta 40) in the cerebral cortex of patients with Alzheimer's disease: A double-labeling immunohistochemical study with antibodies specific for A beta 40 and the A beta that ends at residues alanine42/threonine43 (Abeta 42). *Neurochem. Res.* **22**, 1499–1506 (1997).
71. Attems, J., Jellinger, K., Thal, D. R., and Van Nostrand, W. Review: Sporadic cerebral amyloid angiopathy. *Neuropathol. Appl. Neurobiol.* **37**, 75–93 (2011).
72. Petkova, A. T. *et al.* A structural model for Alzheimer's β-amyloid fibrils based on experimental constraints from solid state NMR. *Proc. Natl. Acad. Sci. USA.* **99**, 16742–16747 (2002).
73. Antzutkin, O. N., Leapman, R. D., Balbach, J. J., and Tycko, R. Supramolecular structural constraints on Alzheimer's β-amyloid fibrils from electron microscopy and solid-state nuclear magnetic resonance. *Biochemistry.* **41**, 15436–15450 (2002).
74. Benzinger, T. L. *et al.* Propagating structure of Alzheimer's beta-amyloid(10-35) is parallel beta-sheet with residues in exact register. *Proc. Natl. Acad. Sci. USA.* **95**, 13407–13412 (1998).

75. Schmidt, M. *et al.* Comparison of Alzheimer Aβ(1–40) and Aβ(1–42) amyloid fibrils reveals similar protofilament structures. *Proc. Natl. Acad. Sci.* **106**, 19813–19818 (2009).
76. Bitan, G. *et al.* Amyloid beta-protein (Abeta) assembly: Abeta 40 and Abeta 42 oligomerize through distinct pathways. *Proc. Natl. Acad. Sci. U S A.* **100**, 330–335 (2003).
77. Bernstein, S. L. *et al.* Amyloid-β protein oligomerization and the importance of tetramers and dodecamers in the aetiology of Alzheimer's disease. *Nat. Chem.* **1**, 326–331 (2009).
78. Kłoniecki, M. *et al.* Ion mobility separation coupled with MS detects two structural states of Alzheimer's disease Aβ 1–40 peptide oligomers. *J. Mol. Biol.* **407**, 110–124 (2011).
79. Kuperstein, I. *et al.* Neurotoxicity of Alzheimer's disease Aβ peptides is induced by small changes in the Aβ42 to Aβ40 ratio. *EMBO J.* **29**, 3408–3420 (2010).
80. Yan, Y. and Wang, C. Aβ40 Protects Non-toxic Aβ42 monomer from aggregation. *J. Mol. Biol.* **369**, 909–916 (2007).
81. Jan, A., Gokce, O., Luthi-Carter, R., and Lashuel, H. A. The ratio of monomeric to aggregated forms of Aβ40 and Aβ42 is an important determinant of amyloid-β aggregation, fibrillogenesis, and toxicity. *J. Biol. Chem.* **283**, 28176–28189 (2008).
82. Murray, M. M. *et al.* Amyloid β protein: Aβ40 inhibits Aβ42 oligomerization. *J. Am. Chem. Soc.* **131**, 6316–6317 (2009).
83. Pauwels, K. *et al.* Structural basis for increased toxicity of pathological Aβ42:Aβ40 ratios in Alzheimer disease. *J. Biol. Chem.* **287**, 5650–5660 (2012).
84. Kim, J. *et al.* Abeta40 inhibits amyloid deposition in vivo. *J. Neurosci.* **27**, 627–633 (2007).
85. Edelhoch, H. Spectroscopic determination of tryptophan and tyrosine in proteins. *Biochemistry.* **6**, 1948–1954 (1967).
86. Santoro, M. M. and Bolen, D. W. Unfolding free energy changes determined by the linear extrapolation method. 1. Unfolding of phenylmethanesulfonyl alpha-chymotrypsin using different denaturants. *Biochemistry.* **27**, 8063–8068 (1988).
87. Kayed, R. *et al.* Common structure of soluble amyloid oligomers implies common mechanism of pathogenesis. *Science.* **300**, 486–489 (2003).
88. Nielsen, L. *et al.* Effect of environmental factors on the kinetics of insulin fibril formation: Elucidation of the molecular mechanism. *Biochemistry.* **40**, 6036–6046 (2001).
89. Kayed, R. *et al.* Fibril specific, conformation dependent antibodies recognize a generic epitope common to amyloid fibrils and fibrillar oligomers that is absent in prefibrillar oligomers. *Mol. Neurodegener.* **2**, 18 (2007).
90. Dunker, A. K., Romero, P., Obradovic, Z., Garner, E. C., and Brown, C. J. Intrinsic protein disorder in complete genomes. *Genome Inform.* **11**, 161–171 (2000).
91. Baumketner, A. *et al.* Structure of the 21–30 fragment of amyloid β-protein. *Protein Sci.* **15**, 1239–1247 (2006).
92. Chen, Y.-R. and Glabe, C. G. Distinct early folding and aggregation properties of Alzheimer amyloid-β peptides Aβ40 and Aβ42: Stable trimer or tetramer formation by Aβ42. *J. Biol. Chem.* **281**, 24414–24422 (2006).
93. Mustafi, S. M., Garai, K., Crick, S. L., Baban, B., and Frieden, C. Substoichiometric inhibition of Aβ1–40 aggregation by a tandem Aβ40–1-Gly8-1-40 peptide. *Biochem. Biophys. Res. Commun.* **397**, 509–512 (2010).

94. Harris, M. E., Hensley, K., Butterfield, D. A., Leedle, R. A., and Carney, J. M. Direct evidence of oxidative injury produced by the Alzheimer's β-amyloid peptide (1–40) in cultured hippocampal neurons. *Exp. Neurol.* **131**, 193–202 (1995).
95. Roccatano, D., Colombo, G., Fioroni, M., and Mark, A. E. Mechanism by which 2,2,2-trifluoroethanol/water mixtures stabilize secondary-structure formation in peptides: A molecular dynamics study. *Proc. Natl. Acad. Sci. USA.* **99**, 12179–12184 (2002).
96. Povey, J. F., Smales, C. M., Hassard, S. J., and Howard, M. J. Comparison of the effects of 2,2,2-trifluoroethanol on peptide and protein structure and function. *J. Struct. Biol.* **157**, 329–338 (2007).
97. Fezoui, Y. and Teplow, D. B. Kinetic studies of amyloid beta-protein fibril assembly. Differential effects of alpha-helix stabilization. *J. Biol. Chem.* **277**, 36948–36954 (2002).
98. Bush, A. I. The metallobiology of Alzheimer's disease. *Trends Neurosci.* **26**, 207–214 (2003).
99. Zatta, P., Drago, D., Bolognin, S., and Sensi, S. L. Alzheimer's disease, metal ions and metal homeostatic therapy. *Trends Pharmacol. Sci.* **30**, 346–355 (2009).
100. Lovell, M. A., Robertson, J. D., Teesdale, W. J., Campbell, J. L., and Markesbery, W. R. Copper, iron and zinc in Alzheimer's disease senile plaques. *J. Neurol. Sci.* **158**, 47–52 (1998).
101. Frederickson, C. J., Koh, J.-Y., and Bush, A. I. The neurobiology of zinc in health and disease. *Nat. Rev. Neurosci.* **6**, 449–462 (2005).
102. Yumoto, S., Kakimi, S., Ohsaki, A., and Ishikawa, A. Demonstration of aluminum in amyloid fibers in the cores of senile plaques in the brains of patients with Alzheimer's disease. *J. Inorg. Biochem.* **103**, 1579–1584 (2009).
103. Tian, Y., Bassit, B., Chau, D., and Li, Y. M. An APP inhibitory domain containing the Flemish mutation residue modulates gamma-secretase activity for Abeta production. *Nat. Struct. Mol. Biol.* **17**, 151–158 (2010).
104. Betts, V. *et al.* Aggregation and catabolism of disease-associated intra-Abeta mutations: Reduced proteolysis of AbetaA21G by neprilysin. *Neurobiol. Dis.* **31**, 442–450 (2008).
105. Roher, A. E. *et al.* Beta-amyloid-(1-42) is a major component of cerebrovascular amyloid deposits: Implications for the pathology of Alzheimer disease. *Proc. Natl. Acad. Sci. U S A.* **90**, 10836–10840 (1993).

Single-Molecule Fluorescence Resonance Energy Transfer Studies of β-Amyloid Clusters in Physiological Solutions

8

Jun Han, Erwen Mei,† Mei-Ping Kung,‡ Hank F. Kung,‡ Jian-Min Yuan§ and Hai-Lung Dai**

**Department of Chemistry, Temple University, Philadelphia, PA 19122*
†Regional Laser and Biotechnology Laboratories,
‡Department of Radiology, University of Pennsylvania, Philadelphia, PA 19104
§Department of Physics, Drexel University, Philadelphia, PA 19104

Formation of oligomers of β-amyloid (Aβ) peptides is a pivotal initial event associated with the pathogenicity of Alzheimer's disease (AD). Using the single-molecule fluorescence resonance energy transfer (smFRET) microscopy technique, we detected the very beginning of the formation of Aβ peptide oligomers in physiological solutions. The existence of oligomers is identified from the FRET between dyes individually bonded to the N terminus of the 40-unit peptides. Several dimers and oligomers identified by the varying FRET efficiency have been detected. The dimer with the highest stability is characterized to have a distance of 43 Å between its two N termini, in an antiparallel structure that is similar to that of the dimer unit (antiparallel hairpin) that has been previously identified in the fibrils.

1. Introduction

Alzheimer's disease (AD) is a neurodegenerative disease affecting millions of Americans.[1] Currently, one in three seniors dies with AD or another form of dementia, and one in every eight older Americans has AD, which is the sixth leading cause of death in the United States[1] *and* the only cause of death among the top 10 in the United States that cannot be either prevented, cured, or slowed. Although the exact mechanisms, which lead to the development of AD, are not fully understood, the prevailing hypothesis of the origin of pathogenicity of AD is the formation of β-amyloid (Aβ) aggregates and plaques in the brain.[2–5] It is suggested that excess Aβ, peptides with 39–43 residue amino acids derived from enzymatic degradation of the Aβ precursor protein (APP), in AD patient's brain coagulate and result in first the production of small oligomers with sizes starting from dimer of the Aβ peptides. These Aβ oligomers, abbreviated as Aβ-derived diffusible ligands (ADDLs), eventually lead to the formation of large Aβ aggregates, fibrils, and plaques. Some reports suggest a direct connection between ADDLs and pathogenicity of AD, i.e., the soluble Aβ peptide oligomers inflict neuronal malfunction prior to the formation of amyloid plaques.[6–11] It is now generally believed that the soluble Aβ oligomers represent the initial toxic species in AD,[12,13] with the dimer being the smallest synaptotoxic species.[14] Understanding how the ADDLs, particularly the dimer, are formed so that compounds and procedures for inhibiting their formation can be designed and tested, therefore, constitutes a viable approach for developing effective treatment of AD.[15,16]

Among the Aβ peptides, β(1–42) is the principal species associated with senile plaque formation,[17] while Aβ(1–40) is 10 times more abundant in cerebrovascular amyloid deposit.[18,19] A variety of experimental methods[15,20–29] has been applied to investigate the initial oligomerization of Aβ peptides. Distinctly different pathways of oligomerization have been elucidated for these two peptides.[20,29–31] It is important to know the rates and mechanisms by which the monomeric peptides coagulate and form oligomeric species and amyloid fibrils. What is the nature of the interactions between the peptides, which as monomers are already partially folded in aqueous solutions,[32] and what molecular properties and environmental factors can be used to influence these interactions? Are any of the Aβ peptides particularly active in ADDL formation? These are among the many fundamental questions concerning the formation of toxic species in AD. Since the dimer is likely the very first formed species in the entire chain reactions leading to fibril, the starting point for answering these questions is to be able to detect the Aβ dimer, measure its structure, and characterize its stability in a physiological solution.

The structure of fibrils of Aβ(1–40) has been proposed based on solid-state nuclear magnetic resonance (NMR) data[33–35] and electron microscopy measurements of Aβ(1–40) fibril dimensions and the mass per length.[36,37] A previous study asserts that Aβ(1–40) fibrils are made of units of dimers parallel with each other with each dimer unit consisting of two peptides in antiparallel hairpin conformation.[33]

In terms of oligomerization, gel filtration chromatography[27,28] and other experimental studies[15,25] have shown that at neutral pH Aβ(1–40) peptides do form dimers. A study of oligomer size distribution by photoinduced cross-linking of unmodified proteins[25] indicated that Aβ(1–40) oligomers consist primarily dimer, trimer, and tetramer species, with the dimer being the most abundant oligomer in the Aβ(1–40) solution.

The primary challenge in characterizing the dimer in a solution is how to identify the dimer among the monomer and oligomers. Single-molecule fluorescence resonance energy transfer (smFRET) has proven effective in locating a molecule labeled with a particular pair of fluorescing chromophores and in measuring the distance between this pair of specific fluorophores.[38,39] This technique can be applied for the identification of the Aβ dimer and provides structural information. In this approach, a donor fluorophore is attached through chemical bonding to a peptide molecule, while a different acceptor fluorophore that can fluoresce following resonant energy transfer from the donor is attached to another peptide molecule. Detection of acceptor fluorescence signal following photoexcitation of the donor indicates that the two peptide molecules are within energy transfer distance and therefore their association as a dimer or higher oligomer. The energy transfer efficiency can be measured for the determination of the distance between the two fluorophores.

Indeed the FRET approach has been applied to analyze amyloid peptides in the *bulk* form where gel filtration chromatography was used to separate oligomers of different sizes and the study concluded that dimerization of Aβ(1–40) with nM to μM concentrations in aqueous Tris-HCl pH 7.4 buffer solution does occur.[15] FRET experiments, conducted with time resolution, on bulk samples of shorter Aβ peptides, Aβ(11–25), have revealed various conformations in their aggregation process.[40] Other studies have used single-molecule imaging to study Aβ(1–42) oligomers interacting with the cellular prion protein[41]; single-molecule fluorescence correlation spectroscopy, in which the diffusivity of the fluorescing object in solution is measured to infer its size, to observe the formation of large oligomers of Aβ(1–40)[22]; and single-molecule quantized photobleaching technique to characterize small oligomers immobilized on coverslip surfaces.[42,43]

In this chapter, we summarize the research on the first identification using the *single-molecule* FRET method of the dimers of Aβ peptides, in this case Aβ(1–40), in physiological solutions. Under single-molecule conditions where the dimer is interrogated one at a time, structural information and stability associated with a particular dimer can be deduced. The most stable dimer under the observation conditions has been identified with its structure characterized in terms of the N-terminal distance.

2. Methodology

The synthetic Aβ(1–40) peptide was labeled at the N terminus either with 5-carboxytetramethylrhodamine (TMR, MW 430) dye as the donor, or with the HiLyte Fluor 647 (HiLyte647, MW 1003) dye as the acceptor. All peptides, including unlabeled Aβ(1–40) were originally dissolved in dimethyl sulfoxide (DMSO) as stock solutions.

It was reported that the Aβ peptides are in random coil conformations when dissolved in DMSO,[44] while in neutral aqueous solution the random coils tend to have more propensity to form β sheets. In all samples under microscopic examination, the DMSO content in samples was no more than 1.5%. A previous study has concluded that there is no significant change in the overall structure of the Aβ(1–40) over the range of DMSO from 0 to 10% in a neutral buffer.[15]

The lyophilized TMR-labeled Aβ(1–40) and HiLyte647-labeled Aβ(1–40) peptides were premixed with an equal molar ratio at 800 μM total peptide concentration in DMSO, followed by sonication to break up any preformed aggregates. For the concentration-dependent experiments, the premixed samples were diluted in phosphate buffered saline (PBS) to reach total peptide concentrations of 5, 8, or 12 μM. These concentrations were chosen since the critical concentration for Aβ(1–40) fibril formation at neutral pH on laboratory time scale has been reported to be *ca.* 10~40 μM.[16] At a specified incubation time a particular sample was then diluted by 100,000-fold to a sub-nM peptide concentration to facilitate single-molecule detection.

The fluorescence signal was detected by an inverted, epi-illumination confocal microscope that uses a sample-scanning stage with closed-loop feedback for accurate positioning of individual molecules.[45] The 514.5 nm line of an argon ion laser with circular polarization was used to excite the sample with near diffraction-limited focus.

The surface of the quartz coverslip (0.17 mm thick) was coated with polyethylene glycol (PEG) to reduce surface adsorption of the peptides.[46] Appropriate combinations of notch (514.5 nm), band-pass, long-pass, and dichroic filters were used to spectrally isolate the fluorescence emitted by the TMR and HiLyte647 dyes for detection by two single-photon counting avalanche diodes separately as channels D and A.

Structural information for the dimer can be obtained from the FRET efficiency. The distance r between the donor and the acceptor fluorophores is related to the FRET efficiency E as

$$E = \frac{R_0^6}{R_0^6 + r^6}, \tag{1}$$

where R_0 is the Förster distance.[47] The natural form of Aβ(1–40) in water is partially folded.[32] Because the fluorescent probe is covalently bonded to the N terminus of the 1 amino acid residue (aspartic acid), the attached dyes are likely to rotate and tilt with high degrees of freedom. It is therefore reasonable to speculate that the orientation constant is set as $2/3$ just like for isotropic, fast rotating transition dipoles.[48] The Förster distance is calculated as 59.4 Å for water as the medium for this set of donor–acceptor dyes. The distance between the two N-terminus-labeled dyes of the two peptides $R_{\text{D-A}}$ in the dimer is related to R_0 and the FRET efficiency E as[47]

$$R_{\text{D-A}} = R_0 \left(\frac{1-E}{E}\right)^{\frac{1}{6}}. \tag{2}$$

The FRET efficiency was reported based on the donor channel intensity I_D^i and the acceptor channel intensity I_A^i as

$$E^i = \frac{I_A^i - \alpha I_D^i}{I_A^i - \alpha I_D^i + \gamma I_D^i}, \tag{3}$$

where, α is the coefficient that counts for the cross talk from the Channel D to Channel A, and the factor γ accounts for the difference between collection efficiencies of the two channels. The cross talk from Channel A to Channel D was found to be negligible by control experiments using only Hilyte647-Aβ(1–40) and unlabeled Aβ(1–40) peptides at an equal molar ratio in PBS. The cross talk coefficient α was determined to be 0.23 ± 0.09 from control experiments using only TMR-Aβ(1–40) and unlabeled Aβ(1–40) peptides at an equimolar ratio in PBS. The following procedure is used to determine the FRET efficiency from the intensity-time FRET spectra such as those shown in Figures 2 and 3 in the following section. First, a threshold T representing noise background is applied to weighing if the fluorescence intensity should be counted as signal: only when the intensity of the ith data point in the acceptor channel $I_A^i \geq T$, it is considered as FRET signal. The magnitude of the threshold T contains contributions from noise as well as leak of donor signal into the acceptor channel and was determined from control experiments using solutions in which no FRET signal is expected.

3. Dimers and Oligomers

It is important to check that the labeling of the peptide with dyes does not change its conformation along its aggregation and polymerization process. According to the solid-state NMR studies,[33] the amino acid residues 1–8 of the peptide Aβ(1–40) are conformationally disordered in the fibril. Because the fluorescent probe is covalently bonded to the N terminus of the 1 amino acid residue (aspartic acid), the attached dyes are most likely to rotate and tilt with high degree of freedom, which suggests that the conformation of the peptide should not be significantly affected by the labeling. Experimentally, previous studies have confirmed that the N-terminal dye labeling of the peptide with either TMR[49] or HiLyte 647 dyes[50,51] does not significantly change its conformation along its aggregation and polymerization process. In this study, we reached the same conclusion by Fourier transform infrared (FTIR) spectroscopy and transmission electron microscopy.

In order for the fluorescence to be detected under single-molecule conditions, the concentration of the species has to be at nM or less. An experiment was conducted at such low concentrations (the total peptide concentration 1 nM with an equimolar ratio between the two labeled peptides). No FRET signal could be detected, indicating that neither dimers nor oligomers exist when incubation peptide concentration is at the nM level or below. In contrast, when the fluorophore-attached peptides were incubated at μM concentrations, which after a certain incubation time were diluted to sub-nM concentrations, ample smFRET signal appeared when this sub-nM solution was examined. Figure 1a shows the fluorescence signals detected in both the donor and acceptor channels for a sample diluted to 0.3 nM from the solution of 8 μM total peptide concentration after 10-h incubation. The recorded fluorescence burst density in the donor channel is much higher than that observed in the acceptor channel, indicating that the monomer is still the dominating species. The clearly observed fluorescence bursts in the acceptor channel (Ch. A), however, indicates that dimer or perhaps even larger oligomers exist in the solution. Each burst represents a species that contains at least a pair of donor and acceptor dyes within FRET distance diffusing through the observation zone. Figure 1b shows an expanded view around the peak labeled with an asterisk in panel (a) to depict discrete single-molecule burst signals.

To characterize the species that gives the FRET bursts, we first determine its corresponding FRET efficiency. The FRET efficiencies of the spectral peaks from a particular sample are expressed in a histogram such as those in Fig. 2.

To determine if the FRET signals arise from the dimer or larger oligomers, we examined how their intensities relate to incubation concentration and time. Intuitively, longer incubation time at a fixed concentration should result in more and larger oligomers. Figure 2 shows the efficiency distribution of the FRET signals detected under single-molecule conditions (diluted to nM) for samples with the total incubation concentration of 8 μM after incubation times ranging from 1 to 96 h. For the sample at the 1 h, which is right after renaturation of the peptides from the denatured state in DMSO, i.e., the high concentration peptide-in-DMSO solution is dissolved in PBS; its single-molecule FRET signals are dominated by high FRET

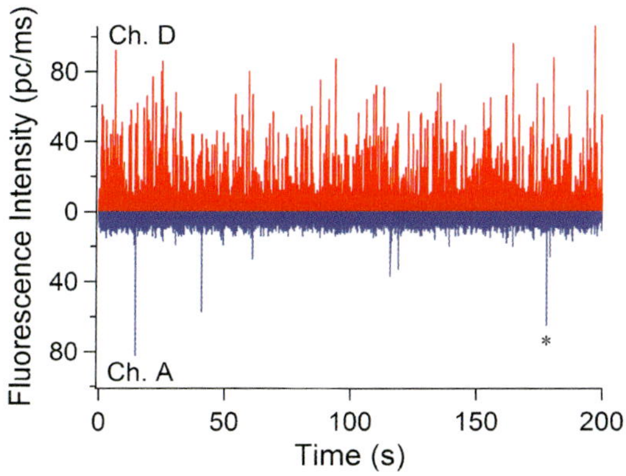

(a)

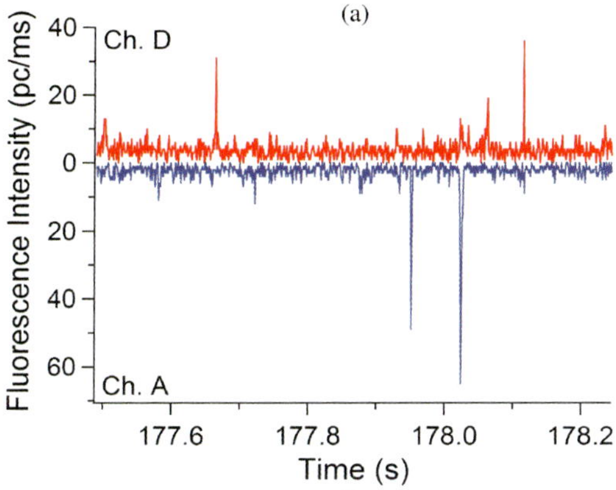

(b)

Figure 1: Fluorescence signals detected in both the donor and acceptor channels as a function of time. (a) Incubation at 8 μM total peptide concentration for 10 h. The total peptide concentration of the diluted sample is 0.3 nM. (b) Expanded view around the peak labeled with an asterisk in panel (a) to show discrete single-molecule burst signals.

efficiency events centered at ~90%. As incubation time proceeds, the FRET signals with efficiencies below 80% begin to appear more abundantly. At 96 h, the strongest FRET band is ~65%.

It is also expected that concentrations of larger oligomers should increase with the peptide concentration at the same incubation time. FRET efficiency histograms for incubation concentrations, 5 and 12 μM, at the same incubation time of 48 h are shown in Fig. 3. For the lowest concentration of 5 μM the dominating FRET band is the one at ~90% efficiency. As concentration increases, the FRET distribution moves to the medium efficiency region. At the 12 μM concentration, the FRET efficiency band becomes broader and is centered at 65%.

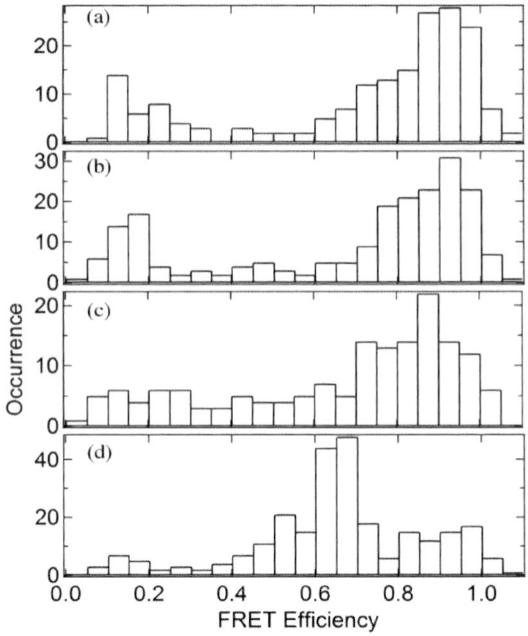

Figure 2: The smFRET efficiency distribution of the fluorescence signals taken for samples, diluted to sub-nM, from a solution initially with 4 μM TMR-Aβ(1–40) and 4 μM HiLyte647-Aβ(1–40) in PBS after incubation times of (a) 1 h, (b) 20 h, (c) 48 h, and (d) 96 h.

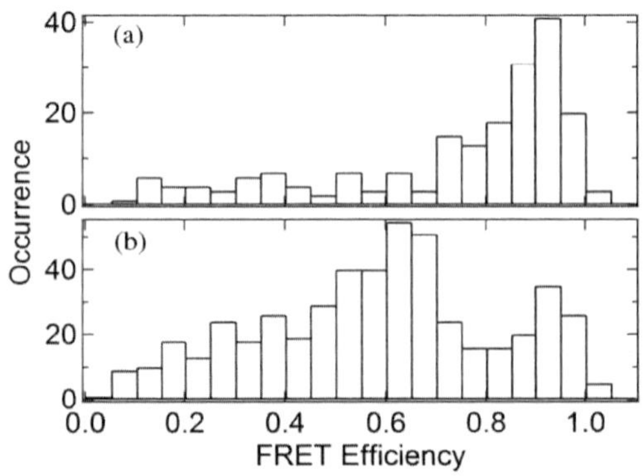

Figure 3: The smFRET efficiency distribution of the fluorescence signals taken for samples diluted to nM concentration from a solution initially with (a) 5 μM and (b) 12 μM total peptide concentration of equal molar mixtures of TMR-Aβ(1–40) and Hilyte647-Aβ(1–40) in PBS after 48-h incubation.

From the observations on the FRET efficiency distribution as a function of incubation concentration and time, we found that the high efficiency (~90%) band appears first, right after the renaturation of the peptide. It is logical to assign this band as FRET from a dimer resulted from the initial step of the polymerization process. Other evidence also supports this assignment. Single-step photobleaching was observed for the acceptor channel fluorescence of the high FRET efficiency bursts followed by a single-step fluorescence intensity jumping in the donor channel. This phenomenon is usually associated with FRET signals from a species containing a single pair of donor and acceptor fluorophores.[52] Furthermore, the dissociation behavior of the oligomers described next is also consistent with this assignment of dimer to the high efficiency band.

At longer incubation times and higher incubation concentrations, mid-range efficiency features appear in the FRET distribution. These features are best assigned to larger oligomers. The corresponding FRET signals in the intensity–time record have in general wider burst width in time, which is defined as the time span of a burst above the threshold T. On average, the width of the oligomers' bursts is 70–400% wider than that of the dimer, which is at the level of the 1-ms time resolution. The assignment of the broader bursts to larger oligomers is based on the expectation that they tend to stay in the observation zone longer.

The corresponding R_{D-A} for the FRET efficiency distribution such as the one in Fig. 2a can be calculated accordingly and expressed in Fig. 4. The $R_{\mathrm{D-A}}$ distribution of the ~90% FRET efficiency band shows a Gaussian distribution from 25–60 Å with the center located at 42.5 Å. This broad distribution in distance may be attributed to the flexibility of the amino acids 1–8 that results in the broad range of the donor–acceptor distance.

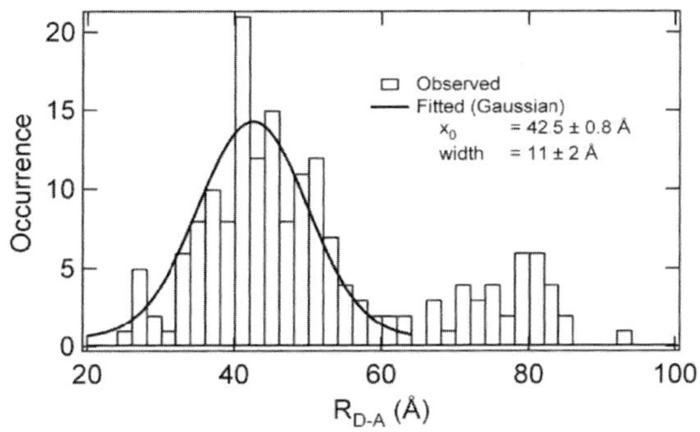

Figure 4: Corresponding $R_{\mathrm{D-A}}$ distribution calculated for Figure 2a. The FRET efficiency band at ~90% corresponds to an $\mathrm{R_{D-A}}$ Gaussian distribution centered at 42.5 Å.

On the basis of a molecular dynamics calculation by Barz and Urbanc, the average N-terminal distance from the center of mass in the Aβ(1–40) dimer is about 1.55 nm,[53] which means that if the two N termini are on the opposite ends of the globular dimer, the average N-termini distance is 31 Å. Considering that the labeled dyes have sizes on the order of 10 Å, this calculated value is in a reasonable agreement with our experimental value. This implies that the dimer has an antiparallel conformation where the two N termini are on the opposite ends, which is in consistence with the structure of the Aβ fibril determined by solid-state NMR that has antiparallel dimers as the building blocks.[33]

The earlier time histograms in Fig. 2 show that there is a small band in the low efficiency region at around 10%. Control experiments indicate that it arises mainly from the leakage of the donor fluorescence into the acceptor channel. The possibility that there exist dimers and oligomers with low FRET efficiency, however, cannot be ruled out. The 90% band corresponds to the dominant component in the dimer and oligomers distribution.

It is important to note that the FRET spectra we have recorded are from samples that were diluted to sub-nM concentration for single-molecule FRET detection. In such low concentrations, oligomers and dimers formed are not in equilibrium with the monomer and they would dissociate. The most abundant dimer detected here, the one with ~90% FRET efficiency, therefore is the last surviving species and the most stable dimer. This deduction suggests that the most stable dimer in the PBS solution has an antiparallel structure same as the dimer unit in the fibril.[33] The stability of the most stable dimer reported here can be characterized by its dissociation rate, which can be measured in a solution where the dimer concentration is in excess of the concentration in equilibrium with the monomer peptide. This condition is created by diluting an incubated solution with high initial dimer/oligomer concentration into a clean saline solution exactly like the ones prepared for single-molecule FRET observation. The dimer dissociation rate can then be measured as the change in time of the number density of the 90% FRET signals following the dilution. The incubated solution may initially contain larger oligomers, which would dissociate and form the dimer. In such situations the number density of the dimer FRET signals may increase first, representing an increase of dimer concentration due to oligomer dissociation, and then followed by a decrease corresponding to dimer dissociation.

We have examined the number of high smFRET efficiency ($\geq$80%) events as a function of time. The sample was prepared from an incubation solution of an equal molar mixture of TMR-Aβ(1–40) and HiLyte647-Aβ(1–40) with 8 μM total peptide concentration. The solution was incubated for 1 h and diluted to 0.4 nM for smFRET detection. The number of events can be converted to concentrations[54] based on the geometric shape of the light cavity of the microscope and the diffusion constant of the dimer estimated from that of the monomer.

As expected, immediately following dilution of the incubated solution to the single-molecule observation concentration, the concentration of the dimer signal increases first and then quickly decreases to the minimum value. The decreasing portion of the intensity can be fit to a single exponential with a rate constant of 0.185 ± 0.007 min^{-1}. This decay of the dimer concentration may be induced by two processes: the adsorption of dimers onto the surface of the PEG-coated quartz coverslip and the dissociation of the dimer. The adsorption rate constant of the Aβ monomer onto the surface of the PEG-coated quartz coverslip was determined to be 0.048 ± 0.003 min^{-1}, using the information from the donor channel where the monomer is the dominant component. Assuming that the diffusion-controlled adsorption rate of the dimer is the same as that of the monomer, we determine the dissociation rate constant of the dimer as 0.14 ± 0.01 min^{-1}. This value corresponds to a lifetime of 7.3 ± 0.4 min for this dimer at the pico-M concentration in the presence of a monomer concentration of 0.1 nM.

Finally, we comment on the FRET efficiency of the larger oligomers. Analysis shows that an oligomer containing more than two dyes will statistically give lower FRET efficiency. For example, statistically, about 40% of the tetramer would have either one or more donor fluorophore with less than one acceptor as its immediate neighbors. These tetramers would have less than 50% FRET efficiency even if the distance between a pair of donor–acceptor dyes warrants high FRET efficiency.

4. Conclusions

Single-molecule FRET detected through a confocal microscope has enabled the identification of formation of dimers and larger oligomers of Aβ(1–40) peptide in physiological solutions. In this experimental approach, the peptide is bonded with one in a pair of donor–acceptor dyes at its N terminus. It is found that FRET signals assignable to dimers and oligomers of the Aβ peptides appear immediately following mixing of the peptides at μM concentrations, but not at sub-nM concentrations. The most abundant and stable dimer that has survived the process of dissolution after the dilution of the incubated solution for single-molecule observation has a high FRET efficiency of ~90%. Accordingly, in the structure of this dimer, the distance between the two N termini is estimated to be 43 Å, which indicates that the structure of this most stable dimer in physiological solutions is such that the two N termini are on the opposite ends of the globular dimer, which is similar to that of the dimer unit (antiparallel hairpin) in the fibrils determined by the solid-state NMR study.[33] Under nonequilibrium conditions this dimer dissociates with a lifetime of 7.3 min. Larger oligomers are also identified at lower FRET efficiencies.

This experimental demonstration for being able to detect the smallest oligomers in physiological solutions under single-molecule conditions provides an approach for testing agents that may inhibit the formation of the Aβ aggregates at its inception. Equally important is that this approach may have the potential to be used as a basis for developing optical technology for early *in vivo* detection of potential AD pathogens.

Acknowledgments

We thank Drs. Feng Gai and Tom Troxler of University of Pennsylvania for the use of the facilities in the NIH-sponsored Regional Laser and Biotechnology Laboratories. We also thank Dr. Brigita Urbanc of Drexel University for stimulating discussions. This work was partially supported by grants from the National Institutes of Health (AG-022559 to HFK and AG-021868 to MPK) and from the National Science Foundation (CHE-1465096 to HLD).

References

1. Alzheimer's Association, 2016 Alzheimer's disease facts and figures. Alzheimer's & Dementia. *J. Alzheimers Assoc.* **12**(4), 459–509 (2016).
2. Yaari, R. and Corey-Bloom, J. Alzheimer's disease. *Semin. Neurol.* **27**(1), 32–41 (2007).
3. Hardy, J. Alzheimer's disease: The amyloid cascade hypothesis: an update and reappraisal. *J. Alzheimers Dis.* **9**(3 Suppl), 151–153 (2006).
4. Roberson, E. D., Scearce-Levie, K., Palop, J. J., Yan, F., Cheng, I. H., Wu, T., and Mucke, L. Reducing endogenous tau ameliorates amyloid beta-induced deficits in an Alzheimer's disease mouse model. *Science.* **316**(5825), 750–754 (2007).
5. Turner, R. S. Alzheimer's disease. *Semin. Neurol.* **26**(5), 499–506 (2006).
6. Klein, W. L. Aβ toxicity in Alzheimer's disease: Globular oligomers (ADDLs) as new vaccine and drug targets. *Neurochem. Int.* **41**(5), 345–352 (2002).
7. Georganopoulou, D. G., Chang, L., Nam, J. M., Thaxton, C. S., Mufson, E. J., Klein, W. L., and Mirkin, C. A. Nanoparticle-based detection in cerebral spinal fluid of a soluble pathogenic biomarker for Alzheimer's disease. *Proc. Natl. Acad. Sci. U.S.A.* **102**(7), 2273–2276 (2005).
8. Hardy, J. and Selkoe, D. J. The amyloid hypothesis of Alzheimer's disease: Progress and problems on the road to therapeutics. *Science.* **297**(5580), 353–356 (2002).

9. Glabe, C. G. Amyloid accumulation and pathogenesis of Alzheimer's disease: Significance of monomeric, oligomeric and fibrillar Abeta. *Subcell. Biochem.* **38**, 167–177 (2005).
10. Lesné, S. E., Sherman, M. A., Grant, M., Kuskowski, M., Schneider, J. A., Bennett, D. A., and Ashe, K. H. Brain amyloid-β oligomers in ageing and Alzheimer's disease. *Brain.* **136**, 1383–1398 (2013).
11. Kostylev, M. A., Kaufman, A. C., Nygaard, H. B., Patel, P., Haas, L. T., Gunther, E. C., and Strittmatter, S. M. Prion-protein-interacting amyloid-β oligomers of high molecular weight are tightly correlated with memory impairment in multiple Alzheimer mouse models. *J. Biol. Chem.* **290**(28), 17415–17438 (2015).
12. Benilova, I., Karran, E., and De Strooper, B. The toxic Abeta oligomer and Alzheimer's disease: An emperor in need of clothes. *Nat. Neurosci.* **15**(3), 349–357 (2012).
13. Yang, T., Hong, S., O'Malley, T., Sperling, R. A., Walsh, D. M., and Selkoe, D. J. New ELISAs with high specificity for soluble oligomers of amyloid β-protein detect natural Aβ oligomers in human brain but not CSF. *Alzheimers Dement.* **9**(2), 99–112 (2013).
14. Shankar, G. M., Li, S., Mehta, T. H., Garcia-Munoz, A., Shepardson, N. E., Smith, I., and Selkoe, D. J. Amyloid-beta protein dimers isolated directly from Alzheimer's brains impair synaptic plasticity and memory. *Nat. Med.* **14**(8), 837 (2008).
15. Garzon-Rodriguez, W., Sepulveda-Becerra, M., Milton, S., and Glabe, C. G. Soluble amyloid Aβ-(1–40) exists as a stable dimer at low concentrations. *J. Biol. Chem.* **272**(34), 21037–21044 (1997).
16. Harper, J. D. and Lansbury, Jr. P. T. Models of amyloid seeding in Alzheimer's disease and scrapie: Mechanistic truths and physiological consequences of the time-dependent solubility of amyloid proteins. *Annu. Rev. Biochem.* **66**, 385–407 (1997).
17. Steven, G. Y. Evidence that Abeta42 is the real culprit in Alzheimer's disease. *Ann. Neurol.* **37**(3), 287–288 (1995).
18. Rogers, J., Cooper, N. R., Webster, S., Schultz, J., McGeer, P. L., Styren, S. D., and Ward, P. Complement activation by beta-amyloid in Alzheimer disease. *Proc. Natl. Acad. Sci. USA.* **89**(21), 10016–10020 (1992).
19. Selkoe, D. J. Translating cell biology into therapeutic advances in Alzheimer's disease. *Nature.* **399**(6738 Suppl), A23–A31 (1999).
20. Bitan, G., Kirkitadze, M. D., Lomakin, A., Vollers, S. S., Benedek, G. B., and Teplow, D. B. Amyloid beta-protein (Abeta) assembly: Abeta40 and Abeta42 oligomerize through distinct pathways. *Proc. Natl. Acad. Sci. U S A.* **100**(1), 330–335 (2003).
21. Bernstein, S. L., Wyttenbach, T., Baumketner, A., Shea, J. E., Bitan, G., Teplow, D. B., and Bowers, M. T. Amyloid beta-protein: Monomer structure and early aggregation states of Abeta42 and its pro19 alloform. *J. Am. Chem. Soc.* **127**(7), 2075–2084 (2005).
22. Tjernberg, L. Q., Pramanik, A., Bjorling, S., Thyberg, P., Thyberg, J., Nordstedt, C., and Rigler, R. Amyloid β-peptide polymerization studied using fluorescence correlation spectroscopy. *Chem. Biol.* **6**(1), 53–62 (1999).
23. Mousseau, N. and Derreumaux, P. Exploring the early steps of amyloid peptide aggregation by computers. *Acc. Chem. Res.* **38**(11), 885–891 (2005).
24. Teplow, D. B., Lazo, N. D., Bitan, G., Bernstein, S., Wyttenbach, T. Bowers, M. T., and Stanley, H. E. Elucidating amyloid-protein folding and assembly: a multidisciplinary approach. *Acc. Chem. Res.* **39**(9), 635–645 (2006).

25. Bitan, G., Lomakin, A., and Teplow, D. B. Amyloid beta-protein oligomerization. prenucleation interactions revealed by photo-induced cross-linking of unmodified proteins. *J. Biol. Chem.* **276**(37), 35176–35184 (2001).
26. Huang, T. H. J., Yang, D. S., Plaskos, N. P., Go, S., Yip, C. M., Fraser, P. E., and Chakrabartty, A. Structural studies of soluble oligomers of the Alzheimer β-amyloid peptide. *J. Mol. Biol.* **297**(1), 73–87 (2000).
27. Hilbich, C., Kisters-Woike, B., Reed, J., Masters, C. L., and Beyreuther, K. Substitutions of hydrophobic amino acids reduce the amyloidogenicity of Alzheimer's disease βA4 peptides. *J. Mol. Biol.* **228**(2), 460–473 (1992).
28. Soreghan, B., Kosmoski, J., and Glabe, C. G. Surfactant properties of Alzheimer's Abeta peptides and the mechanism of amyloid aggregation. *J. Biol. Chem.* **269**(46), 28551–28554 (1994).
29. Chen, Y. R. and Glabe, C. G. Distinct early folding and aggregation properties of Alzheimer amyloid-beta peptides Abeta40 and Abeta42: Stable trimer or tetramer formation by Abeta42. *J. Biol. Chem.* **281**(34), 24414–24422 (2006).
30. Schreck, J. and Yuan, J. M. Statistical mechanical treatments of protein amyloid formation. *Int. J. Mol. Sci.* **14**(9), 17420 (2013).
31. Lee, J., Culyba, E. K., Powers, E. T., and Kelly, J. W. Amyloid-β forms fibrils by nucleated conformational conversion of oligomers. *Nat. Chem. Biol.* **7**(9), 602–609 (2011).
32. Vivekanandan, S., Brender, J. R., Lee, S. Y., and Ramamoorthy, A. A partially folded structure of amyloid-beta(1–40) in an aqueous environment. *Biochem. Biophys. Res. Commun.* **411**(2), 312–316 (2011).
33. Petkova Aneta, T., Yau, W. M., and Tycko, R. Experimental constraints on quaternary structure in Alzheimer's beta-amyloid fibrils. *Biochemistry.* **45**(2), 498–512 (2006).
34. Tycko, R. and Wickner, R. B. Molecular structures of amyloid and prion fibrils: Consensus versus controversy. *Acc. Chem Res.* **46**(7), 1487–1496 (2013).
35. Potapov, A., Yau, W. M., Ghirlando, R., Thurber, K. R., and Tycko, R. Successive stages of amyloid-β self-assembly characterized by solid-state nuclear magnetic resonance with dynamic nuclear polarization. *J. Am. Chem Soc.* **137**(25), 8294–8307 (2015).
36. Antzutkin, O. N., Leapman, R. D., Balbach, J. J., and Tycko, R. Supramolecular structural constraints on Alzheimer's β-amyloid fibrils from electron microscopy and solid-state nuclear magnetic resonance. *Biochemistry.* **41**(51), 15436–15450 (2002).
37. Goldsbury, C. S., Wirtz, S., Muller, S. A., Sunderji, S., Wicki, P., Aebi, U., and Frey, P. Studies on the in vitro assembly of Abeta 1–40: Implications for the search for Abeta fibril formation inhibitors. *J. Struct. Biol.* **130**(2–3), 217–231 (2000).
38. Ha, T., Enderle, T., Ogletree, D. F., Chemla, D. S., Selvin, P. R., and Weiss, S. Probing the interaction between two single molecules: Fluorescence resonance energy transfer between a single donor and a single acceptor. *Proc. Natl. Acad. Sci. USA.* **93**(13), 6264–6268 (1996).
39. Schuler, B., Soranno, A., Hofmann, H., and Nettels, D. Single-molecule FRET spectroscopy and the polymer physics of unfolded and intrinsically disordered proteins. *Annu. Rev. Biophys.* **45**(1), 207–231 (2016).
40. Kim, J. and Lee, M. Observation of multi-step conformation switching in beta-amyloid peptide aggregation by fluorescence resonance energy transfer. *Biochem. Biophys. Res. Commun.* **316**(2), 393–397 (2004).

41. Ganzinger, K. A., Narayan, P., Qamar, S. S., Weimann, L., Ranasinghe, R. T., Aguzzi, A., and Klenerman, D. Single-molecule imaging reveals that small amyloid-β1–42 oligomers interact with the cellular prion protein (PrPC). *Chembiochem.* **15**(17), 2515–2521 (2014).
42. Dukes, K. D., Rodenberg, C. F., and Lammi, R. K. Monitoring the earliest amyloid-β oligomers via quantized photobleaching of dye-labeled peptides. *Anal. Biochem.* **382**(1), 29–34 (2008).
43. Ding, H., Wong, P. T., Lee, E. L., Gafni, A., and Steel, D. G. Determination of the oligomer size of amyloidogenic protein beta-amyloid(1–40) by single-molecule spectroscopy. *Biophys. J.* **97**(3), 912–921 (2009).
44. Shen, C. L. and Murphy, R. M. Solvent effects on self-assembly of beta-amyloid peptide. *Biophys. J.* **69**(2), 640–651 (1995).
45. Mei, E., Sharonov, A., Gao, F., Ferris, H. J., and Hochstrasser, R. M. Anomalously slow diffusion of single molecules near a patterned surface. *J. Phys. Chem. A.* **108**, 7339–7346 (2004).
46. Harris, J. M. and Zalipsky, S. *Poly(ethylene glycol): Chemistry and Biological Applications*, 1st Ed. Washington, DC: American Chemical Society; 1997.
47. J. R. Lakowicz. *Principles of Fluorescence Spectroscopy*, 2nd Ed. New York: Kluwer Academic; 1999.
48. van der Meer, B. W., Coker III, G., and Chen, S. Y. S (eds). *Resonance Energy Transfer: Theory and Data*, New York, NY: VCH Publishers; 1994.
49. Anderson, V. and Webb, W. Transmission electron microscopy characterization of fluorescently labelled amyloid beta 1–40 and alpha-synuclein aggregates. *BMC Biotechnol.* **11**(1), 125 (2011).
50. Johnson, R. D., Schauerte, J. A., Wisser, K. C., Gafni, A., and Steel, D. G. Direct observation of single amyloid-β(1–40) oligomers on live cells: binding and growth at physiological concentrations. *PLoS One.* **6**(8), e23970 (2011).
51. Narayan, P., Orte, A., Clarke, R. W., Bolognesi, B., Hook, S., Ganzinger, K. A., and Klenerman, D. The extracellular chaperone clusterin sequesters oligomeric forms of the amyloid-β(1−40) peptide. *Nat. Struct. Mol. Biol.* **19**(1), 79–83 (2012).
52. Jia, Y., Talaga, D. S., Lau, W. L., Lu, H. S. M., DeGrado, W. F., and Hochstrasser, R. M. Folding dynamics of single GCN-4 peptides by fluorescence resonant energy transfer confocal microscopy. *Chem. Phys.* **247**(1), 69–83 (1999).
53. Barz, B. and Urbanc, B. Dimer formation enhances structural differences between amyloid β-protein (1–40) and (1–42): An explicit-solvent molecular dynamics study. *PLoS One.* **7**(4), e34345 (2012).
54. Eigen, M. and Rigler, R. Sorting single molecules: Application to diagnostics and evolutionary biotechnology. *Proc. Natl. Acad. Sci. USA.* **91**(13), 5740–5747 (1994).

Index